METHODS IN MOLECULAR BIOLOGY

For further volumes:
http://www.springer.com/series/7651

For over 35 years, biological scientists have come to rely on the research protocols and methodologies in the critically acclaimed *Methods in Molecular Biology* series. The series was the first to introduce the step-by-step protocols approach that has become the standard in all biomedical protocol publishing. Each protocol is provided in readily-reproducible step-by-step fashion, opening with an introductory overview, a list of the materials and reagents needed to complete the experiment, and followed by a detailed procedure that is supported with a helpful notes section offering tips and tricks of the trade as well as troubleshooting advice. These hallmark features were introduced by series editor Dr. John Walker and constitute the key ingredient in each and every volume of the *Methods in Molecular Biology* series. Tested and trusted, comprehensive and reliable, all protocols from the series are indexed in PubMed.

Microbial 'Omics

Methods and Protocols

Edited by

Tao Huang

Shanghai Institute of Nutrition and Health, Chinese Academy of Sciences, Shanghai, China

Humana Press

Editor
Tao Huang
Shanghai Institute of Nutrition
and Health
Chinese Academy of Sciences
Shanghai, China

ISSN 1064-3745 ISSN 1940-6029 (electronic)
Methods in Molecular Biology
ISBN 978-1-0716-5263-3 ISBN 978-1-0716-5264-0 (eBook)
https://doi.org/10.1007/978-1-0716-5264-0

This Humana imprint is published by the registered company Springer Science+Business Media, LLC, part of Springer Nature.
The registered company address is: 1 New York Plaza, New York, NY 10004, U.S.A.

Preface

The field of microbial omics has undergone an unprecedented transformation over the past two decades, evolving from a niche area of microbiology to a cornerstone of interdisciplinary research that bridges biology, medicine, ecology, and engineering. What began with the advent of next-generation sequencing (NGS) technologies has blossomed into a holistic discipline that deciphers the complex interplay between microbial communities and their environments, hosts, and industrial systems. Today, microbial omics stands at the intersection of high-throughput data generation, advanced computational analytics, and translational application, offering unprecedented insights into the "invisible majority" that shapes our world.

This book, *Microbial 'Omics: Methods and Protocols*, emerges from a pressing need to synthesize the rapidly expanding toolkit of microbial omics research while providing a structured, accessible guide for both novice and seasoned researchers. It is designed as a comprehensive laboratory companion that navigates the entire research pipeline—from sample collection to data interpretation, and from foundational protocols to cutting-edge applications. Our goal is to address a critical gap in the literature: the lack of a unified resource that integrates traditional experimental methods with modern computational tools and emerging technologies, all while emphasizing reproducibility—a cornerstone of scientific progress.

Section 1: Foundational Overview

The book opens with Chapter 1—a comprehensive introduction that sets the stage for subsequent chapters. This chapter provides a historical context for microbial omics, outlines key methods (from sequencing platforms to bioinformatics tools), and discusses emerging trends and challenges.

Section 2: Core Experimental Methods

The second section focuses on the standardized experimental protocols of microbial omics research. These chapters address the critical, often overlooked steps that determine the success of any omics study—from sample collection to data preprocessing.

Chapter 2 tackles one of the most persistent challenges in microbial omics: contamination control. As microbial communities are highly sensitive to environmental perturbations, this chapter provides detailed protocols for collecting samples from diverse environments while minimizing cross-contamination and preserving community integrity.

Chapter 3 builds on sample collection by guiding readers through the critical steps of cleaning and filtering raw sequencing data. With the exponential growth of sequencing datasets, efficient preprocessing is essential to improve the accuracy of downstream analyses.

Chapter 4 addresses the next challenge: converting short sequencing reads into contiguous sequences (contigs) and predicting protein-coding genes. The quality of metagenomic assemblies directly impacts the ability to identify novel genes and reconstruct metabolic pathways.

Chapter 5 completes the experimental core by moving beyond relative abundance measurements—a limitation of many early omics studies—to absolute quantification. Absolute quantification is particularly critical for clinical and environmental studies, where accurate biomass measurements are essential for linking microbial communities to host health or ecosystem function.

Section 3: Data Analysis Tools

The third section focuses on computational tools that transform processed data into meaningful insights. These chapters bridge the gap between experimental data and biological interpretation, emphasizing practical, reproducible workflows.

Chapter 6 introduces readers to network-based approaches for analyzing microbial interactions. Network analysis has become increasingly important in microbial omics, as it allows researchers to understand the collective behavior of microbial communities.

Chapter 7 represents the intersection of omics and AI. It highlights the transformative potential of AI-driven enzyme discovery for biomanufacturing and medicine.

Section 4: Applied Omics: From Health to Industry

The fourth section is the largest, reflecting the broad applicability of microbial omics. It is divided into three subsections—healthcare, food fermentation, and industrial/environmental applications—each focusing on how omics methods are solving real-world problems.

Healthcare Applications

Microbial omics has revolutionized our understanding of the human microbiome and its role in health and disease. This subsection focuses on host-microbe interactions in cancer and interorgan communication.

Chapter 8 explores the bidirectional communication between the gut and lung microbiomes. The gut-lung axis is a rapidly growing area of research, with implications for both prevention and treatment of inflammatory and infectious diseases.

Chapters 9 and 10 focus on the role of microbes in colorectal cancer (CRC)—one of the most well-studied microbiome-associated diseases. Chapter 9 explores how gut microbial metabolites contribute to CRC development and progression, while Chapter 10 investigates the interplay between Epstein-Barr virus (EBV) infection, epigenetic modifications, and radioresistance in CRC. Together, these chapters highlight the multifaceted role of microbes in cancer, from metabolic dysregulation to immune modulation.

Chapter 11 extends this theme to hepatocellular carcinoma (HCC), focusing on how viral infections (e.g., hepatitis B/C) contribute to liver cancer. This chapter describes omics methods for profiling infection-associated microbial communities, identifying microbial biomarkers for early HCC detection, and developing microbiome-based therapeutics.

Food Fermentation Applications

Fermented foods represent some of the oldest and most widespread human-microbe interactions. This subsection focuses on multispecies symbioses in fermentation processes, a topic of growing interest for food science and industry.

Chapter 12 covers staple fermented foods (e.g., Sauerkraut, Kimchi), describing omics methods for identifying key microbial taxa, characterizing their metabolic contributions (e.g., flavor development, preservation), and optimizing fermentation processes.

Chapter 13 expands to specialty fermentations (e.g., Cheese, Kefir) and nonfood applications (e.g., Fermented Tea, Coffee). This chapter addresses the unique challenges of these systems, such as complex microbial interactions, variable environmental conditions, and industrial scalability.

Industrial and Environmental Applications

Microbial omics are increasingly being used to solve industrial and environmental challenges, from biomanufacturing to bioremediation.

Chapter 14 focuses on the industrial potential of macrofungi (e.g., mushrooms) for polysaccharide production. Polysaccharides from macrofungi have diverse applications in food, pharmaceuticals, and cosmetics, and omics methods are enabling researchers to identify key biosynthetic genes, optimize production, and engineer novel polysaccharides with enhanced properties.

Chapter 15 explores a novel application of microbial omics: the development of functionalized microorganisms for environmental and industrial use. This chapter describes protocols for coating microbes with metal-tannic acid complexes, characterizing their physical and chemical properties, and discussing their applications such as probiotic encapsulation, engineered microbial therapeutics, and microbial-based biosensors.

Chapter 16 represents the pinnacle of applied microbial omics, describing how multi-omics data is used to design and engineer synthetic microbial communities. The integration of omics and synthetic biology is the next frontier of microbial research, enabling researchers to create custom microbes with tailored functions.

Section 5: Cutting-Edge Technologies: AI and Virology

The final section focuses on the latest frontier of microbial omics: the integration of AI and Large Language Models (LLMs) in virology. This section reflects the field's shift toward predictive and data-driven research, highlighting how advanced computational tools are transforming viral genomics and epidemiology.

Chapter 17 focuses on RNA LLMs in viral research. LLMs are being used for a wide range of applications, including viral genome annotation and predicting viral-host interactions. This chapter provides practical guides for using LLMs in viral research, while discussing the challenges of interpretability.

Chapter 18 provides a comprehensive overview of protein language models (PLMs) in viral research. PLMs—such as AlphaFold, ESM, and ProtTrans—use deep learning to predict protein structure, function, and interactions from sequence data alone. This chapter describes how PLMs are being used to orphan protein structure resolution, virosphere-wide functional classification, host factor identification, and therapeutic antibody optimization.

This book would not have been possible without the contributions of the chapter authors who dedicated their time and expertise to its development. I am deeply grateful to them, who generously shared their protocols, insights, and experiences. We also thank the editorial team at Springer for their guidance and support throughout the publication process.

Finally, I would like to dedicate this book to the broader microbial omics community—from the early pioneers who laid the groundwork for the field to the next generation of researchers who will continue to push its boundaries. It is my hope that this book will serve as a valuable resource for years to come, inspiring new discoveries and advancing the field of microbial omics for the benefit of humanity and the planet.

Shanghai Institute of Nutrition and Health
Chinese Academy of Sciences
Shanghai, China

Tao Huang

Contents

Contributors

WEIXU CAI • *Clinical Medical College and Affiliated Hospital of Chengdu University, Chengdu University, Chengdu, Sichuan, China; School of Life Sciences, Department of Life Sciences and Medicine, Northwest University, Xi'an, Shaanxi, China*

XIAO CHANG • *College of Medical Information and Artificial Intelligence, Shandong First Medical University, Jinan, China*

ZHONGHUA CHENG • *Department of Gastroenterology, Shanghai Xuhui Center Hospital, Shanghai, China*

ZHEN FENG • *Department of Gastroenterology, Dahua Hospital of Xuhui District, Shanghai, China*

BAOTONG FU • *Center for Grassland Microbiome, College of Pastoral Agriculture Science and Technology, Lanzhou University, Lanzhou, Gansu, China*

YI-ZHOU GAO • *The Center for Microbes, Development and Health, Shanghai Institute of Immunity and Infection, Chinese Academy of Sciences, Shanghai, China; Department of Environmental Hygiene, School of Public Health, Harbin Medical University, Harbin, China*

JIA-XIN GUO • *The Center for Microbes, Development and Health, Shanghai Institute of Immunity and Infection, Chinese Academy of Sciences, Shanghai, China; Department of Radiation Oncology, Zhejiang Cancer Hospital, Hangzhou, China; Zhejiang Key Laboratory of Particle Radiotherapy Equipment, Hangzhou, China*

HUAWEN HAN • *Center for Grassland Microbiome, College of Pastoral Agriculture Science and Technology, Lanzhou University, Lanzhou, Gansu, China*

TAO HUANG • *Department of Artificial Intelligence and Digital Health, CAS Engineering Laboratory for Nutrition, Shanghai Institute of Nutrition and Health, University of Chinese Academy of Sciences, Chinese Academy of Sciences, Shanghai, China; Bio-Med Big Data Center, CAS Key Laboratory of Computational Biology, Shanghai Institute of Nutrition and Health, University of Chinese Academy of Sciences, Chinese Academy of Sciences, Shanghai, China*

JI-HANG JIANG • *State Key Laboratory of Microbial Diversity and Innovative Utilization, Institute of Microbiology, Chinese Academy of Sciences, Beijing, China*

BO LI • *Guangdong Basic Research Center of Excellence for Ecological Security and Green Development, School of Ecology, Environment and Resources, Guangdong University of Technology, Guangzhou, China*

JIUFENG LI • *Department of Environmental Health, School of Public Health, NHC Key Laboratory of Health Technology Assessment, Key Laboratory of Public Health Safety of the Ministry of Education, Fudan University, Shanghai, China*

YIXUE LI • *Department of Artificial Intelligence and Digital Health, CAS Engineering Laboratory for Nutrition, Shanghai Institute of Nutrition and Health, University of Chinese Academy of Sciences, Chinese Academy of Sciences, Shanghai, China; Bio-Med Big Data Center, CAS Key Laboratory of Computational Biology, Shanghai Institute of Nutrition and Health, University of Chinese Academy of Sciences, Chinese Academy of Sciences, Shanghai, China; Guangzhou National Laboratory, Guangzhou, China; GMU-GIBH Joint School of Life Sciences, The Guangdong-Hong Kong-Macau Joint Laboratory for Cell Fate Regulation and Diseases, Guangzhou Laboratory, Guangzhou Medical*

University, Guangzhou, China; Key Laboratory of Systems Health Science of Zhejiang Province, School of Life Science, Hangzhou Institute for Advanced Study, University of Chinese Academy of Sciences, Hangzhou, China

YUANQI LI • *Tumor Biological Diagnosis and Treatment Center, the Third Affiliated Hospital of Soochow University, Changzhou, China; Jiangsu Engineering Research Center for Tumor Immunotherapy, Changzhou, China; Institute of Cell Therapy, Soochow University, Changzhou, China*

DESHUI LIU • *Beijing Life Science Academy, Beijing, China*

JIAJUN LIU • *Department of Artificial Intelligence and Digital Health, CAS Engineering Laboratory for Nutrition, Shanghai Institute of Nutrition and Health, University of Chinese Academy of Sciences, Chinese Academy of Sciences, Shanghai, China; Bio-Med Big Data Center, CAS Key Laboratory of Computational Biology, Shanghai Institute of Nutrition and Health, University of Chinese Academy of Sciences, Chinese Academy of Sciences, Shanghai, China*

CIMIN LONG • *Laboratory of Animal Nutritional Physiology and Metabolic Process, Key Laboratory of Agro-Ecological Processes in Subtropical Region, National Engineering Laboratory for Pollution Control and Waste Utilization in Livestock and Poultry Production, Institute of Subtropical Agriculture, Chinese Academy of Sciences, Changsha, China*

HUILONG LUO • *Guangzhou Municipal and Guangdong Provincial Key Laboratory of Molecular Target and Clinical Pharmacology, the NMPA and State Key Laboratory of Respiratory Disease, School of Pharmaceutical Sciences and the Second Affiliated Hospital, Guangzhou Medical University, Guangzhou, China*

LINGXIN LUO • *Department of Artificial Intelligence and Digital Health, CAS Engineering Laboratory for Nutrition, Shanghai Institute of Nutrition and Health, University of Chinese Academy of Sciences, Chinese Academy of Sciences, Shanghai, China; Bio-Med Big Data Center, CAS Key Laboratory of Computational Biology, Shanghai Institute of Nutrition and Health, University of Chinese Academy of Sciences, Chinese Academy of Sciences, Shanghai, China; School of Life Sciences, Shanghai University, Shanghai, China*

XING LUO • *State Key Laboratory of Microbial Diversity and Innovative Utilization, Institute of Microbiology, Chinese Academy of Sciences, Beijing, China; University of Chinese Academy of Sciences, Beijing, China*

QIYU MENG • *Guangdong Basic Research Center of Excellence for Ecological Security and Green Development, School of Ecology, Environment and Resources, Guangdong University of Technology, Guangzhou, China*

BICHEN PENG • *College of Medical Information and Artificial Intelligence, Shandong First Medical University, Jinan, China*

ZHENG-DONG QI • *Clinical Medical College and Affiliated Hospital of Chengdu University, Chengdu University, Chengdu, Sichuan, China; School of Life Sciences, Department of Life Sciences and Medicine, Northwest University, Xi'an, Shaanxi, China*

YAOWEI QIN • *Shanghai Pudong Hospital and Department of Anatomy and Histoembryology, School of Basic Medical Sciences, Fudan University, Shanghai, China*

YUAN QIN • *Guangzhou Municipal and Guangdong Provincial Key Laboratory of Molecular Target and Clinical Pharmacology, the NMPA and State Key Laboratory of Respiratory Disease, School of Pharmaceutical Sciences and the Second Affiliated Hospital, Guangzhou Medical University, Guangzhou, China*

WEIWEI SHEN • *School of Life Sciences, Shanghai University, Shanghai, China; Department of Artificial Intelligence and Digital Health, CAS engineering Laboratory for Nutrition, Shanghai Institute of Nutrition and Health, University of Chinese Academy of Sciences, Chinese Academy of Sciences, Shanghai, China; Bio-Med Big Data Center, CAS Key Laboratory of Computational Biology, Shanghai Institute of Nutrition and Health, University of Chinese Academy of Sciences, Chinese Academy of Sciences, Shanghai, China*

QIAN SONG • *Shandong Provincial Key Laboratory of Precision Oncology, Shandong Cancer Hospital and Institute, Shandong First Medical University, Shandong Academy of Medical Sciences, Jinan, Shandong, China*

FAN WANG • *Qingdao Institute of Bioenergy and Bioprocess Technology, Chinese Academy of Sciences, Qingdao, China; Shandong Energy Institute, Qingdao, China; Qingdao New Energy Shandong Laboratory, Qingdao, China*

LELI WANG • *Laboratory of Animal Nutritional Physiology and Metabolic Process, Key Laboratory of Agro-Ecological Processes in Subtropical Region, National Engineering Laboratory for Pollution Control and Waste Utilization in Livestock and Poultry Production, Institute of Subtropical Agriculture, Chinese Academy of Sciences, Changsha, China*

QIAN WANG • *Tobacco Research Institute, Chinese Academy of Agricultural Sciences, Qingdao, China*

ZUOYUN WANG • *Shanghai Pudong Hospital and Department of Anatomy and Histoembryology, School of Basic Medical Sciences, Fudan University, Shanghai, China*

DAI-XU WEI • *Clinical Medical College and Affiliated Hospital of Chengdu University, Chengdu University, Chengdu, Sichuan, China*

DANFENG WEN • *Department of Environmental Health, School of Public Health, NHC Key Laboratory of Health Technology Assessment, Key Laboratory of Public Health Safety of the Ministry of Education, Fudan University, Shanghai, China*

HONGMING XIA • *Center for Grassland Microbiome, College of Pastoral Agriculture Science and Technology, Lanzhou University, Lanzhou, Gansu, China*

JIYING XU • *Guangdong Basic Research Center of Excellence for Ecological Security and Green Development, School of Ecology, Environment and Resources, Guangdong University of Technology, Guangzhou, China*

CHAO YAN • *Laboratory of Animal Nutritional Physiology and Metabolic Process, Key Laboratory of Agro-Ecological Processes in Subtropical Region, National Engineering Laboratory for Pollution Control and Waste Utilization in Livestock and Poultry Production, Institute of Subtropical Agriculture, Chinese Academy of Sciences, Changsha, China; Shenzhen Branch, Guangdong Laboratory of Lingnan Modern Agriculture, Key Laboratory of Livestock and Poultry Multi-Omics of MARA, Agricultural Genomics Institute at Shenzhen, Chinese Academy of Agricultural Sciences, Shenzhen, China*

XUE YANG • *Guangdong Basic Research Center of Excellence for Ecological Security and Green Development, School of Ecology, Environment and Resources, Guangdong University of Technology, Guangzhou, China*

QIDONG YIN • *School of Civil Engineering, Sun Yat-Sen University, Zhuhai, China*

YULONG YIN • *Laboratory of Animal Nutritional Physiology and Metabolic Process, Key Laboratory of Agro-Ecological Processes in Subtropical Region, National Engineering Laboratory for Pollution Control and Waste Utilization in Livestock and Poultry Production, Institute of Subtropical Agriculture, Chinese Academy of Sciences, Changsha, China*

Zhimeng Yuan • *Clinical Medical College and Affiliated Hospital of Chengdu University, Chengdu University, Chengdu, Sichuan, China*

Feng Zhang • *Laboratory of Animal Nutritional Physiology and Metabolic Process, Key Laboratory of Agro-Ecological Processes in Subtropical Region, National Engineering Laboratory for Pollution Control and Waste Utilization in Livestock and Poultry Production, Institute of Subtropical Agriculture, Chinese Academy of Sciences, Changsha, China*

Haibo Zhang • *Qingdao Institute of Bioenergy and Bioprocess Technology, Chinese Academy of Sciences, Qingdao, China; Shandong Energy Institute, Qingdao, China; Qingdao New Energy Shandong Laboratory, Qingdao, China*

Jiahao Zhang • *Clinical Medical College and Affiliated Hospital of Chengdu University, Chengdu University, Chengdu, Sichuan, China*

Yuemei Zhang • *Guangzhou Municipal and Guangdong Provincial Key Laboratory of Molecular Target and Clinical Pharmacology, the NMPA and State Key Laboratory of Respiratory Disease, School of Pharmaceutical Sciences and the Second Affiliated Hospital, Guangzhou Medical University, Guangzhou, China*

Tongyi Zhao • *Guangdong Basic Research Center of Excellence for Ecological Security and Green Development, School of Ecology, Environment and Resources, Guangdong University of Technology, Guangzhou, China*

Chunxiu Zhou • *Center for Grassland Microbiome, College of Pastoral Agriculture Science and Technology, Lanzhou University, Lanzhou, Gansu, China*

Li-Wei Zhou • *State Key Laboratory of Microbial Diversity and Innovative Utilization, Institute of Microbiology, Chinese Academy of Sciences, Beijing, China*

You Zhou • *Tumor Biological Diagnosis and Treatment Center, the Third Affiliated Hospital of Soochow University, Changzhou, China; Jiangsu Engineering Research Center for Tumor Immunotherapy, Changzhou, China; Institute of Cell Therapy, Soochow University, Changzhou, China*

Yuping Zhou • *Department of Gastroenterology, Institute of Digestive Disease, the First Affiliated Hospital of Ningbo University, Ningbo Key Laboratory of Translational Medicine Research on Gastroenterology and Hepatology, Ningbo, China*

Liucun Zhu • *School of Life Sciences, Shanghai University, Shanghai, China*

Yongcheng Zhu • *Guangzhou Municipal and Guangdong Provincial Key Laboratory of Molecular Target and Clinical Pharmacology, the NMPA and State Key Laboratory of Respiratory Disease, School of Pharmaceutical Sciences and the Second Affiliated Hospital, Guangzhou Medical University, Guangzhou, China*

Yanting Zou • *Guangzhou eDNAmega Biotechnology Co., Ltd, Guangzhou, China*

Chapter 1

A Brief Review of Microbial Omics: Methods and Perspectives

Chao Yan, Feng Zhang, Cimin Long, Yulong Yin, and Leli Wang

Abstract

Microbial omics has progressed from isolated genomic analyses into a comprehensive, integrated multi-omics framework, profoundly advancing our understanding of microbial complexity and functionality. This mini-review systematically outlines the core technologies within microbial omics—including genomics, transcriptomics, proteomics, and metabolomics—by introducing their fundamental principles, common experimental workflows, and state-of-the-art bioinformatic strategies. We particularly highlight the emergence of single-cell microbial omics as a transformative methodology that resolves molecular and functional heterogeneity within communities, enabling the identification of rare taxa, strain-level microdiversity, and specialized functional roles that are obscured in bulk analyses. Furthermore, we discuss how artificial intelligence (AI)-driven tools are revolutionizing the interpretation of high-dimensional omics data, uncovering latent biological patterns, improving predictive modeling of microbial behavior, and facilitating the translation of microbiome insights into clinical and environmental applications. The review concludes by comparing the strengths, limitations, and optimal use cases of each omics layer and single-cell approach while also addressing ongoing technical challenges and future directions in the field.

Key words Microbiota, Microbial omics, Methods, Single-cell microbial cell artificial intelligence

1 Introduction

In the seventeenth century, Antony van Leeuwenhoek, known as "the father of microbiology," was the first to discover bacteria. Since then, the understanding of the microbial world, comprising bacteria, viruses, fungi, and protozoa, has evolved dramatically. This microbial world is complex and vital, influencing a wide range of biological processes, and is linked to various health conditions, such as obesity, inflammatory bowel disease, and neurological disorders [1]. The intricate cross-talk between microbiota and host significantly affects the host's metabolism, immune response, and endocrine systems through precise molecular mechanisms [2, 3]. Importantly, metabolites serve as crucial intermediaries in these processes. For instance, microbiota contribute to the

Tao Huang (ed.), *Microbial 'Omics: Methods and Protocols*, Methods in Molecular Biology, vol. 3033,
https://doi.org/10.1007/978-1-0716-5264-0_1,

production of essential metabolites, such as short-chain fatty acids (SCFAs), indoles, and vitamins [4], all of which play significant roles in maintaining host functions and health [5]. The field of microbiology has undergone a remarkable transformation, shifting from culture-based methods to sequencing technologies, greatly advancing microbial research and its applications. With the rapid development of sequencing technologies and omics (e.g., genomics, transcriptomics, proteomics, and metabolomics, Table 1), there has been a significant advancement in the study and understanding of microorganisms [6, 7]. Recent advancements in single-cell transcriptomics (scRNA-seq) and spatial omics technologies have enabled researchers to explore the microbial community at unprecedented resolution [8]. Additionally, the integration of artificial intelligence (AI) is beginning to impact microbiome research, offering powerful new avenues for analyzing and interpreting complex microbial data [9]. This review aims to provide researchers with a comprehensive methodological guide, helping them select the most appropriate technologies and understand the key aspects of experimental design.

2 Metagenomic Sequencing

2.1 16S rRNA Gene Sequencing

16S rRNA gene sequencing is a fundamental method for analyzing complex microbiomes and is the most widely used marker gene for profiling bacterial communities [10]. The 16S rRNA contains both variable and conserved regions, with the conserved regions being targeted for amplification using specific primers, enabling the classification of bacteria. The amplified fragments are then sequenced using high-throughput sequencing platforms, such as Illumina or PacBio.

The analysis of 16S rRNA gene sequencing involves several key steps using various software tools, including QIIME 2 [11], MOTHUR, DADA2, VSEARCH [12], and USEARCH [13]. The microbial analysis process includes merging paired-end reads, trimming primers, quality filtering, dereplication, clustering sequences into OTUs (operational taxonomic units) or ASVs (amplicon sequence variants), discarding chimeras, and assigning taxonomy [14]. The resulting sequences are compared to reference databases (e.g., SILVA, Greengenes, and RDP) to assign taxonomy and assess the microbial composition of the sample. Alpha and beta diversity analyses are performed to evaluate within-sample richness and between-sample differences, respectively [15]. Additionally, microbial abundance and functional profiles are assessed, with Phylogenetic Investigation of Communities by Reconstruction of Unobserved States (PICRUSt2) being used to predict the metagenomic functional content based on marker gene data. Overall, 16S rRNA gene sequencing remains a powerful and cost-effective tool

Table 1
Comparison of microbiome omics technologies

Method	Analysis focus	Advantages	Disadvantages
16S rRNA gene sequencing	Microbial species and relative abundance, based on 16S rRNA gene sequences	Simple, fast, and cost-effective High throughput Suitable for bacteria and archaea Can analyze community composition	Limited to bacteria and archaea Cannot differentiate strains within the same species Low resolution, only species-level information
Metagenomics	Complete genomic information of microbial communities	Comprehensive analysis of microbial genomes Reveals functional genes and metabolic potential Advantageous for unculturable microbes	Large data volume, requiring significant computational resources and storage Difficult assembly and incomplete genome data
Metatranscriptomics	RNA expression (mRNA) of microbial communities	Reveals active gene expression and metabolic activity Reflects microbial response to environmental changes Provides dynamic gene expression data	RNA instability can affect data quality Low abundance gene expression is hard to detect Complexities in post-transcriptional modifications
Metaproteomics	Protein content of microbial communities	Provides direct functional information, revealing microbial activities High resolution for protein and enzyme functions-reveals functional networks	Complex data analysis, requiring high-quality databases Difficult to detect low-abundance proteins Challenging mass spectrometry data interpretation
Metabolomics	Metabolites of microbial communities	Metabolites directly reflect microbial metabolic states, with high biological relevance Reveals microbial metabolic pathways and physiological status	Complex and diverse metabolites, data interpretation is challenging Sample preparation is difficult, susceptible to contamination Quantification limited by equipment and technology

(continued)

Table 1
(continued)

Method	Analysis focus	Advantages	Disadvantages
Single-cell microbial omics	Single-cell genomic, transcriptomic, or proteomic information	High resolution, reveals heterogeneity within microbial communities Can study rare populations or specific cell types Suitable for single-cell analysis	Technically complex and costly Single cell data may have high noise and errors Data processing and analysis are challenging

for studying microbial communities [16], but it is often complemented by other omics approaches, such as shotgun metagenomics, for a more comprehensive understanding of microbial diversity and function.

2.2 Metagenome

Shotgun metagenomics enables direct analysis of the complete genomic DNA and diverse microbial communities extracted from environmental or clinical samples [17]. Unlike 16S rRNA gene sequencing—which targets a specific phylogenetic marker—metagenomics captures all DNA present, thereby offering a broader perspective on microbial community composition, including bacteria, archaea, fungi, and viruses. Commonly employed high-throughput sequencing platforms for such analyses include Illumina, PacBio, and Nanopore technologies.

A standard metagenomic analysis workflow typically consists of the following steps: quality control and filtering of raw sequencing data using tools such as Fastp [18] or Trimmomatic [19] to remove low-quality sequences and adapters; High-quality reads are aligned to the host reference genome (e.g., using Bowtie2 [20]) to remove host-derived sequences. The resulting high-quality, host-depleted reads can be analyzed using three main analytical strategies [15, 21]:

1. Taxonomic profiling (read-based): For rapid and accurate estimation of microbial community composition at the species or strain level, quality-filtered reads were directly classified using either MetaPhlAn 4 [22] or the Kraken 2 [23]/Bracken [24] pipeline. Subsequently, the functional annotation of the communities was inferred from the metagenomic reads using HUMAnN 4 [25].
2. Functional profiling (assembly-based): Clean reads are assembled with either MEGAHIT [26] or metaSPAdes [27]. The quality of the assembled contigs was assessed with QUAST [28]. Genes were predicted from the contigs using MetaProdigal [29], and a nonredundant gene catalog was constructed

with CD-HIT [30]. The abundance of these genes was quantified with Salmon. Functional annotation was performed by aligning the predicted gene sequences against databases including the Kyoto Encyclopedia of Genes and Genomes (KEGG), Clusters of Orthologous Groups (COG), and the Carbohydrate-Active Enzymes (CAZy) database using eggNOG-mapper, thereby identifying the metabolic potentials and functional pathways of the microbial community.

3. De novo assembly and genome reconstruction (Binning): To reconstruct metagenome-assembled genomes (MAGs), the Subheading 2.2(2) assembled contigs were processed using the "binning" and "bin_refinement" modules in MetaWRAP [31]. The resulting MAGs were dereplicated to address species redundancy using dRep [32], and their abundance across samples was calculated with CoverM. MAG quality (completeness and contamination) was evaluated with CheckM2 [33], and taxonomic classification was conducted using GTDB-Tk [34]. Functional annotation of the MAGs was performed using eggNOG-mapper [35], DIAMOND [36], and RGI [37].

Metagenomics enables the direct identification of genes encoding critical functions—such as antibiotic resistance, virulence factors, and key metabolic pathways—thereby offering deep insights into the functional potential and gene repertoire of microbial communities. By providing a comprehensive, cultivation-independent view of both microbial diversity and functional capacity, metagenomics is frequently integrated with other omics approaches, including metatranscriptomics, metaproteomics, or metabolomics, to build a more complete and dynamic understanding of microbial community behavior, activity, and ecological roles.

3 Microbial Transcriptomics

Metatranscriptomics involves the direct sequencing and analysis of the total community RNA pool (predominantly mRNA, along with rRNA and noncoding RNAs) extracted from collected samples [38]. This approach captures a real-time snapshot of gene expression across the entire microbiome. In contrast to metagenomics, which reveals the genomic potential of a community, metatranscriptomics reveals which genes are actually transcribed at the time of sampling. It thereby provides critical insights into real-time functional activity, metabolic responses, and regulatory dynamics of microbial communities, including bacteria, archaea, fungi, and viruses [17].

A standard metatranscriptomic analysis workflow typically comprises the following steps: total RNA is extracted using kits optimized for microbial communities, followed by DNase I treatment and ribosomal RNA depletion. cDNA synthesis is performed using random hexamer primers. Stranded libraries are prepared with the Illumina Stranded Total RNA Prep kit, while unstranded libraries are constructed using the NEBNext Ultra II Directional RNA Library Prep kit. Sequencing is typically performed on Illumina, PacBio Iso-Seq, or Oxford Nanopore.

High-quality reads are processed using Fastp or Trimmomatic for adapter trimming and quality filtering. Ribosomal RNA reads are depleted using SortMeRNA [39], RiboDetector, or KneadData, retaining primarily mRNA and noncoding RNA. Non-rRNA reads are mapped to the host reference genome to deplete host RNA using HISAT2 [40] or STAR [41]. The resulting nonhost reads are then subjected to metatranscriptomic analysis with three complementary strategies [42]:

1. Taxonomic profiling (mapping-based): The active microbial community is characterized by analyzing non-rRNA reads, which is achieved through like tools Kraken 2/Bracken or MetaPhlAn 4, which help identify transcriptionally active organisms. The functional annotation of quality-controlled RNA was conducted using HUMAnN 4.
2. Functional profiling (assembly-based): Non-rRNA reads are assembled into transcript contigs using Trinity [43] and rnaSPAdes [44] (for short reads), or the IsoSeq pipeline [45] (for PacBio full-length reads). Putative protein-coding genes were predicted from the resulting high-confidence transcript sequences using TransDecoder. Functional annotation of the nonredundant protein catalog, derived from the predicted coding sequences, was performed using MMseqs2 [46] against specialized databases, including KEGG, CAZy, eggNOG, NR, and SWISS-PROT, to elucidate the functional potential of the microbial community. Gene- and transcript-level expression quantification was performed by mapping quality-controlled, non-rRNA reads back to the assembled transcriptome using Salmon, which accounts for read mapping uncertainty and improves quantification accuracy. Alternatively, Kallisto can be employed for rapid pseudoalignment-based quantification. The resulting counts for downstream differential expression analysis with DESeq2. This assembly-based workflow enables the discovery of novel transcripts, accurate isoform-level quantification, and comprehensive functional annotation, particularly valuable in highly diverse or poorly characterized microbial communities.
3. De novo transcriptome assembly and annotation (Assembly-based): The Subheading 3(2) assembled contigs were collapsed at 95% identity using CD-HIT-EST or Corset. Taxonomic

origin and relative transcript abundance were assigned via MMseqs2 or by lowest common ancestor (LCA) classification of DIAMOND. The final transcript set was functionally annotated using the Trinotate [47] and eggNOG-mapper, supplemented by sensitive DIAMOND searches against UniRef90, Swiss-Prot, KEGG, CAZ, MEROPS, VFDB, and CARD to characterize expressed metabolic pathways, carbohydrate-active enzymes, virulence factors, and antimicrobial resistance genes in the active microbiota.

Metatranscriptomics enables the direct identification of actively expressed genes involved in critical processes—such as antibiotic resistance, virulence factor production, nutrient cycling, and stress responses—under specific environmental or physiological conditions. By revealing not just what microbes are capable of but what they are actually doing, metatranscriptomics is frequently integrated with shotgun metagenomics, metaproteomics, and metabolomics in multi-omics frameworks to provide a comprehensive, dynamic understanding of microbial community function, regulation, and ecosystem-level interactions.

4 Microbial Proteomics and Metabolomics

4.1 Metaproteome

Metaproteomics enables direct identification and quantification of the entire set of proteins expressed by a microbial community at a given time and condition, providing a functional readout that bridges genomic potential (metagenomics) and transcriptional activity (metatranscriptomics) [48]. Unlike DNA- or RNA-based approaches, metaproteomics captures posttranslational modifications, protein stability, and actual enzymatic machinery, thus revealing the true biochemical activities of bacteria, archaea, fungi, and—in some cases—viruses within complex environmental, host-associated, or clinical samples [49].

A typical metaproteomic workflow encompasses sample preparation, LC-MS/MS analysis, and downstream bioinformatic processing to characterize protein expression within complex microbial communities. The protocol typically begins with mechanical cell lysis, protein isolation, enzymatic digestion of proteins into peptides, liquid chromatography-tandem mass spectrometry (LC-MS/MS) in data-dependent (DDA) or data-independent acquisition (DIA) mode [50]. Taxonomic annotation of identified metaproteins was performed using three complementary strategies [51]:

1. Reference database annotation: peptides were searched against NCBI-nr, UniProt, or UniRef using MaxQuant or MetaLab, directly inheriting taxonomic lineage from the best-matching reference sequences. In addition, the complete tryptic peptide list was submitted to UniPept [52], and Peptonizer2000 [53],

which rapidly returned taxonomic distribution via the LCA algorithm. Although typically limited to genus-level or higher resolution due to peptide conservation and database bias, this approach provided a fast, independent validation of the dominant active taxa.

2. Shotgun metagenome-directed database annotation: A custom nonredundant protein database was constructed from the same samples' shotgun metagenomic assemblies (read assembly into contigs using MEGAHIT, or metaSPAdes; contigs predicted with Prodigal; clustered via MMseqs2; and high-quality MAGs classified by GTDB-Tk). Peptides were re-searched against this database with MetaProteomeAnalyzer or MaxQuant, markedly increasing identification rates and taxonomic specificity.
3. Amplicon-directed database annotation: The process typically involves inferring microbial community composition using amplicon analysis pipelines (e.g., QIIME 2 or mothur) with reference databases, such as SILVA or Greengenes; selecting dominant taxa to generate a list of reference organisms; and retrieving the corresponding proteomes of these organisms from public databases, like UniProt or NCBI, to compile a custom protein sequence database. To reduce redundancy, clustered databases like UniRef90 or UniRef50 can be used, though this may further compromise taxonomic resolution.

Final taxonomic assignments were integrated from all three strategies, with priority given to sample-matched metagenomic hits, yielding robust and high-confidence linkage of expressed proteins to active community members. Metaproteomics directly identifies actively functioning enzymes and pathways—such as antibiotic resistance determinants, virulence factors, substrate-degradation machinery, and central carbon metabolism—under real physiological conditions [54]. Because proteins represent the ultimate molecular actors of microbial communities, metaproteomics is increasingly integrated with metagenomics, metatranscriptomics, and metabolomics in multi-omics studies to deliver a system-level understanding of microbial activity, regulation, and ecological impact in diverse habitats.

4.2 Metabolome

Microbial metabolomics enables direct detection and quantification of the complete set of small-molecule metabolites produced and modified by the collective metabolic activity of a microbiome at a specific time and condition [55]. In contrast to metagenomics (genomic feature), metatranscriptomics (transcriptional activity), and metaproteomics (protein expression), metabolomics captures the end products and substrates of biochemical reactions, thereby providing direct insights into the physiological status and functional activity of microbial communities and even host-microbe crosstalk.

A standard microbial metabolomic workflow typically comprises the following steps: metabolites are extracted from biomass or liquid samples (e.g., feces or gut contents) using cold biphasic solvents (e.g., methanol/chloroform/water) or single-phase 80% methanol to simultaneously quench metabolic activity and precipitate proteins. The resulting extracts are typically split for parallel analysis on complementary analytical platforms:

1. Hydrophilic interaction liquid chromatography-mass spectrometry (HILIC-MS/MS) for polar metabolites (e.g., amino acids, organic acids, and nucleotides).
2. Reverse-phase C18 LC-MS/MS for lipids, bile acids, and semi-polar secondary metabolites.
3. Gas chromatography-mass spectrometry (GC-MS) for volatile compounds and central carbon primary metabolites.

Analyses were performed on high-resolution instruments, including Thermo Orbitrap Exploris 480, Q-Exactive series, Sciex TripleTOF 6600, or Bruker timsTOF Pro, which are commonly operated in DDA or DIA acquisition mode, frequently with rapid positive/negative ionization switching to maximize metabolite coverage [55].

Raw data are processed using Compound Discoverer, XCMS, MZmine 3, or MS-DIAL, enabling feature detection, retention-time alignment, and gap filling [55]. Metabolite identification is performed at the metabolomics standards initiative (MSI) level 1 (using authentic chemical standards), level 2 (spectral matching against public databases like GNPS, MoNA, Metlin, or HMDB), or level 3 (annotation by molecular family or compound class). Quantification is typically based on peak area with correction using isotope-labeled internal standards when available.

When integrated with metagenomic data, metabolites can be mapped to microbial pathways using tools such as mummichog2, MetaboAnalyst, or KEGG/MetaCyc modules reconstructed from matched MAGs, helping to infer active microbial functions (e.g., SCFAs production, bile acid transformation). Taxon-specific metabolic contributions may be estimated using correlation-based methods or specialized tools like MelonnPan and tax4met.

Because metabolites represent the ultimate downstream output of cellular regulatory processes, metabolomics is increasingly integrated with metagenomics, metatranscriptomics, and metaproteomics in multi-omics frameworks to identify specific biomarker molecules associated with structural and metabolic states. Notably, its approaches are targeted and untargeted, relying on the research objectives and questions. Metabolomics directly reveals key bioactive molecules—such as SCFAs, neurotransmitters, quorum-sensing signals, virulence-associated metabolites, and antibiotic resistance–related intermediates—that mediate microbial community behavior, host-microbiota interactions, and ecosystem functioning.

5 Single-Cell Microbial Omics

Single-cell microbial omics resolves genomic, transcriptomic, and epigenetic heterogeneity of individual cells from complex communities, revealing rare taxa, strain-level microdiversity, and functional specialization hidden in bulk analyses [56]. Thus, it provides critical insight into the cellular mechanisms governing microbial adaptation and ecosystem function. However, scRNA-seq has made significant contributions to mammalian biology, but applying these methodologies to microbes presents several challenges. These include lower mRNA content, shorter half-lives of mRNA, lack of poly(A) tails, and the rigid structure of microbial cell walls, which hinders their lysis using protocols designed for mammalian cells [8, 57]. To address these challenges, several methods for microbial scRNA-seq have been developed in recent years. These methods can be broadly categorized into two groups:

1. Sequencing-based approaches. Microbial scRNA-seq techniques involve fixing microbes to prevent RNA degradation, enzymatically digesting cell walls for permeabilization, and using random priming or gene-specific probes for mRNA capture. Strategies like Cas9 nuclease or RNase H-based methods help reduce rRNA contamination. Methods like PETRI-seq, microSPLiT, BaSSSH-seq, and mscRNA-seq use combinatorial indexing for high-throughput, scalable profiling. Probe-based techniques (e.g., ProBac-seq [58]) and droplet-based methods (e.g., smRandom-seq [59], BacDrop) offer efficient mRNA capture, with potential for cross-kingdom profiling in microbes and fungi.
2. Fluorescence in situ hybridization (FISH) based techniques. Microscopic methods like FISH offer single-cell resolution and spatial data, enabling analysis from microbial communities to subcellular mRNA localization. Unlike scRNA-seq, FISH can be applied to fixed cells, biofilms, or host tissues. Technologies like parseqFISH and MERFISH allow gene expression profiling and visualization of spatial heterogeneity in biofilms, as well as the subcellular localization of specific genes. However, bacterial mRNA's high transcript density poses challenges for transcriptome-wide combinatorial barcoding. Expansion microscopy and bacterial-MERFISH help address this by expanding bacteria for better resolution.

So far, no single protocol works optimally for all microbes, especially for complex, polymicrobial communities. Developing robust, universal wet-lab and bioinformatic workflows is an area of active development. Here, a brief example is presented to introduce the analytical method [60]:

1. Reference pangenome construction: A microbial pangenome database was constructed by integrating assembled high-quality MAGs.
2. Sample collection: Microbial cells were fixed with 4% paraformaldehyde.
3. Library preparation: Reverse transcription was performed with pre-barcoded random primers and dA tailing. Single-cell droplets were generated and encapsulated for amplification and extension using a microfluidic platform, followed by sequencing.
4. Microbiome scRNA-seq upstream analysis: Data quality control and filtering were performed using Trimmomatic. Gene abundance was calculated, and high-quality single cells were screened by aligning clean reads to the microbial pangenome using BWA [61] and BEDTools [62], creating the single-cell gene expression matrix. Clean reads were classified by Kraken 2 against a customized genome OTUs (gOTUs) database derived from microbial pangenome, with cells assigned to seven phylogenetic levels (domain, phylum, class, order, family, genus, and species). The normalized read counts for each taxonomy were estimated using Bracken.
5. Microbiome scRNA-seq downstream analysis: The single-cell gene expression matrix was imported into Seurat for further analysis. Cell clustering and functional cluster identification were performed, followed by interaction analysis and pathway functional activity analysis. Pseudo-time analysis was conducted to investigate cell state transitions, and enrichment analysis was used to assess the functional significance of differentially expressed genes.

Single-cell microbial omics directly reveals rare biosphere members, intracellular viruses, plasmid distribution, metabolic specialization (e.g., nitrogen-fixing versus non-fixing cells within the same species), phage-host interactions, and functional heterogeneity driving community resilience or dysbiosis. Increasingly integrated with bulk metagenomics, metatranscriptomics, and stable-isotope probing, single-cell approaches are transforming our understanding of microbial ecophysiology, evolution, and functional roles in natural and host-associated ecosystems.

6 Artificial Intelligence and Microbiome Research

Artificial intelligence (AI) is currently driving a transformative revolution in science and is poised to profoundly reshape microbiome research in the coming years [9]. As Georg K. Gerber aptly framed in his perspective, "AI in microbiome research: Where have we been, and where are we going?"—the field stands at a pivotal inflection point, with emerging machine-learning (ML) and deep-learning approaches promising to unlock new levels of insight into microbial ecology, function, and host-microbiome interactions.

Traditional methods of microbial identification often depend on statistical techniques, such as t-tests, ANOVA, and network analysis algorithms, to identify differentially abundant taxa or disrupted network modules. Recently, AI-based approaches have become increasingly valuable for microbiologists, offering powerful tools for analyzing complex biological data [63]. These techniques are typically divided into unsupervised learning and supervised learning, each serving different purposes in microbiological research.

1. Unsupervised learning: Methods like principal component analysis (PCA) and clustering help identify patterns in high-dimensional microbial data, such as distinguishing microbial communities or functional groups. Traditional machine learning methods, like regression, random forests, support vector machines (SVMs), and gradient boosting, are also used for tasks, such as predicting microbial behaviors or classifying species.
2. Supervised learning: Methods are applied when known outcomes are available, allowing predictive modeling and classification. Neural networks, including multilayer perceptrons (MLP), convolutional neural networks (CNN), and recurrent neural networks (RNN), can model complex, nonlinear relationships in microbial data, from gene expression to disease prediction. Graph convolutional networks (GCN) are particularly useful for analyzing microbial interaction networks, while autoencoders help with dimensionality reduction and feature extraction in high-dimensional datasets. These methods are especially useful in applications, such as microbiome analysis, disease prediction, and antimicrobial resistance studies.

The integration of ML into microbiology holds great promise for advancing microbiome research, disease diagnostics, and drug discovery. By analyzing microbial community structures and their relationships with health outcomes, ML can identify disease biomarkers and enhance personalized medicine [64]. AI-driven approaches, such as neural networks and clustering algorithms, are increasingly used to uncover patterns in microbiome data, facilitating the identification of a core microbiome signature that can serve as an indicator of health [65]. Additionally, fecal microbial load, a major determinant of gut microbiome variation, plays a key role in disease associations and highlights the need for AI-driven approaches to better understand these complex relationships and improve health outcomes [66]. For example, the synergy between AI and probiotics is emerging as a promising avenue for health promotion and industrial innovation [67]. However, challenges remain, including issues with data quality, overfitting, and model interpretability. Future advancements in ML will focus on integrating multi-omics data, improving model transparency, and utilizing real-time clinical data to develop more accurate, microbiome-based treatments.

7 Conclusion and Prospection

The study of microbiomes has rapidly evolved from traditional culture-based techniques to sophisticated sequencing technologies (Table 2), enabling comprehensive insights into microbial diversity, function, and their dynamic interactions with hosts. Advances in metagenomics, transcriptomics, proteomics, and metabolomics have provided an unprecedented understanding of microbial communities and their roles in health and disease.

Single-cell microbial omics has emerged as a transformative approach to unravel the molecular and functional diversity within microbial communities, offering deeper insights into rare taxa, strain-level variations, and functional specializations. When integrated with AI-driven techniques, such as machine learning and deep learning, microbiome research is entering a new era, with AI promising to revolutionize the way we analyze complex biological data, uncover hidden patterns, and predict microbial behaviors and health outcomes.

The growing interest in host-microbiota interactions highlights the importance of understanding how the gut microbiome influences host physiology [73], with emerging methodologies such as genome-wide association studies (GWAS), microbial composition, scRNA-seq of host cells [74], and spatial host-microbiome sequencing (SHM-seq), providing critical insights into these complex relationships [75]. Additionally, recent advancements in the study of host-microbiota interactions, particularly in contexts like intestinal stem cell homeostasis [76] and spatial microbiome profiling [77], underscore the complexity of the microbiome's influence on host biology. These developments offer valuable insights into the role of microbial communities in maintaining tissue integrity and regulating host immune and metabolic functions.

Table 2
Key software tools commonly used in microbial bioinformatics analysis

Analysis step	Software	Function	Refs
Quality control & preprocessing	Trimmomatic	Fast read trimming and adapter removal tool	[19]
	Fastp	Ultrafast all-in-one read quality control, adapter trimming and filtering tool	[18]
Rrna filtering	SortMeRNA	Ultrafast rrna read filtering and taxonomic assignment tool	[39]
Read-based taxonomic profiling	Kraken 2	Fast taxonomic classification of metagenomic reads	[23]
	Bracken	Accurate taxonomic abundance estimation from kraken classification results	[24]
	MetaPhlAn 4	Taxonomic profiling from metagenomic data	[22]
De novo assembly	MEGAHIT	A fast and scalable metagenome assembler	[26]
	metaSPAdes	Strain-level metagenomic assembler	[27]
Metatranscriptome assembly	Trinity	Leading de novo RNA-seq assembler for transcriptomes without reference	[43]
	rnaSPAdes	De novo RNA-seq assembler optimized for isoforms and strain-level variants	[44]
Read mapping & coverage	BWA	Ultra-fast, memory-efficient short-read aligner for DNA-seq	[61]
	STAR	Ultra-fast, splice-aware RNA-seq aligner with high sensitivity	[41]
	CoverM	Fast, accurate read coverage calculator for genomes and mags	[68]
Transcript quantification	Kallisto	Ultra-fast, lightweight pseudoalignment for RNA-seq quantification	[69]
	Salmon	Ultra-fast, accurate transcript quantification from RNA-seq (alignment-free)	[70]
Assembly quality assessment	QUAST	Genome assembly, metagenome assembly	[28]
Binning refinement & dereplication	dRep	Fast genome dereplication and MAG quality comparison tool	[32]
	MetaWRAP	All-in-one metagenomics pipeline for QC, assembly, binning, profiling, and annotation	[31]
Mag taxonomy	GTDB-Tk	Taxonomic classification of mags using GTDB reference tree	[34]
Gene prediction	Prodigal	Fast, accurate prokaryotic gene predictor	[71]
Functional annotation	eggNOG-mapper	Fast protein annotation via orthology and functional transfer	[35]
	DIAMOND	Ultrafast protein alignment tool (BLASTX/BLASTP alternative)	[36]
Pathway & functional profiling	HUMAnN 4	Fast metagenomic functional profiling and pathway abundance quantification tool	[25]

(continued)

Table 2 (continued)

Analysis step	Software	Function	Refs
Sequence clustering	CD-HIT	Fast sequence clustering and redundancy removal tool	[30]
	MMseqs2	Ultrafast sequence search and clustering toolkit (sensitive alternative to BLAST/CD-HIT)	[46]
16S amplicon analysis	QIIME2	Microbial community analysis, diversity analysis, taxonomic profiling	[11]
	VSEARCH	Microbial ecology analysis, community composition, diversity analysis	[12]
	USEARCH	A fast, open-source search and clustering engine for microbial analysis	[13]
	PICRUSt2	Predicting the functional potential of the microbiome from marker gene sequences	[72]

Despite the considerable potential, several hurdles remain, including challenges with data quality, model interpretability, and the integration of multi-omics data for a more comprehensive understanding of host-microbiome dynamics. As AI continues to advance, its integration with other omics technologies will drive further breakthroughs in microbial ecology, host-microbiome interactions, and personalized medicine. Ultimately, the continued development of more robust, universal experimental workflows, coupled with AI-driven analyses, holds the key to unlocking novel therapeutic strategies and innovations in microbiome-based health promotion.

Conflict of Interest

This work was supported by the National Natural Science Foundation of China (Grant No. 32472943) and the National Science and Technology Major Project on Biological Breeding (Grant No. 2023ZD040430103). The authors have no conflicts of interest to declare that are relevant to the content of this chapter.

References

1. Fan Y, Pedersen O (2021) Gut microbiota in human metabolic health and disease. Nat Rev Microbiol 19(1):55–71. https://doi.org/10.1038/s41579-020-0433-9
2. Morais LH, Schreiber HL, Mazmanian SK (2021) The gut microbiota-brain axis in behaviour and brain disorders. Nat Rev Microbiol 19(4):241–255. https://doi.org/10.1038/s41579-020-00460-0
3. Wang L, Liu Q, Chen Y, Zheng X, Wang C, Qi Y, Dong Y, Xiao Y, Chen C, Chen T, Huang Q, Zhai Z, Long C, Yang H, Li J, Wang L, Zhang G, Liao P, Liu Y-X, Huang P, Huang J, Wang Q, Chu H, Yin J, Yin Y (2022) Antioxidant potential of Pediococcus pentosaceus strains from the sow milk bacterial collection in weaned piglets. Microbiome 10(1). https://doi.org/10.1186/s40168-022-01278-z

4. Krautkramer KA, Fan J, Backhed F (2021) Gut microbial metabolites as multi-kingdom intermediates. Nat Rev Microbiol 19(2):77–94. https://doi.org/10.1038/s41579-020-0438-4
5. Chen S, Luo S, Yan C (2021) Gut microbiota implications for health and welfare in farm animals: a review. Animals (Basel) 12(1):93. https://doi.org/10.3390/ani12010093
6. Duan D, Wang M, Han J, Li M, Wang Z, Zhou S, Xin W, Li X (2024) Advances in multi-omics integrated analysis methods based on the gut microbiome and their applications. Front Microbiol 15:1509117. https://doi.org/10.3389/fmicb.2024.1509117
7. Kumari P, Prakash P, Yadav S, Saran V (2022) Microbiome analysis: an emerging forensic investigative tool. Forensic Sci Int 340: 111462. https://doi.org/10.1016/j.forsciint.2022.111462
8. Pountain AW, Yanai I (2025) Dissecting microbial communities with single-cell transcriptome analysis. Science 389(6764):eadp6252. https://doi.org/10.1126/science.adp6252
9. Gerber GK (2024) AI in microbiome research: where have we been,where are we going? Cell Host Microbe 32(8):1230–1234. https://doi.org/10.1016/j.chom.2024.07.021
10. Ward DM, Weller R, Bateson MM (1990) 16S rRNA sequences reveal numerous uncultured microorganisms in a natural community. Nature 345(6270):63–65. https://doi.org/10.1038/345063a0
11. Bolyen E, Rideout JR, Dillon MR, Bokulich NA, Abnet CC, Al-Ghalith GA, Alexander H, Alm EJ, Arumugam M, Asnicar F, Bai Y, Bisanz JE, Bittinger K, Brejnrod A, Brislawn CJ, Brown CT, Callahan BJ, Caraballo-Rodriguez AM, Chase J, Cope EK, Da Silva R, Diener C, Dorrestein PC, Douglas GM, Durall DM, Duvallet C, Edwardson CF, Ernst M, Estaki M, Fouquier J, Gauglitz JM, Gibbons SM, Gibson DL, Gonzalez A, Gorlick K, Guo J, Hillmann B, Holmes S, Holste H, Huttenhower C, Huttley GA, Janssen S, Jarmusch AK, Jiang L, Kaehler BD, Kang KB, Keefe CR, Keim P, Kelley ST, Knights D, Koester I, Kosciolek T, Kreps J, Langille MGI, Lee J, Ley R, Liu YX, Loftfield E, Lozupone C, Maher M, Marotz C, Martin BD, McDonald D, McIver LJ, Melnik AV, Metcalf JL, Morgan SC, Morton JT, Naimey AT, Navas-Molina JA, Nothias LF, Orchanian SB, Pearson T, Peoples SL, Petras D, Preuss ML, Pruesse E, Rasmussen LB, Rivers A, Robeson MS, 2nd, Rosenthal P, Segata N, Shaffer M, Shiffer A, Sinha R, Song SJ, Spear JR, Swafford AD, Thompson LR, Torres PJ, Trinh P, Tripathi A, Turnbaugh PJ, Ul-Hasan S, van der Hooft JJJ, Vargas F, Vazquez-Baeza Y, Vogtmann E, von Hippel M, Walters W, Wan Y, Wang M, Warren J, Weber KC, Williamson CHD, Willis AD, Xu ZZ, Zaneveld JR, Zhang Y, Zhu Q, Knight R, Caporaso JG (2019) Reproducible, interactive, scalable and extensible microbiome data science using QIIME 2. Nat Biotechnol 37 (8):852–857. doi:https://doi.org/10.1038/s41587-019-0209-9
12. Rognes T, Flouri T, Nichols B, Quince C, Mahe F (2016) VSEARCH: a versatile open source tool for metagenomics. PeerJ 4:e2584. https://doi.org/10.7717/peerj.2584
13. Edgar RC (2013) UPARSE: highly accurate OTU sequences from microbial amplicon reads. Nat Methods 10(10):996–998. https://doi.org/10.1038/nmeth.2604
14. Nguyen NP, Warnow T, Pop M, White B (2016) A perspective on 16S rRNA operational taxonomic unit clustering using sequence similarity. NPJ Biofilms Microbiomes 2:16004. https://doi.org/10.1038/npjbiofilms.2016.4
15. Liu YX, Qin Y, Chen T, Lu M, Qian X, Guo X, Bai Y (2021) A practical guide to amplicon and metagenomic analysis of microbiome data. Protein Cell 12(5):315–330. https://doi.org/10.1007/s13238-020-00724-8
16. Johnson JS, Spakowicz DJ, Hong BY, Petersen LM, Demkowicz P, Chen L, Leopold SR, Hanson BM, Agresta HO, Gerstein M, Sodergren E, Weinstock GM (2019) Evaluation of 16S rRNA gene sequencing for species and strain-level microbiome analysis. Nat Commun 10(1):5029. https://doi.org/10.1038/s41467-019-13036-1
17. Franzosa EA, McIver LJ, Rahnavard G, Thompson LR, Schirmer M, Weingart G, Lipson KS, Knight R, Caporaso JG, Segata N, Huttenhower C (2018) Species-level functional profiling of metagenomes and metatranscriptomes. Nat Methods 15(11):962–968. https://doi.org/10.1038/s41592-018-0176-y
18. Chen S (2023) Ultrafast one-pass FASTQ data preprocessing, quality control, and deduplication using Fastp. iMeta 2(2):e107. https://doi.org/10.1002/imt2.107
19. Bolger AM, Lohse M, Usadel B (2014) Trimmomatic: a flexible trimmer for Illumina sequence data. Bioinformatics 30(15): 2114–2120. https://doi.org/10.1093/bioinformatics/btu170
20. Langmead B, Salzberg SL (2012) Fast gapped-read alignment with Bowtie 2. Nat Methods 9(4):357–359. https://doi.org/10.1038/nmeth.1923

21. Bai D, Chen T, Xun J, Ma C, Luo H, Yang H, Cao C, Cao X, Cui J, Deng YP, Deng Z, Dong W, Dong W, Du J, Fang Q, Fang W, Fang Y, Fu F, Fu M, Fu YT, Gao H, Ge J, Gong Q, Gu L, Guo P, Guo Y, Hai T, Liu H, He J, He ZY, Hou H, Huang C, Ji S, Jiang C, Jiang GL, Jiang L, Jin LN, Kan Y, Kang D, Kou J, Lam KL, Li C, Li C, Li F, Li L, Li M, Li X, Li Y, Li ZT, Liang J, Lin Y, Liu C, Liu D, Liu F, Liu J, Liu T, Liu T, Liu X, Liu Y, Liu B, Liu M, Lou W, Luan Y, Luo Y, Lv H, Ma T, Mai Z, Mo J, Niu D, Pan Z, Qi H, Shi Z, Song C, Sun F, Sun Y, Tian S, Wan X, Wang G, Wang H, Wang H, Wang H, Wang J, Wang J, Wang K, Wang L, Wang SK, Wang X, Wang Y, Xiao Z, Xing H, Xu Y, Yan SY, Yang L, Yang S, Yang Y, Yao X, Yousuf S, Yu H, Lei Y, Yuan Z, Zeng M, Zhang C, Zhang C, Zhang H, Zhang J, Zhang N, Zhang T, Zhang YB, Zhang Y, Zhang Z, Zhou M, Zhou Y, Zhu C, Zhu L, Zhu Y, Zhu Z, Zou H, Zuo A, Dong W, Wen T, Chen S, Li G, Gao Y, Liu YX (2025) Easy metagenome: a user-friendly and flexible pipeline for shotgun metagenomic analysis in microbiome research. iMeta 4(1):e70001. https://doi.org/10.1002/imt2.70001

22. Blanco-Miguez A, Beghini F, Cumbo F, McIver LJ, Thompson KN, Zolfo M, Manghi P, Dubois L, Huang KD, Thomas AM, Nickols WA, Piccinno G, Piperni E, Puncochar M, Valles-Colomer M, Tett A, Giordano F, Davies R, Wolf J, Berry SE, Spector TD, Franzosa EA, Pasolli E, Asnicar F, Huttenhower C, Segata N (2023) Extending and improving metagenomic taxonomic profiling with uncharacterized species using MetaPhlAn 4. Nat Biotechnol 41(11):1633–1644. https://doi.org/10.1038/s41587-023-01688-w

23. Wood DE, Lu J, Langmead B (2019) Improved metagenomic analysis with Kraken 2. Genome Biol 20(1):257. https://doi.org/10.1186/s13059-019-1891-0

24. Lu J, Breitwieser FP, Thielen P, Salzberg SL (2017) Bracken: estimating species abundance in metagenomics data. Peer J Comput Sci 3. https://doi.org/10.7717/peerj-cs.104

25. Beghini F, McIver LJ, Blanco-Miguez A, Dubois L, Asnicar F, Maharjan S, Mailyan A, Manghi P, Scholz M, Thomas AM, Valles-Colomer M, Weingart G, Zhang Y, Zolfo M, Huttenhower C, Franzosa EA, Segata N (2021) Integrating taxonomic, functional, and strain-level profiling of diverse microbial communities with bioBakery 3. elife 10. https://doi.org/10.7554/eLife.65088

26. Li D, Luo R, Liu CM, Leung CM, Ting HF, Sadakane K, Yamashita H, Lam TW (2016) MEGAHIT v1. 0: a fast and scalable metagenome assembler driven by advanced methodologies and community practices. Methods 102: 3–11. https://doi.org/10.1016/j.ymeth.2016.02.020

27. Prjibelski A, Antipov D, Meleshko D, Lapidus A, Korobeynikov A (2020) Using SPAdes De Novo assembler. Curr Protoc Bioinform 70(1):e102. https://doi.org/10.1002/cpbi.102

28. Gurevich A, Saveliev V, Vyahhi N, Tesler G (2013) QUAST: quality assessment tool for genome assemblies. Bioinformatics 29(8): 1072–1075. https://doi.org/10.1093/bioinformatics/btt086

29. Hyatt D, LoCascio PF, Hauser LJ, Uberbacher EC (2012) Gene and translation initiation site prediction in metagenomic sequences. Bioinformatics 28(17):2223–2230. https://doi.org/10.1093/bioinformatics/bts429

30. Fu L, Niu B, Zhu Z, Wu S, Li W (2012) CD-HIT: accelerated for clustering the next-generation sequencing data. Bioinformatics 28(23):3150–3152. https://doi.org/10.1093/bioinformatics/bts565

31. Uritskiy GV, DiRuggiero J, Taylor J (2018) MetaWRAP-a flexible pipeline for genome-resolved metagenomic data analysis. Microbiome 6(1):158. https://doi.org/10.1186/s40168-018-0541-1

32. Olm MR, Brown CT, Brooks B, Banfield JF (2017) dRep: a tool for fast and accurate genomic comparisons that enables improved genome recovery from metagenomes through de-replication. ISME J 11(12):2864–2868. https://doi.org/10.1038/ismej.2017.126

33. Chklovski A, Parks DH, Woodcroft BJ, Tyson GW (2023) CheckM2: a rapid, scalable and accurate tool for assessing microbial genome quality using machine learning. Nat Methods 20(8):1203–1212. https://doi.org/10.1038/s41592-023-01940-w

34. Parks DH, Chuvochina M, Waite DW, Rinke C, Skarshewski A, Chaumeil PA, Hugenholtz P (2018) A standardized bacterial taxonomy based on genome phylogeny substantially revises the tree of life. Nat Biotechnol 36(10): 996–1004. https://doi.org/10.1038/nbt.4229

35. Cantalapiedra CP, Hernandez-Plaza A, Letunic I, Bork P, Huerta-Cepas J (2021) eggNOG-mapper v2: functional annotation, orthology assignments, and domain prediction at the metagenomic scale. Mol Biol Evol

38(12):5825–5829. https://doi.org/10.1093/molbev/msab293

36. Buchfink B, Reuter K, Drost HG (2021) Sensitive protein alignments at tree-of-life scale using DIAMOND. Nat Methods 18(4): 366–368. https://doi.org/10.1038/s41592-021-01101-x
37. Alcock BP, Huynh W, Chalil R, Smith KW, Raphenya Amogelang R, Wlodarski MA, Edalatmand A, Petkau A, Syed SA, Tsang KK, Baker SJC, Dave M, McCarthy Madeline C, Mukiri KM, Nasir JA, Golbon B, Imtiaz H, Jiang X, Kaur K, Kwong M, Liang ZC, Niu KC, Shan P, Yang JYJ, Gray Kristen L, Hoad GR, Jia B, Bhando T, Carfrae Lindsey A, Farha Maya A, French S, Gordzevich R, Rachwalski K, Tu Megan M, Bordeleau E, Dooley D, Griffiths E, Zubyk HL, Brown ED, Maguire F, Beiko Robert G, Hsiao WWL, Brinkman FSL, Van Domselaar G, McArthur AG (2023) CARD 2023: expanded curation, support for machine learning, and resistome prediction at the comprehensive antibiotic resistance database. Nucleic Acids Res 51(D1):D690–D699. https://doi.org/10.1093/nar/gkac920
38. Bashiardes S, Zilberman-Schapira G, Elinav E (2016) Use of metatranscriptomics in microbiome research. Bioinform Biol Insights 10: 19–25. https://doi.org/10.4137/BBI.S34610
39. Kopylova E, Noe L, Touzet H (2012) SortMeRNA: fast and accurate filtering of ribosomal RNAs in metatranscriptomic data. Bioinformatics 28(24):3211–3217. https://doi.org/10.1093/bioinformatics/bts611
40. Kim D, Paggi JM, Park C, Bennett C, Salzberg SL (2019) Graph-based genome alignment and genotyping with HISAT2 and HISAT-genotype. Nat Biotechnol 37(8):907–915. https://doi.org/10.1038/s41587-019-0201-4
41. Dobin A, Davis CA, Schlesinger F, Drenkow J, Zaleski C, Jha S, Batut P, Chaisson M, Gingeras TR (2013) STAR: ultrafast universal RNA-seq aligner. Bioinformatics 29(1):15–21. https://doi.org/10.1093/bioinformatics/bts635
42. Zhang Y, Thompson KN, Branck T, Yan Y, Nguyen LH, Franzosa EA, Huttenhower C (2021) Metatranscriptomics for the human microbiome and microbial community functional profiling. Annu Rev Biomed Data Sci 4: 279–311. https://doi.org/10.1146/annurev-biodatasci-031121-103035
43. Haas BJ, Papanicolaou A, Yassour M, Grabherr M, Blood PD, Bowden J, Couger MB, Eccles D, Li B, Lieber M, MacManes MD, Ott M, Orvis J, Pochet N, Strozzi F, Weeks N, Westerman R, William T, Dewey CN, Henschel R, LeDuc RD, Friedman N, Regev A (2013) De novo transcript sequence reconstruction from RNA-seq using the Trinity platform for reference generation and analysis. Nat Protoc 8(8):1494–1512. https://doi.org/10.1038/nprot.2013.084
44. Bushmanova E, Antipov D, Lapidus A, Prjibelski AD (2019) rnaSPAdes: a de novo transcriptome assembler and its application to RNA-Seq data. Gigascience 8(9). https://doi.org/10.1093/gigascience/giz100
45. Marx V (2023) Method of the year: long-read sequencing. Nat Methods 20(1):6–11. https://doi.org/10.1038/s41592-022-01730-w
46. Kallenborn F, Chacon A, Hundt C, Sirelkhatim H, Didi K, Cha S, Dallago C, Mirdita M, Schmidt B, Steinegger M (2025) GPU-accelerated homology search with MMseqs2. Nat Methods 22(10):2024–2027. https://doi.org/10.1038/s41592-025-02819-8
47. Bryant DM, Johnson K, DiTommaso T, Tickle T, Couger MB, Payzin-Dogru D, Lee TJ, Leigh ND, Kuo TH, Davis FG, Bateman J, Bryant S, Guzikowski AR, Tsai SL, Coyne S, Ye WW, Freeman RM Jr, Peshkin L, Tabin CJ, Regev A, Haas BJ, Whited JL (2017) A tissue-mapped axolotl De Novo transcriptome enables identification of limb regeneration factors. Cell Rep 18(3):762–776. https://doi.org/10.1016/j.celrep.2016.12.063
48. Valdes-Mas R, Leshem A, Zheng D, Cohen Y, Kern L, Zmora N, He Y, Katina C, Eliyahu-Miller S, Yosef-Hevroni T, Richman L, Raykhel B, Allswang S, Better R, Shmueli M, Saftien A, Cullin N, Slamovitz F, Ciocan D, Ouyang KS, Mor U, Dori-Bachash M, Molina S, Levin Y, Atarashi K, Jona G, Puschhof J, Harmelin A, Stettner N, Chen M, Suez J, Honda K, Lieb W, Bang C, Kori M, Maharshak N, Merbl Y, Shibolet O, Halpern Z, Shouval DS, Shamir R, Franke A, Abdeen SK, Shapiro H, Savidor A, Elinav E (2025) Metagenome-informed metaproteomics of the human gut microbiome, host, and dietary exposome uncovers signatures of health and inflammatory bowel disease. Cell 188(4): 1062–1083 e1036. https://doi.org/10.1016/j.cell.2024.12.016
49. Tanca A, Palomba A, Fraumene C, Pagnozzi D, Manghina V, Deligios M, Muth T, Rapp E, Martens L, Addis MF, Uzzau S (2016) The impact of sequence database choice on metaproteomic results in gut

microbiota studies. Microbiome 4(1):51. https://doi.org/10.1186/s40168-016-0196-8

50. Gomez-Varela D, Xian F, Grundtner S, Sondermann JR, Carta G, Schmidt M (2023) Increasing taxonomic and functional characterization of host-microbiome interactions by DIA-PASEF metaproteomics. Front Microbiol 14:1258703. https://doi.org/10.3389/fmicb.2023.1258703
51. Nebauer DJ, Pearson LA, Neilan BA (2024) Critical steps in an environmental metaproteomics workflow. Environ Microbiol 26(5): e16637. https://doi.org/10.1111/1462-2920.16637
52. Gurdeep Singh R, Tanca A, Palomba A, Van der Jeugt F, Verschaffelt P, Uzzau S, Martens L, Dawyndt P, Mesuere B (2019) Unipept 4.0: functional analysis of metaproteome data. J Proteome Res 18(2):606–615. https://doi.org/10.1021/acs.jproteome.8b00716
53. Holstein T, Verschaffelt P, Van Den Bossche T, Van de Vyver S, Martens L, Mesuere B, Muth T (2025) https://doi.org/10.1101/2024.05.20.594958
54. Van Den Bossche T, Armengaud J, Benndorf D, Blakeley-Ruiz JA, Brauer M, Cheng K, Creskey M, Figeys D, Grenga L, Griffin TJ, Henry C, Hettich RL, Holstein T, Jagtap PD, Jehmlich N, Kim J, Kleiner M, Kunath BJ, Malliet X, Martens L, Mehta S, Mesuere B, Ning Z, Tanca A, Uzzau S, Verschaffelt P, Wang J, Wilmes P, Zhang X, Zhang X, Li L, Metaproteomics I (2025) The microbiologist's guide to metaproteomics. Imeta 4(3):e70031. https://doi.org/10.1002/imt2.70031
55. Gonzalez-Riano C, Dudzik D, Garcia A, Gil-de-la-Fuente A, Gradillas A, Godzien J, Lopez-Gonzalvez A, Rey-Stolle F, Rojo D, Ruperez FJ, Saiz J, Barbas C (2020) Recent developments along the analytical process for metabolomics workflows. Anal Chem 92(1): 203–226. https://doi.org/10.1021/acs.analchem.9b04553
56. Zhang Y, Gao J, Huang Y, Wang J (2018) Recent developments in single-cell RNA-Seq of microorganisms. Biophys J 115(2): 173–180. https://doi.org/10.1016/j.bpj.2018.06.008
57. Imdahl F, Saliba AE (2020) Advances and challenges in single-cell RNA-seq of microbial communities. Curr Opin Microbiol 57:102–110. https://doi.org/10.1016/j.mib.2020.10.001
58. McNulty R, Sritharan D, Pahng SH, Meisch JP, Liu S, Brennan MA, Saxer G, Hormoz S, Rosenthal AZ (2023) Probe-based bacterial single-cell RNA sequencing predicts toxin regulation. Nat Microbiol 8(5):934–945. https://doi.org/10.1038/s41564-023-01348-4
59. Xu Z, Wang Y, Cai W, Chen Y, Wang Y (2025) Single microorganism RNA sequencing of microbiomes using smRandom-Seq. Nat Protoc. https://doi.org/10.1038/s41596-025-01181-5
60. Jia M, Zhu S, Xue M-Y, Chen H, Xu J, Song M, Tang Y, Liu X, Tao Y, Zhang T, Liu J-X, Wang Y, Sun H-Z (2024) Single-cell transcriptomics across 2,534 microbial species reveals functional heterogeneity in the rumen microbiome. Nat Microbiol 9(7):1884–1898. https://doi.org/10.1038/s41564-024-01723-9
61. Li H, Durbin R (2010) Fast and accurate long-read alignment with Burrows–Wheeler transform. Bioinformatics 26(5):589–595. https://doi.org/10.1093/bioinformatics/btp698
62. Quinlan AR, Hall IM (2010) BEDTools: a flexible suite of utilities for comparing genomic features. Bioinformatics 26(6):841–842. https://doi.org/10.1093/bioinformatics/btq033
63. Asnicar F, Thomas AM, Passerini A, Waldron L, Segata N (2024) Machine learning for microbiologists. Nat Rev Microbiol 22(4): 191–205. https://doi.org/10.1038/s41579-023-00984-1
64. Xiao L, Zhao F (2024) Exploring the frontier of microbiome biomarker discovery with artificial intelligence. Natl Sci Rev 11(11):nwae325. https://doi.org/10.1093/nsr/nwae325
65. Wu G, Xu T, Zhao N, Lam YY, Ding X, Wei D, Fan J, Shi Y, Li X, Li M, Ji S, Wang X, Fu H, Zhang F, Shi Y, Zhang C, Peng Y, Zhao L (2024) A core microbiome signature as an indicator of health. Cell 187(23):6550–6565. e6511. https://doi.org/10.1016/j.cell.2024.09.019
66. Nishijima S, Stankevic E, Aasmets O, Schmidt TSB, Nagata N, Keller MI, Ferretti P, Juel HB, Fullam A, Robbani SM, Schudoma C, Hansen JK, Holm LA, Israelsen M, Schierwagen R, Torp N, Telzerow A, Hercog R, Kandels S, Hazenbrink DHM, Arumugam M, Bendtsen F, Brons C, Fonvig CE, Holm JC, Nielsen T, Pedersen JS, Thiele MS, Trebicka J, Org E, Krag A, Hansen T, Kuhn M, Bork P, Galaxy, MicrobLiver C (2025) Fecal microbial load is a major determinant of gut microbiomevariation and a confounder for disease associations. Cell 188(1):222–236. e215. https://doi.org/10.1016/j.cell.2024.10.022
67. Han X, Liu Q, Li Y, Zhang M, Liu K, Kwok L-Y, Zhang H, Zhang W (2025) Synergizing

artificial intelligence and probiotics: a comprehensive review of emerging applications in health promotion and industrial innovation. Trends Food Sci Technol 159:104938. https://doi.org/10.1016/j.tifs.2025.104938

68. Aroney STN, Newell RJP, Nissen JN, Camargo AP, Tyson GW, Woodcroft BJ (2025) CoverM: read alignment statistics for metagenomics. Bioinformatics 41(4). https://doi.org/10.1093/bioinformatics/btaf147
69. Bray NL, Pimentel H, Melsted P, Pachter L (2016) Near-optimal probabilistic RNA-seq quantification. Nat Biotechnol 34(5): 525–527. https://doi.org/10.1038/nbt.3519
70. Patro R, Duggal G, Love MI, Irizarry RA, Kingsford C (2017) Salmon provides fast and bias-aware quantification of transcript expression. Nat Methods 14(4):417–419. https://doi.org/10.1038/nmeth.4197
71. Hyatt D, Chen GL, Locascio PF, Land ML, Larimer FW, Hauser LJ (2010) Prodigal: prokaryotic gene recognition and translation initiation site identification. BMC Bioinform 11: 119. https://doi.org/10.1186/1471-2105-11-119
72. Douglas GM, Maffei VJ, Zaneveld JR, Yurgel SN, Brown JR, Taylor CM, Huttenhower C, Langille MGI (2020) PICRUSt2 for prediction of metagenome functions. Nat Biotechnol 38(6):685–688. https://doi.org/10.1038/s41587-020-0548-6
73. Sonnert ND, Rosen CE, Ghazi AR, Franzosa EA, Duncan-Lowey B, Gonzalez-Hernandez JA, Huck JD, Yang Y, Dai Y, Rice TA, Nguyen MT, Song D, Cao Y, Martin AL, Bielecka AA, Fischer S, Guan C, Oh J, Huttenhower C, Ring AM, Palm NW (2024) A host–microbiota interactome reveals extensive transkingdom connectivity. Nature 628(8006):171–179. https://doi.org/10.1038/s41586-024-07162-0
74. Li J, Ma Y, Cao Y, Zheng G, Ren Q, Chen C, Zhu Q, Zhou Y, Lu Y, Zhang Y, Deng C, Chen WH, Su J (2025) Integrating microbial GWAS and single-cell transcriptomics reveals associations between host cell populations and the gut microbiome. Nat Microbiol 10(5): 1210–1226. https://doi.org/10.1038/s41564-025-01978-w
75. Lotstedt B, Strazar M, Xavier R, Regev A, Vickovic S (2024) Spatial host–microbiome sequencing reveals niches in the mouse gut. Nat Biotechnol 42(9):1394–1403. https://doi.org/10.1038/s41587-023-01988-1
76. Wu H, Mu C, Xu L, Yu K, Shen L, Zhu W (2024) Host-microbiota interaction in intestinal stem cell homeostasis. Gut Microbes 16(1): 2353399. https://doi.org/10.1080/19490976.2024.2353399
77. Shen J, Liang W, Zhao R, Chen Y, Liu Y, Cheng W, Chai T, Zhang Y, Chen S, Liu J, Chen X, Deng Y, Zhang Z, Huang Y, Yang H, Pang L, Qiu Q, Deng H, Pan S, Wang L, Ye J, Luo W, Jiang X, Huang X, Li W, Leung EL, Zhang L, Huang L, Yang Z, Chen R, Mei J, Yue Z, Wei H, Karsten K, Han L, Fang X (2025) Cross-tissue multi-omics analyses reveal the gut microbiota's absence impacts organ morphology, immune homeostasis, bile acid and lipid metabolism. iMeta 4(1):e272. https://doi.org/10.1002/imt2.272

Chapter 2

Standardized Protocols for Environmental Sample Collection: Minimizing Contamination and Preserving Microbial Community Integrity

Bo Li, Tongyi Zhao, Jiying Xu, Qiyu Meng, Qidong Yin, and Yanting Zou

Abstract

Accurate microbial community assessment begins with reliable sample collection. Environmental matrices, such as soil, water, and human-associated habitats, each present unique challenges that can introduce contamination or alternative microbial structures in samples. This chapter provides an overview of standardized sampling strategies for these environments, emphasizing principles for minimizing external contamination and preserving the integrity of microbial communities. Key considerations in field practice, sample handling, preservation, and transport are summarized, along with common pitfalls and practical solutions.

Key words Environmental sampling, Microbiome, Contamination control, Sample preservation, Community structure

1 Introduction

High-quality microbiome research begins with the acquisition of samples that accurately reflect the in situ microbial community. The sampling stage is widely recognized as one of the most susceptible steps to bias in microbial ecology and metagenomics workflows [1]. Even minor deviations, such as inconsistent sampling depth, surface contamination, or prolonged storage before preservation, can introduce compositional shifts that are larger than biological differences among samples [2]. With the growing sensitivity of high-throughput sequencing and multi-omics technologies, the influence of improper sampling has become even more pronounced.

Tao Huang (ed.), *Microbial 'Omics: Methods and Protocols*, Methods in Molecular Biology, vol. 3033,
https://doi.org/10.1007/978-1-0716-5264-0_2,

Environmental matrices differ significantly in structure and microbial distribution. Soil is highly heterogeneous, containing strong inhibitors such as humic acids and often showing microscale variation in microbial abundance [3]. Aquatic environments typically feature low microbial biomass, making samples prone to contamination from equipment, filtration systems, or laboratory air [4]. Human-associated microbiomes introduce additional challenges, including ethical considerations, host-derived contaminants, and the need for strict standardization of swabbing or collection procedures [5]. Despite these differences, all sampling protocols share a central objective: minimizing external contamination and preserving the microbial community as close to its original state as possible.

To ensure sample integrity, researchers have developed a variety of field and laboratory workflows, including the use of field blanks, single-use sterile tools, rapid nucleic-acid stabilization reagents, and controlled-temperature transport systems [6, 7]. These practices have become essential components of modern microbiome research and are now integrated into widely adopted standards for environmental and clinical sampling.

This chapter provides a comprehensive overview of sampling principles and protocols tailored to soil, water, and human-associated environments. It highlights contamination risks, practical mitigation strategies, and preservation approaches that maintain community stability from the moment of collection through transport and storage. By synthesizing commonly adopted practices and frequently encountered pitfalls, the chapter aims to guide researchers in generating reliable, reproducible, and comparable microbiome datasets that can support robust ecological and molecular analyses.

2 Materials

All materials listed below are required for sampling microbial communities from soil, water, and human-associated environments. Use sterile, nuclease-free consumables and ensure all equipment is cleaned according to institutional biosafety guidelines. Prepare sufficient materials for all sampling events, including additional units to accommodate field blanks and replicate sampling. Always wear appropriate personal protective equipment (PPE) during handling and transportation of samples.

2.1 General Principles for Minimizing Contamination

An overview of the standardized field workflow, highlighting key contamination control steps from preparation through transport, is shown in Fig. 1.

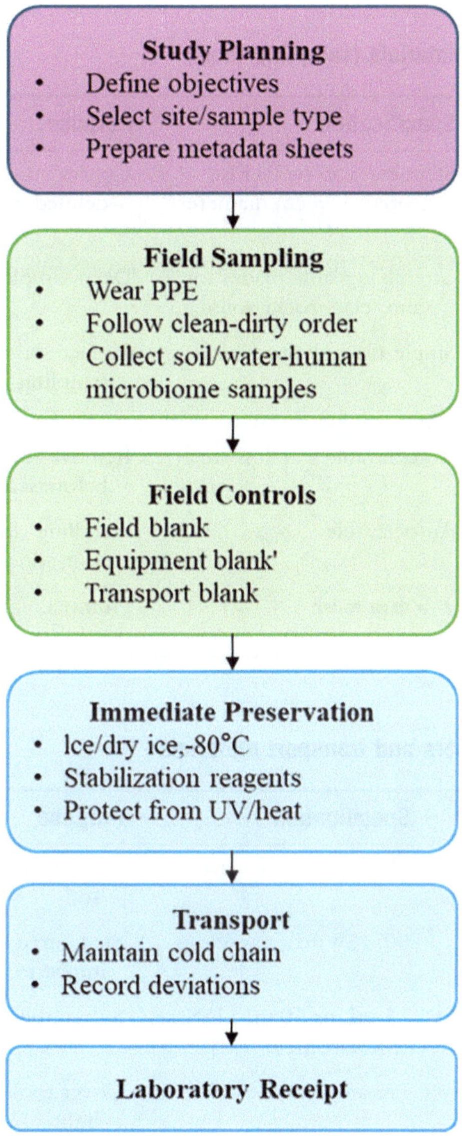

Fig. 1 Workflow of field sampling

2.2 Soil Sampling Materials

Soil sampling for microbiome analysis requires tools that can accurately capture spatially heterogeneous substrates while minimizing environmental and operator-derived contamination. Because soils differ widely in texture, moisture, organic content, and microbial biomass, materials must be selected to match the sampling strategy—whether surface soil, subsurface cores, rhizosphere samples, or composite plots. This section categorizes soil sampling materials into four functional groups: (1) sampling tools, (2) sample containers and transport materials, (3) field measurement and documentation tools, and (4) PPE and decontamination supplies. This

Table 1
Soil sampling materials (sampling tools)

Item	Specification	Purpose	Notes
Soil corer	Stainless steel or PTFE-coated; 1–5 cm diameter	Collect intact cores at defined depth	PTFE reduces contamination and soil adhesion
Soil auger	Interchangeable blades for sand/clay/rocky soils	Deep sampling (>30 cm)	Match auger type to soil texture
Sterile scoops/spatulas	Single-use polypropylene	Surface soil or composite sampling	Replace after each sample
Hand trowel	Autoclavable stainless steel	Remove surface debris before sampling	Clean between sites
Sterile forceps	Autoclavable	Handling rhizosphere or root-adhered soil	Avoid touching non-sterile surfaces
Soil sieve (optional)	2–5 mm mesh	Homogenize bulk soil	Not recommended for low-biomass samples

Table 2
Sample containers and transport materials

Item	Specification	Purpose	Notes
Whirl-Pak® sampling bags	Sterile, 55–1500 mL	Collect bulk or homogenized soil	Leak-proof, lightweight
Polypropylene jars	50–250 mL, screw-cap	Structured sampling & moisture retention	Higher mechanical protection
Sterile centrifuge tubes	15 mL or 50 mL, DNase/RNase-free	Subsampling or stabilization	Label before field use
Aluminum foil	Pre-autoclaved	Cover tools, protect from light	Also creates sterile surfaces
Transport cooler	Insulated with ice packs	Maintain low temperature during transport	Minimize temperature fluctuations

structured layout helps researchers select appropriate materials for their sampling design and ensures reproducibility across sites and studies. The categorized materials required for soil microbiome sampling, including tools, containers, field documentation equipment, and decontamination supplies, are summarized in Tables 1, 2, 3, and 4.

2.3 Water Sampling Materials

Water sampling for microbiome analysis presents unique challenges because aquatic environments often contain low microbial biomass, heterogeneous particle loads, and strong physical-chemical

Table 3
Field measurement and documentation tools

Item	Specification	Purpose	Notes
GPS-enabled field tablet/GPS unit	Waterproof; ±3 m accuracy	Geospatial metadata	Essential for mapping
Soil moisture meter	Probe-based digital meter	Measure soil water content	Record immediately after sampling
Portable pH meter/pH strips	Field-compatible; pH range 3–10	Assess soil pH in situ	Avoid prolonged probe exposure
Environmental meter (optional)	Measures temperature/ humidity	Supporting metadata	Important for time-sensitive sampling
Field notebook/ metadata sheet	Waterproof paper or tablet	Record sample details	Redundant documentation recommended
Permanent marker	Solvent-resistant	Labeling tubes/ bags	Use on both bag and tag

Table 4
PPE and decontamination supplies

Item	Specification	Purpose	Notes
Powder-free nitrile gloves	Single-use	Prevent operator contamination	Change between samples
Disposable boot covers	Waterproof; single-use	Avoid cross-site contamination	Required in agricultural sites
10% bleach solution	Freshly prepared	Decontaminate corers/tools	Follow with ethanol (corrosion risk)
70% ethanol	Molecular biology grade	Remove bleach residue & speed drying	Avoid pooling on metal
Sterile water	Ultrapure	Final rinse for tools; equipment blanks	Keep sealed until use
Biohazard/autoclavable waste bags	Heavy-duty	Collect contaminated disposables	Comply with local regulations

gradients. These factors make water samples particularly susceptible to contamination from pumps, tubing, filtration devices, or airborne particles, especially in oligotrophic lakes, drinking water systems, and open ocean environments [8]. Collecting sufficient volumes is essential for recovering adequate DNA for sequencing, and filtration-based concentration remains the dominant method in environmental microbiology and eDNA studies [9]. Depth-specific samplers such as Niskin bottles are required for stratified water bodies, where microbial communities vary significantly across

Table 5
Water sampling devices

Item	Specification	Purpose	Notes
HDPE/PP sampling bottles	Sterile, 500 mL to 2 L	Grab water collection	Low DNA adsorption; durable
Peristaltic pump	Portable; sterile tubing	Depth-specific sampling	Avoid backflow contamination
Submersible pump	Battery powered	High-volume sampling	Clean exterior between sites
Niskin/Van Dorn sampler	Acrylic/PVC; 1–10 L	Stratified water column sampling	Standard for marine layers [12] (Galand et al. 2010)
Sterile syringes	20–60 mL	Manual sampling or filtration	Single-use only
Sampling pole	Glass/acrylic	Shoreline collection	Avoid disturbing sediment

thermoclines and oxygen gradients [10]. Environmental metadata—including pH, temperature, dissolved oxygen, conductivity, and turbidity—must be recorded to support ecological interpretation [11]. Table 2 summarizes the categorized materials required for water microbiome sampling. The materials required for water microbiome sampling, including sampling devices, filtration systems, preservation supplies, field measurement instruments, and personal protective equipment, are summarized in Tables 5, 6, 7, 8, and 9.

2.4 Human Microbiome Sampling Materials

Human-associated microbiome sampling requires specialized materials designed to capture microbial communities from distinct anatomical sites while minimizing contamination from host cells, environmental surfaces, and operators. Because biomass levels vary substantially among body sites—ranging from high-biomass stool samples to extremely low-biomass skin and nasal samples—sampling materials must be selected based on microbial load and site accessibility [18]. Sterile, individually wrapped swabs are the standard for mucosal and skin sampling, whereas saliva and stool require containers capable of preventing leakage and ensuring biosafety. Human sampling materials must also comply with ethical, biosafety, and clinical standards, including the use of PPE, specimen bags, and labeling systems that maintain sample integrity and traceability [19]. Following tables summarize the categorized materials required for sampling the human microbiome. The categorized materials required for human-associated microbiome sampling, including sampling devices, containers, documentation tools, personal protective equipment, and compliance materials, are summarized in Tables 10, 11, 12, 13, and 14.

Table 6
Filtration and biomass boncentration

Item	Specification	Purpose	Notes
Disposable filtration units	0.22 μm or 0.45 μm pore size	Concentration of microbial biomass from water samples	0.22 μm captures most bacteria and archaea; 0.45 μm reduces clogging in turbid waters [13]
Inline capsule filters	0.2–1.0 μm; sterile, single-use	High-volume microbiome sampling (5 to 20+ liters)	Suitable when large volumes are required to capture low biomass
Filter membranes	Polyethersulfone (PES), cellulose nitrate (CN), polycarbonate (PC); 25–47 mm	Capture microbial cells for DNA extraction	PES has lowest DNA adsorption, widely used in aquatic microbiome studies [14]
Filter holders	Single-use plastic or autoclavable polycarbonate	Supports membrane filtration in field or lab	Avoid metal holders that retain residual DNA
Vacuum pump/ hand vacuum	Portable	Filtration-driven concentration	Hand pumps advantageous in remote sites
Sterile filter forceps	Stainless steel or plastic	Transfer membranes into sterile tubes	Avoid touching membrane surface

Table 7
Field preservation and transport materials

Item	Specification	Purpose	Notes
Sterile tubes	15 mL, 50 mL	Store filters	Label before field deployment
Whirl-Pak bags	Sterile	Temporary storage	Useful for small volumes
Cryovials	1.5–5 mL	Filter storage	O-ring seals prevent leaks
Insulated coolers	With ice/dry ice	Temperature control	Avoid repeated thawing [15] (Lauber et al. 2010)
Secondary containment bags	Biohazard-marked	Transport safety	Mandatory for wastewater

Table 8
Field measurement and documentation tools

Item	Specification	Purpose	Notes
Multiparameter water quality meter	Temp, pH, DO, EC, turbidity	Environmental metadata	Calibrate daily [16] (Gans et al. 2019)
Thermometer/temp logger	Digital	Real-time temp	Required for metabolic models
Turbidity meter	0–1000 NTU	Particle load	Affects filtration rate
GPS-enabled tablet	Waterproof; high accuracy	Metadata logging	Use standardized forms
Field notebook/sheets	Waterproof	Observations	Maintain redundancy
Waterproof markers	Permanent	Label bottles/ tubes	Avoid smearing

Table 9
Personal protective equipment and decontamination materials

Item	Specification	Purpose	Notes
Nitrile gloves	Powder-free	Prevent contamination	Replace between samples [17]
Surgical masks	Disposable	Reduce aerosols	Especially important for low-biomass water
10% bleach	Fresh	Decontaminate tools	Follow with ethanol
70% ethanol	Molecular grade	Remove bleach residue	Speeds drying
Sterile water	Ultrapure	Equipment blanks	Keep sealed

3 Methods

3.1 Soil Sampling Procedure

This protocol provides detailed stepwise instructions for standardized soil microbiome sampling. Soil often contains natural PCR inhibitors such as humic acids, making consistent sampling and proper documentation important (*see* **Note 1**). All steps should be performed using sterile, DNA-free materials and follow institutional biosafety guidelines.

3.1.1 Preparation

1. Sterilize all soil corers, spatulas, trowels, and forceps by autoclaving at 121 °C for 20 min or wiping thoroughly with 10% bleach followed by 70% ethanol.
2. Pre-label all sampling tubes and bags with:
   - Sample ID.
   - Site code.
   - Depth (e.g., 0–5 cm, 10–20 cm).

Table 10
Sampling devices

Item	Specification	Purpose	Notes
Flocked swabs	Sterile nylon or polyester; individually wrapped	Oral, nasal, skin, and urogenital microbiome sampling	High microbial capture efficiency [20]
Foam or rayon swabs	Sterile, soft-tipped	Sensitive mucosal sites	Lower biomass recovery than flocked swabs
Saliva collection kits	Funnel-based or integrated stabilization tube	Collect unstimulated saliva for microbiome profiling	Avoid eating/drinking 30 min before sampling [21]
Stool collection kits	Leak-proof container + sampling scoop	Gut microbiome sampling	Must meet biosafety transport regulations
Rectal swabs (optional)	Sterile, flocked	Alternative to stool collection	Useful in clinical settings [22]
Tongue depressors	Wooden or plastic; sterile	Aid in oral sampling	Prevents swab contamination
Sterile cryovials	1.5–5 mL; screw-cap with O-ring	Long-term storage of swabs or aliquots	Suitable for −80 °C storage

Table 11
Sample containers and transport materials

Item	Specification	Purpose	Notes
Screw-cap tubes	2–15 mL; sterile	Place swabs or small-volume samples	Label tubes before sampling
Saliva tubes	5–10 mL; leak-proof	Saliva storage	Keep upright during transport
Stool containers	30–120 mL; sterile, wide-mouth	Stool collection	Must be shatterproof and leak-proof
Biohazard specimen bags	Two-layer, sealable	Clinical-grade sample transport	Required for all human specimens
Cryoboxes	Polypropylene	Organize multiple cryovials	Maintain tube orientation
Secondary containment	Hard cases with absorbent lining	Regulatory compliance	Required for transport between facilities

Table 12
Field Measurement and documentation tools

Item	Specification	Purpose	Notes
Metadata sheets/ digital forms	Standardized templates	Record diet, medications, time of last cleaning, body site	Essential for reproducibility [23]
Temperature loggers	±0.5 °C accuracy	Monitor sample temperature	Useful for multisite studies
Barcode labels	Waterproof, freezer-safe	Sample traceability	Avoid handwritten labels when possible
Permanent markers	Solvent-resistant	Backup labeling	Use on tube side & cap

Table 13
PPE and decontamination supplies

Item	Specification	Purpose	Notes
Nitrile gloves	Powder-free, single-use	Avoid operator-derived contamination	Replace between each participant [24]
Surgical masks	Disposable	Prevent exhaled contamination	Especially important for oral/nasal sampling
Disposable gowns/lab coats	Fluid-resistant	Protect operator and sample	Required in clinical settings
Alcohol wipes	70% isopropanol	Clean surfaces before sampling	Do not clean skin before sampling unless protocol requires
70% ethanol	Molecular-grade	Decontaminate equipment	Avoid overspray onto participants
Bleach wipes (optional)	Low-splash formulations	Clean non-contact surfaces	Not used on sampling sites
Sharps container (if needed)	Clinical-grade	Dispose of tongue depressors or sharps	Follow biosafety regulations

- Replicate number.
- Date & time.

3. Prepare a sampling map or GPS grid and identify all sampling points.
   - For transect sampling: mark positions at 5–10 m intervals.
   - For quadrat sampling: mark 1 m × 1 m or 5 m × 5 m grids.
4. Wear powder-free nitrile gloves. Change gloves before each sample.

Table 14
Safety, ethics, and compliance materials

Item	Specification	Purpose	Notes
Informed consent forms	IRB-approved	Ethical compliance	Mandatory for human studies
Sample ID sheets	Standardized	Maintain subject anonymity	Separate from metadata
Participant instruction leaflets	Plain-language	Standardized pre-sampling behavior	Avoid food/drink/brushing (oral)
Biohazard labels	Adhesive	Mark containers and boxes	Required for transport & storage
PPE disposal bags	Autoclavable	Waste management	Use per institutional guidelines

5. Prepare an insulated cooler with 4–6 frozen ice packs for temporary preservation.
6. Bring sufficient sterile 50 mL tubes for subsamples (minimum 3 per sampling point) (*see* **Note 10**).

3.1.2 Surface Soil Sampling (0–5 cm)

1. Remove visible litter such as leaves or stones without disturbing the soil surface.
2. Insert a sterile spatula into the soil at a 45° angle, collecting the top 0–5 cm layer.
3. Collect at least 10–20 g of soil (roughly 2–3 spatula scoops) into a sterile bag.
4. Avoid touching the inside of the bag or tube with tools or gloves.
5. Expel excess air and seal immediately.
6. Place the sample into the cooler.
7. Record the following metadata:
   (a) GPS coordinates (±3 m).
   (b) Surface temperature.
   (c) Vegetation type.
   (d) Soil color and texture (manual classification).
   (e) Recent weather conditions (past 48 h).

3.1.3 Subsurface Soil Coring (5–30 cm or Deeper)

1. Insert the sterile soil corer vertically into the ground to the target depth (e.g., 10–20 cm).
2. Rotate the corer clockwise three full turns to cut through compacted layers.

3. Gently pull the corer up in one motion to prevent mixing of soil horizons.
4. Remove the liner or open the corer window and extrude the soil core onto sterile foil (*see* **Note 2**).
5. Identify the required depth segment (e.g., 10–15 cm) using a sterile ruler (*see* **Note 2**).
6. Use a sterile spatula to subsample 5–10 g from the center of the soil core (avoid outer edges, which contact the metal tube).
7. Transfer soil into a sterile tube or bag.
8. Seal immediately.
9. Decontaminate the corer with bleach → ethanol → sterile water before the next sampling point.

3.1.4 Composite Soil Sampling

1. Define a sampling area (e.g., 10 m × 10 m).
2. Select 5–10 random points within the area using a random number table or GPS.
3. At each point, collect 2–3 g of surface soil (0–5 cm).
4. Combine all subsamples (total 10–30 g) into a single sterile Whirl-Pak bag.
5. Seal and homogenize by gently massaging the outside of the bag for 10–15 s.
6. Avoid overmixing to prevent heating or oxygen exposure.
7. Aliquot homogenized soil into replicate 50 mL tubes if needed.

3.1.5 Rhizosphere Soil Sampling

1. Excavate the root system using a sterile trowel, removing ~10 cm of surrounding soil.
2. Shake the plant gently to detach loosely adhering bulk soil.
3. Using sterile forceps or a gloved hand, collect soil firmly attached to fine roots (the rhizoplane zone).
4. Transfer collected rhizosphere soil (~2 to 5 g) to a sterile tube.
5. If collecting root sections, cut 1–2 cm pieces using sterile scissors.
6. Place samples into a cooler immediately.

3.1.6 Sample Preservation and Transport

1. Keep all soil samples at 4 °C during fieldwork.
2. Transfer to the laboratory within 4–6 h whenever possible.
3. For long trips (>8 h), add dry ice to maintain a low temperature.
4. Record temperature fluctuations using a portable temperature logger.

5. Upon arrival, store samples at:
   - −20 °C for short-term storage (<1 week).
   - −80 °C for long-term preservation.
6. Avoid freeze-thaw cycles (*see* **Note 3**).
7. Complete all chain-of-custody documentation, including:
   - Time of sampling.
   - Time of cooling.
   - Time of laboratory arrival.
   - Condition of sample containers.

3.2 Water Sampling Procedure

The following protocol describes standardized procedures for collecting water samples for microbiome analysis across freshwater, marine, riverine, and wastewater environments. All steps must be performed using sterile materials and handled in a manner that minimizes environmental and operator-derived contamination (*see* **Note 4**). Adjust sample volume and filtration strategy based on expected biomass.

3.2.1 Preparation

1. Verify that all sampling bottles (HDPE/PP), filters, tubing, and pumps are sterile and packaged intact.
2. Pre-label all bottles, filter tubes, and cryovials with sample ID, site code, depth, replicate number, and time (*see* **Note 8**).
3. Prepare a metadata form including parameters for temperature, pH, dissolved oxygen, conductivity, turbidity, weather conditions, and GPS coordinates (*see* **Note 8**).
4. Calibrate the multiparameter water quality meter according to manufacturer instructions (typically single-point or multipoint calibration).
5. Assemble an insulated cooler with at least 4–6 frozen ice packs.
6. Wear powder-free nitrile gloves. Replace gloves before each sampling.
7. Prepare sterile field blanks (sterile bottled water) and keep sealed until use (*see* **Note 4**).

3.2.2 Grab Sampling (Direct Water Collection)

Suitable for lakes, reservoirs, rivers, and nearshore marine environments.

1. Stand facing upstream or upwind to avoid disturbing the water before sampling (*see* **Note 9**).
2. Submerge the sterile sampling bottle to a depth of 20–30 cm, avoiding surface films.
3. Fill the bottle completely, leaving 0.5–1 cm headspace to allow gentle mixing.

4. Seal the bottle immediately without touching the inside of the cap.
5. In fast-flowing rivers, collect water as swiftly as possible to avoid sediment resuspension (*see* **Note 9**).
6. Store the bottle in a cooler immediately after sampling.
7. Record water temperature and physical parameters at the same site and time.

3.2.3 Subsurface or Stratified Water Sampling (Niskin/Van Dorn Sampler)

Used for thermally or chemically stratified lakes, deep marine sampling, reservoirs, and depth-specific microbiome studies.

1. Deploy the sampler to the desired depth using a calibrated line or winch (*see* **Note 8**).
2. Trigger the sampler to close at the target depth without disturbing the surrounding water.
3. Retrieve the sampler smoothly to prevent mixing water layers.
4. Transfer water into sterile bottles through the sampler spigot.
5. Avoid allowing water to flow over your hands or clothing to prevent contamination.
6. Process samples immediately or place them in a cooler.

3.2.4 Pump-Based Sampling (Peristaltic/Submersible Pump)

Used when large volumes (2–20 L) are required due to low biomass.

1. Attach sterile tubing to the pump outlet. Keep tubing in sterile packaging until the moment of use.
2. Prime the pump with ~200 mL of sample water to remove residual air.
3. Position the tubing inlet at least 10–20 cm below the surface, avoiding sediment disturbance.
4. Pump water at a consistent flow rate (typically 0.5–1.5 L/min) into sterile containers or directly into filter units.
5. If collecting into containers, fill them completely and cap immediately.
6. Between sampling sites, flush the pump tubing with 500–1000 mL of the new site's water. If contamination risk is high (e.g., very low-biomass water), replace the inlet tubing segment that directly contacts the water column (*see* **Note 18**).

3.2.5 Filtration and Biomass Concentration

Filtration is the central step for concentrating microbial biomass from aquatic environments, particularly in low-biomass systems, such as oligotrophic lakes, groundwater, and oceanic surface waters. Because microbial cells are often sparse and unevenly distributed in water, filtration efficiency directly determines downstream DNA yield, taxonomic richness, and detection sensitivity.

All filtration steps should be conducted using sterile, DNA-free materials, avoiding direct hand or surface contact with filters, tubing, or filter housings.

1. Prepare filtration equipment, including disposable filtration units, inline capsule filters, or membrane holder assemblies. Ensure all components that contact water are sterile and remain sealed until use (*see* **Note 5**).
2. Select the appropriate pore size based on expected biomass and turbidity.
   - *0.22 μm membranes* are recommended for free-living bacteria and archaea.
   - *0.45 μm membranes* reduce clogging in turbid waters but incorporate more particle-attached communities.

     Using inconsistent pore sizes between sites introduces structural artifacts (*see* **Note 6**).
3. Choose membrane material carefully. Polyethersulfone (PES) membranes typically provide higher DNA recovery and lower adsorption than cellulose nitrate or PVDF, particularly in low-biomass environments (*see* **Note** 7).
4. Set up the filtration system according to site conditions:
   - *Gravity or vacuum filtration* for small volumes (50 mL to 1 L).
   - *Peristaltic pump + inline capsule filter* for large-volume sampling (2–20 L).
   - *Hand-powered vacuum pumps* for remote fieldwork.

     Prevent backflow or contact between tubing and non-sterile surfaces.
5. *Filter the sample immediately after collection* to minimize biomass loss or compositional shifts. Avoid allowing filters to dry completely, as drying leads to membrane cracking and cell rupture (*see* **Note 16**).
6. *Transfer filters* using sterile forceps into pre-labeled sterile tubes or cryovials. Do not fold or excessively compress the membrane, which may mechanically lyse cells.
7. *Record filtration metadata*, including sample volume, filtration time, type of membrane, pore size, and any clogging events. These parameters influence biomass recovery and must be reported for reproducibility.
8. *Store membranes* on ice immediately after filtration and transfer to −20 °C or −80 °C upon return to the laboratory (*see* **Note 16**). Avoid freeze-thaw cycles.

Filtration thus plays a decisive role in the recovery of microbial communities from aquatic systems. Careful selection of pore size, membrane chemistry, and filtration method—combined with precise metadata documentation—ensures the resulting microbial profiles accurately reflect in situ community structure.

3.3 Human Microbiome Sampling Procedure

This protocol provides standardized procedures for the collection of human-associated microbiome samples, including oral, saliva, nasal, skin, and stool specimens. All procedures must comply with ethical approvals, informed consent, and biosafety regulations. Because many body sites contain low microbial biomass, contamination control is essential throughout the process.

3.3.1 Preparation

1. Confirm that informed consent has been obtained for each participant.
2. Prepare sterile sampling materials (flocked swabs, saliva tubes, stool containers, cryovials, gloves, masks).
3. Pre-label all sample tubes with subject ID, body site, date, and time.
4. Prepare metadata sheets including recent diet, medication use, oral hygiene or skin cleaning time, and sampling conditions.
5. Wear powder-free nitrile gloves and a surgical mask.
6. Change gloves *before sampling each new participant.*
7. Ensure all tools are laid out on a clean disposable workspace (bench pad or sterile drape).

3.3.2 Oral Microbiome Sampling (Swab Method)

1. Ask the participant to avoid eating, drinking, chewing gum, or brushing teeth for at least 30 min before sampling.
2. Open the flocked swab package without touching the tip (*see* **Note 11**).
3. Insert the swab into the oral cavity and gently scrape:
   - Buccal mucosa: 10–15 strokes per side.
   - Tongue dorsum: 10–15 strokes.
4. Maintain consistent pressure and speed across participants.
5. Immediately place the swab head into a sterile tube and break the shaft at the breakpoint.
6. Seal the tube tightly.
7. Store the sample on ice immediately.

3.3.3 Saliva Microbiome Sampling

1. Instruct the participant to pool saliva naturally in the mouth for 1–2 min (*see* **Note 12**).
2. Ask the participant to spit 2–5 mL of unstimulated saliva into the sterile collection tube.
3. Avoid coughing, throat clearing, or nasal mucus contamination.

4. Cap the tube tightly and invert gently 2–3 times to homogenize.
5. Place the sample in a cooler.

3.3.4 Nasal Microbiome Sampling

1. Open the sterile flocked swab package.
2. Insert the swab 1–1.5 cm into the anterior nares.
3. Rotate the swab against the nasal wall 5–10 times.
4. Repeat in the other nostril using the same swab unless protocol requires bilateral duplicates.
5. Break the swab into a sterile tube.
6. Seal immediately and place on ice.

3.3.5 Skin Microbiome Sampling

1. Identify the target skin site (e.g., forearm, forehead, dorsal hand).
2. Do not clean the skin prior to sampling unless studying post-cleaning effects.
3. Moisten the sterile swab with 100 μL sterile buffer (optional but improves biomass recovery).
4. Rub the swab across a 4 cm × 4 cm square area with moderate pressure for 20–30 s.
5. Rotate the swab during sampling to maximize surface contact.
6. Insert the swab head into a sterile tube and break the shaft.
7. Store on ice.

3.3.6 Stool Microbiome Sampling

1. Provide the participant with a sterile stool collection kit.
2. Instruct the participant to collect the sample without urine contamination.
3. Using the sterile scoop attached to the lid, collect 1–2 g of stool from the internal portion of the sample (avoid outer layer exposure to air) (*see* **Note 13**).
4. Place stool into the sterile container and seal tightly.
5. Immediately transfer an aliquot (~200 to 300 mg) into a sterile cryovial if subsampling in the field.
6. Store in a cooler until transport.

3.3.7 Rectal Swab (Optional Alternative to Stool)

1. Ask the participant to lie on their side in a comfortable position (*see* **Note 14**).
2. Insert a sterile flocked swab 2–3 cm past the anal verge.
3. Rotate gently for 10–15 s.
4. Withdraw the swab and place it into a sterile tube.
5. Seal and store on ice.

3.3.8 Negative Controls (Low-Biomass Specific)

1. Prepare an unopened sterile swab as a field blank.
2. Expose the swab to room air for 5–10 s at the sampling location.
3. Process the blank exactly as a real sample.
4. Store on ice.

3.3.9 Sample Preservation and Transport

1. Place all samples on ice immediately after collection (*see* **Note 15**).
2. Maintain transport temperature between 2 and 8 °C.
3. Deliver samples to the laboratory within 4–6 h when possible.
4. For longer transport times (>8 h), use dry ice or insulated shipping containers.
5. Upon laboratory arrival, store samples at:
6. 4 °C if processing within 24 h.
7. −20 °C for short-term storage (≤1 week).
8. −80 °C for long-term preservation.
9. Record sample temperature, timing, and chain-of-custody details.
10. Avoid freeze-thaw cycles at all stages.

4 Notes

1. Humic acids, polyphenols, and metal ions in soils inhibit polymerases and suppress amplification of specific clades. This inhibition is taxon-specific and can alter relative abundances, producing artificial community shifts even before sequencing. Recording soil type, organic content, and color helps interpret inhibitor-driven bias [25].
2. Anaerobic soil or sediment layers undergo rapid community shifts when exposed to oxygen. Even brief oxidation can lead to selective loss of strict anaerobes and enrichment of facultative aerobes, destroying in situ community structure. Using gas-tight containers and minimizing exposure is critical in studies targeting anoxic zones [26].
3. Freeze-thaw cycles preferentially lyse Gram-negative bacteria while leaving Gram-positive taxa intact, causing artificial enrichment of the latter. DNA released from lysed cells is also subject to nuclease degradation, compounding the bias. Aliquoting samples prevents taxonomic distortions [27].
4. Extraction kits, PCR reagents, and even ultrapure water contain trace microbial DNA from environmental manufacturing sources. These reagent contaminants contribute consistent

taxa—especially Actinobacteria and Proteobacteria—that may be misinterpreted as biological signals. Parallel reagent blanks are required to distinguish background DNA from true low-biomass taxa [28].

5. DNA adheres strongly to many common plastics (PP, PE) and to metal corners or tools, particularly when biomass is low. This leads to the selective loss of rare or free-living taxa and underestimates true diversity. Using low-DNA-binding plastics and minimizing transfer steps improves taxonomic recovery [29]. Low-biomass samples, such as groundwater, oligotrophic lakes, ocean water, and skin swabs, are easily dominated by contaminant DNA. Reagent DNA or operator-derived aerosol contamination can exceed natural microbial loads and generate entirely artificial communities. Including field blanks and extraction blanks is essential to identify these background signals [30].
6. Filtration pore size determines which microbial fractions are captured: 0.22 μm retains free-living bacteria and archaea, while larger pores incorporate particle-attached cells and debris. Comparing data collected with different pore sizes introduces major structural artifacts and should be avoided [31].
7. Filter chemistry affects microbial retention and DNA recovery. PES membranes show significantly higher recovery efficiency and lower DNA adsorption than cellulose nitrate or PVDF, especially in low-biomass water. Using inconsistent membrane types across sites introduces recovery bias [32].
8. Thermal and oxygen stratification create sharp vertical microbial gradients in lakes, reservoirs, and oceans. Mixing layers during sampling—or failing to record precise depth—destroys the ecological signal and prevents accurate interpretation of community structure [33].
9. Biofilms on sediments and hard surfaces harbor dense, distinct microbial communities. Disturbing shorelines or resuspending sediments introduces these biofilm taxa into water samples, overwhelming planktonic communities and inflating apparent richness [34].
10. Operator differences—including sampling pressure, swabbing technique, tool handling, and experience—can create batch effects larger than biological effects. Multisite studies must standardize procedures and train personnel to reduce interoperator variability. Technical variation attributable to sampling teams is a recognized confounder in microbiome studies [35].

11. Swab material has a strong influence on DNA yield from human surfaces. Flocked swabs provide substantially higher biomass recovery than cotton or rayon by maximizing the release of adherent microbial cells. Inconsistent swab types introduce artificial differences between subjects [36].
12. Human microbiomes respond rapidly to short-term behaviors such as eating, brushing teeth, washing skin, applying cosmetics, or even brief physical activity. These behavioral effects may exceed treatment effects in magnitude. Standardizing pre-sampling behavior is therefore essential for reproducibility [37].
13. The external surface of stool is oxygen-exposed and microbiologically distinct from internal anaerobic regions. Sampling from the surface produces profiles that do not represent the gut lumen. Only internal portions should be collected to ensure accurate community characterization [38].
14. Rectal swabs sample mucosa-associated communities, which differ systematically from stool communities in both composition and diversity. They should not be used interchangeably with stool-derived microbiomes and should be interpreted as distinct ecological niches [39].
15. Nucleic acid stabilizers differ in lysis kinetics and chemical selectivity. Some formulations preferentially lyse Gram-negative cells or degrade RNA rapidly, altering taxonomic representation before extraction. Buffer selection must match downstream DNA/RNA workflows to prevent selective loss [40].
16. Filtering water samples to dryness causes membrane cracking and mechanical rupture of microbial cells. This not only releases intracellular DNA but also disrupts size-fractionation fidelity by mixing particle-attached and free-living taxa. Stopping filtration before dryness preserves cellular integrity [41].
17. Resuspension of benthic sediments introduces orders of magnitude more biomass into water samples, masking natural planktonic signatures. Sediment-associated taxa are typically much more diverse and abundant than water-column microbes, leading to artificial richness inflation [42].
18. Biofilms can accumulate inside tubing used for water pumping systems. Replacing only the inlet tubing—rather than the entire assembly—adequately removes risk of cross-contamination while minimizing waste. Flushing with site water further reduces carryover and is standard practice in environmental monitoring [43].

Conflict of Interest

The authors have no conflicts of interest to declare that are relevant to the content of this chapter.

References

1. Castle SC, Samac DA, Sadowsky MJ, Rosen CJ, Gutknecht JL, Kinkel LL (2019) Impacts of sampling design on estimates of microbial community diversity and composition in agricultural soils. Microb Ecol 78(3):753–763
2. Zeng EY (2023) Microplastic contamination in aquatic environments: an emerging matter of environmental urgency. Elsevier
3. Wilpiszeski RL, Aufrecht JA, Retterer ST, Sullivan MB, Graham DE, Pierce EM, Zablocki OD, Palumbo AV, Elias DA (2019) Soil aggregate microbial communities: towards understanding microbiome interactions at biologically relevant scales. Appl Environ Microbiol 85(14):e00324–e00319. https://doi.org/10.1128/AEM.00324-19
4. Deiner K, Bik HM, Mächler E, Seymour M, Lacoursière-Roussel A, Altermatt F, Creer S, Bista I, Lodge DM, de Vere N, Pfrender ME, Bernatchez L (2017) Environmental DNA metabarcoding: transforming how we survey animal and plant communities. Mol Ecol 26(21):5872–5895. https://doi.org/10.1111/mec.14350
5. Gorzelak MA, Gill SK, Tasnim N, Ahmadi-Vand Z, Jay M, Gibson DL (2015) Methods for improving human gut microbiome data by reducing variability through sample processing and storage of stool. PLoS One 10(8): e0134802
6. Williamson EM, Hammer TJ, Hogendoorn K, Eisenhofer R (2025) Blanking on blanks: few insect microbiota studies control for contaminants. MBio 16(4):e02658–e02624
7. Rissanen AJ, Kurhela E, Aho T, Oittinen T, Tiirola M (2010) Storage of environmental samples for guaranteeing nucleic acid yields for molecular microbiological studies. Appl Microbiol Biotechnol 88(4):977–984
8. Bhateria R, Jain D (2016) Water quality assessment of lake water: a review. Sustain Water Resour Manag 2(2):161–173
9. Bruce K, Blackman RC, Bourlat SJ, Hellström M, Bakker J, Bista I, Bohmann K, Bouchez A, Brys R, Clark K (2021) A practical guide to DNA-based methods for biodiversity assessment. Pensoft Advanced Books
10. Fortunato CS, Eiler A, Herfort L, Needoba JA, Peterson TD, Crump BC (2013) Determining indicator taxa across spatial and seasonal gradients in the Columbia River coastal margin. ISME J 7(10):1899–1911
11. Dwivedi B, Xue B, Lundin D, Edwards RA, Breitbart M (2013) A bioinformatic analysis of ribonucleotide reductase genes in phage genomes and metagenomes. BMC Evol Biol 13(1):33
12. Noble A, Tuit C, Maney J, Wait AJEF (2020) A review of marine water sampling methods for trace metals. Environ Forensics 21(3–4): 267–290
13. Mull B, Hill VR (2012) Recovery of diverse microbes in high turbidity surface water samples using dead-end ultrafiltration. J Microbiol Methods 91(3):429–433
14. Li J, Huang F, Ma Q, Guo W, Feng K, Huang T, Cai YD (2023) Identification of genes related to immune enhancement caused by heterologous ChAdOx1-BNT162b2 vaccines in lymphocytes at single-cell resolution with machine learning methods. Front Immunol 14:1131051. https://doi.org/10.3389/fimmu.2023.1131051
15. Hunt CJJTM (2019) Technical considerations in the freezing, low-temperature storage and thawing of stem cells for cellular therapies. Hemotherapy 46(3):134–150
16. Alam AU, Clyne D, Deen MJJS (2021) A low-cost multi-parameter water quality monitoring system. Sensors 21(11):3775
17. Esmizadeh E, Chang B, Jubinville D, Seto C, Ojogbo E, Tzoganakis C, Mekonnen TH (2021) Stability of nitrile and vinyl latex gloves under repeated disinfection cycles. Mater Today Sustain 11:100067
18. Shi Y, Wang G, Lau HC-H, Yu J (2022) Metagenomic sequencing for microbial DNA in human samples: emerging technological advances. Int J Mol Sci 23(4):2181
19. Harriss D, MacSween A, Atkinson G (2017) Standards for ethics in sport and exercise science research: 2018 update. Int J Sports Med 38(14):1126–1131
20. Jansson L, Akel Y, Eriksson R, Lavander M, Hedman J (2020) Impact of swab material on microbial surface sampling. J Microbiol Methods 176:106006

21. Ng TK, Udeh-Momoh C, Lim MA, Gleerup HS, Leifert W, Ajalo C, Ashton N, Zetterberg H, Rissman RA, Winston CN (2025) Guidelines for the standardization of pre-analytical variables for salivary biomarker studies in Alzheimer's disease research: an updated review and consensus of the Salivary Biomarkers for Dementia Research Working Group. Alzheimer's & Dementia 21(2): e14420
22. Bassis CM, Moore NM, Lolans K, Seekatz AM, Weinstein RA, Young VB, Hayden MK (2017) Comparison of stool versus rectal swab samples and storage conditions on bacterial community profiles. BMC Microbiol 17(1):78
23. Miller SJ (2022) Metadata for digital collections. American Library Association
24. Sawyer J, Bennett A (2006) Comparing the level of dexterity offered by latex and nitrile SafeSkin gloves. Ann Occup Hyg 50(3): 289–296
25. Lear G, Dickie I, Banks J, Boyer S, Buckley HL, Buckley TR, Cruickshank R, Dopheide A, Handley KM, Hermans S (2018) Methods for the extraction, storage, amplification and sequencing of DNA from environmental samples. N Z J Ecol 42(1):10–50A
26. Papkovsky DB, Dmitriev RI (2013) Biological detection by optical oxygen sensing. Chem Soc Rev 42(22):8700–8732
27. Ganda E, Beck KL, Haiminen N, Silverman JD, Kawas B, Cronk BD, Anderson RR, Goodman LB, Wiedmann M (2021) DNA extraction and host depletion methods significantly impact and potentially bias bacterial detection in a biological fluid. Msystems 6(3). https://doi.org/10.1128/msystems.00619-00621
28. Salter SJ, Cox MJ, Turek EM, Calus ST, Cookson WO, Moffatt MF, Turner P, Parkhill J, Loman NJ, Walker AW (2014) Reagent and laboratory contamination can critically impact sequence-based microbiome analyses. BMC Biol 12(1):87
29. Bala S, Garg D, Thirumalesh BV, Sharma M, Sridhar K, Inbaraj BS, Tripathi M (2022) Recent strategies for bioremediation of emerging pollutants: a review for a green and sustainable environment. Toxics 10(8):484
30. Adinolfi M, Aglieri Rinella G, Albrecht E, Bellunato T, Benson S, Blake T, Blanks C, Brisbane S, Brook N, Calvi M (2013) Performance of the LHCb RICH detector at the LHC. Eur Phys J C 73(5):1–17
31. Bouxsein ML, Boyd SK, Christiansen BA, Guldberg RE, Jepsen KJ, Müller R (2010) Guidelines for assessment of bone microstructure in rodents using micro–computed tomography. J Bone Miner Res 25(7):1468–1486
32. Nearing JT, Comeau AM, Langille MG (2021) Identifying biases and their potential solutions in human microbiome studies. Microbiome 9(1):113
33. Cook ER, Kairiukstis LA (2013) Methods of dendrochronology: applications in the environmental sciences. Springer Science & Business Media, Dordrecht
34. Munn CB (2019) Marine microbiology: ecology & applications. CRC Press, Milton
35. Poussin C, Sierro N, Boué S, Battey J, Scotti E, Belcastro V, Peitsch MC, Ivanov NV, Hoeng J (2018) Interrogating the microbiome: experimental and computational considerations in support of study reproducibility. Drug Discov Today 23(9):1644–1657
36. Lee RA, Herigon JC, Benedetti A, Pollock NR, Denkinger CM (2021) Performance of saliva, oropharyngeal swabs, and nasal swabs for SARS-CoV-2 molecular detection: a systematic review and meta-analysis. J Clin Microbiol 59(5). https://doi.org/10.1128/jcm.02881-02820
37. Taff CC, Schoenle LA, Vitousek MN (2018) The repeatability of glucocorticoids: a review and meta-analysis. Gen Comp Endocrinol 260: 136–145
38. Ramette A (2009) Quantitative community fingerprinting methods for estimating the abundance of operational taxonomic units in natural microbial communities. Appl Environ Microbiol 75(8):2495–2505
39. Newton DP, Ho P-Y, Huang KC (2024) The network structure of cross-feeding impacts microbial community diversity under growth-inhibiting stresses.2024.2010. 2029.620519
40. Shi H, Zhou Y, Jia E, Pan M, Bai Y, Ge Q (2021) Bias in RNA-seq library preparation: current challenges and solutions. BioMed Res Int 2021(1):6647597
41. Chen T, Acker JP, Eroglu A, Cheley S, Bayley H, Fowler A, Toner M (2001) Beneficial effect of intracellular trehalose on the membrane integrity of dried mammalian cells. Cryobiology 43(2):168–181
42. Cavaco MA, St. Louis VL, Engel K, St. Pierre KA, Schiff SL, Stibal M, Neufeld JD (2019) Freshwater microbial community diversity in a rapidly changing High Arctic watershed. FEMS Microbiol Ecol 95(11):fiz161
43. Di Modugno M, Gioia A, Gorgoglione A, Iacobellis V, La Forgia G, Piccinni AF, Ranieri E (2015) Build-up/wash-off monitoring and assessment for sustainable management of first flush in an urban area. Sustainability 7(5): 5050–5070

Chapter 3

Metagenomic Data Preprocessing and Quality Control

Bo Li, Jiying Xu, Tongyi Zhao, Xue Yang, Qidong Yin, and Yanting Zou

Abstract

Accurate metagenomic analysis relies on clean and well-processed sequencing reads. This chapter presents a concise four-step workflow for preprocessing: raw data assessment, adapter and quality filtering, host DNA removal, and final clean-read evaluation. By standardizing these essential procedures, researchers can minimize contamination, reduce technical bias, and ensure reliable inputs for assembly and downstream metagenomic analyses.

Key words Metagenomic data, Quality control, Read trimming, Host removal

1 Introduction

High-throughput metagenomic sequencing enables comprehensive characterization of microbial communities, but the accuracy of downstream analyses depends critically on the integrity of the raw sequencing reads. Unlike single-genome datasets, metagenomic samples contain complex mixtures of microbial, viral, plasmid-derived, and often host-derived DNA, creating multiple layers of technical noise. Common artifacts, including sequencing adapters, low-quality bases, chimeras, and environmental or reagent contaminants, can substantially distort taxonomic, functional, and genome-resolved outputs if not properly removed [1].

The importance of data preprocessing arises from its direct influence on nearly every stage of metagenomic analysis. Low-quality or untrimmed reads may inflate species richness, reduce assembly contiguity, introduce errors into functional annotation, and impede the recovery of high-quality metagenome-assembled genomes (MAGs) [2]. Host DNA, especially in human- and plant-associated samples, can dominate libraries and obscure microbial signals, while laboratory reagent contaminants introduce characteristic taxa that may be misinterpreted as biologically meaningful—particularly in low-biomass samples [3, 4].

Tao Huang (ed.), *Microbial 'Omics: Methods and Protocols*, Methods in Molecular Biology, vol. 3033,
https://doi.org/10.1007/978-1-0716-5264-0_3,

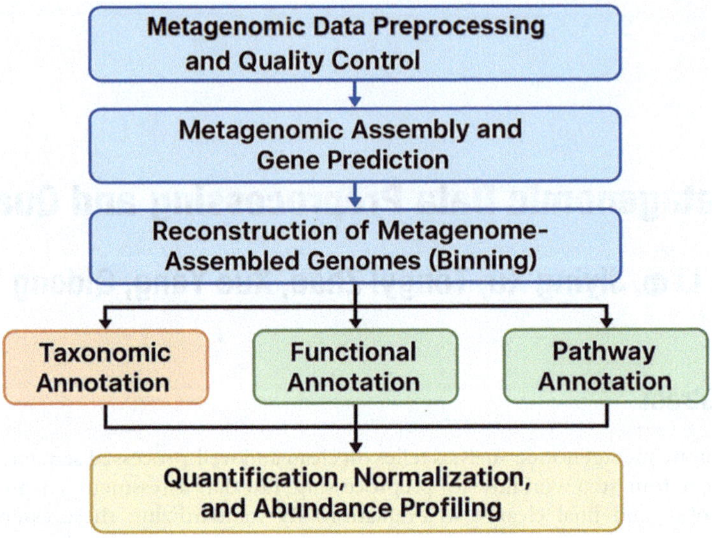

Fig. 1 Typical workflow of metagenomic data analysis for microbiome

Over the past decade, preprocessing has evolved from simple quality trimming into a structured, multistep workflow. Foundational tools such as FastQC [5] and MultiQC [6] established standardized reporting, while Cutadapt [7], Trimmomatic [8], and fastp [9] provided robust trimming capabilities. Alignment tools including Bowtie2 and Minimap2 enabled efficient host-read depletion, and the increasing awareness of reagent contaminants highlighted the need for negative controls and stringent filtering practices [10]. These developments collectively shaped the modern view that rigorous preprocessing is indispensable for reliable metagenomic inference.

Although preprocessing strategies vary among studies, a four-step structure—comprising raw data assessment, adapter and quality filtering, host read depletion, and final clean-read evaluation—has become a widely used and effective framework in metagenomic research. This chapter provides an overview of each step, emphasizing practical considerations, widely used tools, and key decision points. By adopting standardized preprocessing practices, researchers can enhance reproducibility, minimize technical bias, and ensure that downstream analyses accurately reflect the biological composition of microbial communities. A typical end-to-end workflow for metagenomic data analysis is summarized in Fig. 1.

2 Raw Data Assessment

The FASTQ format is the standard file type for storing raw sequencing reads and their base-level quality scores. Each record consists of four lines—read identifier, nucleotide sequence, a

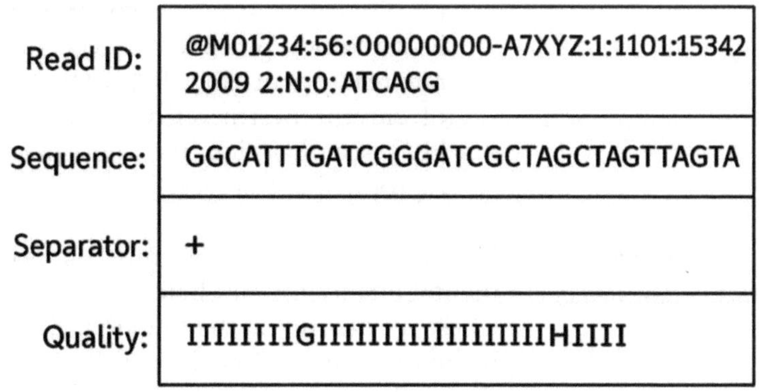

Fig. 2 An illustration of fastq format

2.1 FASTQ Format and Associated Metadata

separator line, and ASCII-encoded Phred quality scores. This integrated structure enables direct evaluation of sequencing accuracy and is the foundation of all raw data quality assessment.

1. *Structure of FASTQ and the Meaning of Phred Scores*

   The quality string in FASTQ encodes Phred scores (*Q*), a logarithmic measure of base-calling accuracy, where $Q = -10 \log_{10} P(error)$. Modern Illumina systems encode Phred scores with an ASCII offset of 33. Understanding this mapping is essential because quality-control tools interpret these scores to generate diagnostic metrics such as per-base quality plots, error rate estimations, and identification of problematic cycles. The basic four-line structure of a FASTQ record and its associated quality string are illustrated in Fig. 2.

2. *Information Contained in FASTQ Headers*

   Read identifiers typically include flowcell ID, lane number, tile coordinates, read direction (R1/R2), and sample barcodes. These metadata allow detection of platform-specific artifacts—for example, tile failures, lane-specific batch effects, or incorrect pairing during demultiplexing. Abnormalities in a subset of reads (e.g., periodic drops in quality or lane-biased GC shifts) are often traceable to these header components.

3. *Sample- and Library-Level Metadata Embedded in FASTQ*

   Many workflows incorporate unique molecular identifiers (UMIs), dual indices, or barcode sequences into read headers. These elements support the detection of barcode hopping, index cross-talk, and reagent-derived contamination—issues especially problematic in low-biomass metagenomes [11]. Correct interpretation of these metadata ensures that the sequenced reads can be confidently attributed to the correct biological sample.

4. *Metadata Relevance for Downstream QC and Reproducibility*

   FASTQ metadata determines how subsequent QC steps are applied. For instance, adapter detection requires knowing the library preparation kit (e.g., Nextera vs. TruSeq), while host-removal pipelines rely on correctly paired reads and accurate read orientation. Reporting FASTQ-level metadata is part of current best-practice guidelines for metagenomic studies [12], emphasizing its importance in ensuring transparent, reproducible analyses.

2.2 Key Quality Metrics in Raw Reads

Raw metagenomic reads must be evaluated using several fundamental quality metrics before any trimming or filtering is applied. These metrics reveal the underlying accuracy of sequencing, detect technical artifacts, and help determine how aggressively the data should be preprocessed. The following indicators constitute the core of raw-read quality assessment across metagenomic workflows.

1. *Per-Base Phred Quality Scores*

   Phred quality scores estimate the probability of incorrect base calls; they typically decline toward the 3′ end of reads due to reduced signal intensity in later sequencing cycles. Quality-control tools such as FastQC generate per-base quality plots that help identify anomalous patterns—including periodic fluctuations, abrupt drops, or tile-linked artifacts—which may arise from flowcell imperfections, overclustered lanes, or platform-specific chemistry issues [13]. Consistently low-quality tails indicate the need for 3′ end trimming, while widespread quality deterioration may suggest library preparation failure.

2. *GC Content Distribution*

   GC content is one of the most sensitive indicators of contamination or misprepared libraries. In microbial metagenomics, the expected GC distribution is typically broad yet unimodal, reflecting diverse taxa. Deviations—such as shoulder peaks, sharp bimodality, or pronounced shifts—can indicate reagent contaminants, co-sequenced host DNA, or unintended amplification biases [14]. For low-biomass samples, unexpected GC peaks often signal exogenous DNA from kits or airborne contaminants [15]. GC anomalies are therefore used both for contamination detection and to assess the biological plausibility of the sample.

3. *Frequency of Ambiguous Bases (N Content)*

   Ambiguous bases ("*N*") arise when the sequencer cannot confidently assign a nucleotide at a specific cycle. Elevated *N* frequencies typically indicate signal dropout, lane-specific issues, or chemical interference during sequencing. While occasional Ns are expected, sustained spikes—especially at identical cycles across reads—suggest systematic instrument failure or poor-quality library inserts. Samples with >1–2% Ns often require aggressive filtering or resequencing, as high *N* rates compromise mapping accuracy and assembly continuity.

4. *Overrepresented and Duplicate Sequences*

   FastQC identifies sequences present at disproportionately high frequencies, including:

   - Residual adapters.
   - PCR duplicates.
   - Primer sequences.
   - PhiX control reads.
   - Reagent-derived microbial contaminants ("kitome").
   - Repetitive artifacts from tagmentation or short-fragment libraries.

   In metagenomic datasets, overrepresented sequences must be interpreted carefully. PCR duplicates may reflect low input biomass, while short-fragment overrepresentation indicates degraded DNA or suboptimal library size selection [16]. Persistent detection of the same microbial species across unrelated samples—especially *Ralstonia*, *Cutibacterium*, *Escherichia*, or *Pseudomonas*—is a hallmark of background contamination [17]. Identifying these sequences early prevents misinterpretation of environmental, clinical, or host-associated signals.

5. *Read Length Distribution and Complexity*

   Raw-read length distributions provide insight into the success of library construction. A tight distribution indicates consistent fragment size selection, while widespread variability suggests library degradation or incomplete tagmentation. Read complexity metrics—such as the proportion of low-complexity sequences, homopolymers, or repetitive motifs—further help diagnose amplification biases or platform-specific artifacts [18]. Low-complexity reads frequently originate from failed library preparations and can severely impair assembly and taxonomic classification.

Together, these quality metrics form the diagnostic foundation of metagenomic preprocessing. Evaluating them before filtering ensures that trimming parameters are data-driven, contamination is detected early, and downstream analyses—assembly, profiling, or functional annotation—are based on reliable, biologically meaningful reads.

2.3 Identifying Adapter Contamination

Adapter contamination is one of the most frequent technical artifacts observed in raw metagenomic reads. It typically arises when the DNA insert is shorter than the read length, causing the sequencer to read into the library adapter sequence ("read-through"). This problem is especially common in environmental and clinical metagenomes, where DNA is often fragmented or size selection is inconsistent.

In quality-control reports, adapter contamination exhibits several recognizable patterns. FastQC's *Adapter Content* module displays a characteristic rise in adapter-derived k-mers toward the 3′ end of reads. The *Overrepresented Sequences* module often lists fragments matching Illumina TruSeq or Nextera adapters, while unusual convergence of A/T/G/C composition in late cycles may indicate partial adapter read-through. These signatures should be interpreted together, as they strengthen the evidence for adapter presence even when the exact adapter sequence is unknown [19].

Identifying adapters early is essential because untrimmed adapter sequences distort multiple downstream steps. They bias k-mer distributions used in taxonomic classification, fragment contigs during assembly, introduce spurious open reading frames, and inflate duplicate read counts—problems that are difficult to correct later in the analysis. For this reason, most preprocessing pipelines rely on FastQC for initial detection, followed by automated confirmation using tools such as fastp (auto-detection), Cutadapt (user-defined adapters), or AdapterRemoval (optimized for short-insert libraries).

In practice, accurate adapter identification depends on having correct library metadata. Different library preparation kits use distinct adapter motifs (e.g., Illumina TruSeq, Nextera XT, dual-index barcodes), and misidentifying the adapter sequence can lead to incomplete trimming. Therefore, adapter screening should always be performed before any quality filtering, host depletion, or merging steps to ensure that the raw data reflect a true biological signal rather than technical artifacts.

2.4 Tools and Usage

A range of computational tools supports the assessment of raw metagenomic read quality. These tools do not modify data; instead, they generate diagnostic summaries revealing technical artifacts such as base-calling errors, GC bias, adapter signatures, or contamination. Effective use of these tools allows researchers to determine whether sequencing output is trustworthy and guides the choice of trimming and filtering strategies in later steps. Because no single tool captures all aspects of quality, raw-read assessment typically relies on a small set of complementary programs.

To evaluate read-level accuracy and structural artifacts, tools such as FastQC and fastp generate modular reports summarizing base quality, sequence composition, duplication rates, and overrepresented sequences. Multisample projects additionally benefit from MultiQC, which aggregates all QC outputs into a single summary, enabling detection of lane-specific or batch-related issues. To assess biological plausibility, contamination-screening tools such as FastQ Screen [20] or Kraken2 [21] (used with a small QC database) can quickly reveal host-derived reads, cross-sample contamination, or reagent-associated taxa. For long-read sequencing data, dedicated diagnostics such as NanoPlot, NanoPack [22] and LongQC [23]

Table 1
Summary of common QC tools

Tool	Primary purpose	Assessment focus
FastQC	Single-sample QC	Base quality, GC, adapters, duplication
fastp (QC-only)	Fast QC reporting	Quality trends, duplication, auto adapter detection
SeqKit/seqtk	Quick statistics	Counts, GC%, *N* content
MultiQC	Multisample summary	Batch effects, cross-sample variation
FastQ Screen	Contamination screening	Host/cross-sample contamination
Kraken2 (QC DB)	Taxonomic plausibility	Unexpected taxa, kitome contaminants
NanoPlot/NanoPack LongQC	Long-read QC	Read-length, error patterns

capture read-length distributions and platform-specific error signatures. Together, these tools provide a comprehensive quality portrait before any trimming or filtering is performed. Commonly used assemblers for short-read, long-read, and hybrid metagenomic data, together with their key features, strengths, and typical use cases, are summarized in Table 1.

1. *Read-Level Diagnostic Tools*

   Used to examine per-base quality, GC content, *N* frequency, and adapter signatures.

   - *FastQC*—Standard single-sample QC; modular panels.
   - *fastp(QC Mode)*—Fast generation of HTML QC reports with automatic adapter detection.
   - *SeqKit/seqtk*—Quick statistics for sanity checks.

2. *Multisample QC Summary*

   Used to compare samples and detect batch effects.

   - *MultiQC*—Aggregates QC outputs, highlights global trends and anomalies.

3. *Contamination-Screening Tools*

   Used to detect host DNA, cross-sample contamination, and kitome taxa.

   - *FastQ Screen*—Maps reads to multiple reference genomes.
   - *Kraken2/Centrifuge(QC database)*—Rapid detection of unexpected taxa.

4. *Long-Read QC Tools*

   Used for Nanopore/PacBio data, where read-length and error patterns differ.

- *NanoPlot/NanoQC*—Read-length and quality summaries.
- *LongQC*—Long-read complexity and error-profile diagnostics.

Basic Usage Examples
FastQC—generate quality reports:

```
fastqc sample_R1.fastq.gz sample_R2.fastq.gz -o
qc_reports/
```

MultiQC—aggregate FastQC results:

```
multiqc qc_reports/ -o multiqc_summary/
```

fastp—QC-only report:

```
fastp \
-i sample_R1.fastq.gz \
-I sample_R2.fastq.gz \
-h sample_fastp.html \
-j sample_fastp.json
```

FastQ Screen—detect contamination:

```
fastq_screen sample_R1.fastq.gz --outdir screen_results/
```

Kraken2 (QC database)—quick taxonomic screen:

```
kraken2 --db qc_mini_db sample.fastq.gz --report sample.
qc.report
```

SeqKit—quick sanity check:

```
seqkit stats sample.fastq.gz
```

3 Adapter Removal and Quality Filtering

3.1 Rationale and Principles

Once raw metagenomic reads have been assessed and deemed technically consistent, the next step is to correct identifiable artifacts that compromise downstream analyses. This stage—adapter removal and quality filtering—forms the core of metagenomic quality control. Its rationale is rooted in the reality that high-throughput sequencing platforms, especially Illumina systems, frequently generate artificial signals: short-insert libraries produce adapter read-through events; terminal bases show reduced base-calling accuracy; and random low-quality regions accumulate across cycles. Without timely correction, these artifacts propagate into assembly fragmentation, false taxonomic assignments, inflated error rates in variant calling, and biased estimates of community diversity [24].

Adapter sequences pose the most immediate risk because they introduce synthetic k-mers that do not exist in any biological genome. Even small amounts of untrimmed adapter can distort k-mer spectra, mislead de Bruijn graph assemblers, and inflate duplication levels detected in QC diagnostics. Similarly, low-quality bases—particularly at the 3′ ends of reads—introduce nucleotide miscalls that mimic biological variation, causing spurious SNPs or false detection of low-abundance taxa. Importantly, aggressive quality filtering carries its own risks: excessive trimming shortens reads, reduces effective coverage, and disproportionately removes data belonging to low-abundance organisms. Therefore, the guiding principle of this stage is not to maximize the amount of trimmed data but to remove the minimum sequence necessary to eliminate technical noise while retaining maximal biological information. Consistency across samples is equally vital; applying different trimming thresholds introduces artificial differences in read length, coverage, or diversity, which can incorrectly inflate beta-diversity or distort ecological interpretations.

3.2 Adapter Removal

Adapter removal targets synthetic oligonucleotides incorporated during library preparation whose presence in reads is a consequence of short insert sizes or read-through during paired-end sequencing. In Illumina workflows, the most common pattern is 3′ read-through: when the actual DNA fragment is shorter than the read length, sequencing extends into the adapter, generating hybrid reads composed of biological sequence followed by synthetic motifs such as AGATCGGAAGAGC for Nextera or TruSeq kits [25]. These adapter-derived sequences are non-genomic, nonrandom, and strongly distort k-mer-based analyses. Their presence also masks true fragment boundaries, complicating downstream merging, mapping, or assembly.

Accurate identification of adapters begins in the assessment stage: FastQC and MultiQC typically reveal elevated adapter content, abnormal 3′ sequence composition, or overrepresented sequences that match known adapters. In this step, the goal is to remove these segments reliably and without excessive loss of biological sequence. Two strategies dominate modern practice. The first is explicit adapter trimming, in which the exact adapter sequence is provided to a tool such as Cutadapt or Atropos. These tools locate adapter-read overlaps based on alignment algorithms that account for sequencing errors, allowing controlled specification of minimum overlap length, maximum mismatch rate, and anchored versus unanchored trimming. The second is automatic adapter detection, implemented in tools like Trimmomatic or fastp, which infer adapter signatures directly from data, useful when library metadata are incomplete.

A key consideration is parameter calibration. Overly permissive overlap thresholds may inadvertently trim true biological sequence, especially in regions with low complexity; conversely, overly strict thresholds miss partial adapters, leaving synthetic k-mers in the dataset. Error-rate parameters must reflect expected sequencing accuracy: allowing too many mismatches may yield false-positive trimming, whereas setting them too low fails to capture real adapter contamination. In paired-end data, synchronized trimming ensures that both reads maintain consistency after adapter removal. Ultimately, adapter trimming must balance sensitivity and specificity, removing obvious contamination while minimizing biological sequence loss.

3.3 Quality Filtering

Quality filtering aims to remove or trim regions where base-calling confidence is low. In Illumina sequencing, terminal cycles—particularly the 3′ end—exhibit systematic quality decay due to phasing, signal bleed, and reduced cluster density over time [26]. Low-quality regions manifest as reduced per-base Phred scores, elevated substitution rates, and increased ambiguity calls (Ns). If retained, such regions introduce random noise that inflates variant calling errors, generates chimeric contigs during assembly, and causes taxonomic classifiers to misassign reads, especially at low abundance.

Filtering strategies fall broadly into three categories. Base-level trimming removes low-quality segments from read termini until a threshold is reached. Sliding-window trimming evaluates quality over a short window (e.g., 4–10 bases) and trims when the window average falls below a defined Phred score. This adaptive strategy is widely used because it avoids trimming individual sporadic low-quality positions while still removing extended low-Q regions. Finally, minimum length filtering ensures that excessively short reads—often lacking discriminatory genomic content—do not enter downstream analysis. In metagenomics, setting this threshold requires caution: shorter reads may still contain meaningful taxonomic or functional signal; however, extremely short fragments (<50 bp) often contribute more noise than information.

One of the greatest challenges is avoiding over-filtering. Because microbial communities contain taxa with very low abundance or variable GC content, aggressive trimming disproportionately affects certain lineages, causing ecological signal loss. Differences in trimming parameters between samples—not biological variation—can become the dominant source of beta-diversity. Therefore, parameters must be standardized across all samples in a project. The objective is not to maximize stringency but to ensure balanced retention of sequence information while removing regions that would unquestionably pollute downstream computations.

3.4 Tools and Practical Usage

A small set of trimming and filtering tools dominates contemporary metagenomic workflows. Although their underlying algorithms differ, they all aim to remove adapter-derived sequences and low-quality regions while preserving as much biological signal as possible. In practice, the choice of tool is less important than the consistent and well-documented use of parameters across all samples in a study. This section provides a unified overview of the most commonly used tools—Cutadapt, Trimmomatic, and fastp—followed by practical usage examples suitable for both small-scale and large-scale projects. Typical parameter choices for short-read metagenomic assembly are summarized in Table 2.

Cutadapt is the most widely used tool for precise adapter removal because it allows explicit specification of adapter motifs, mismatch tolerance, anchoring, and minimum overlap. Developed originally for small-RNA applications, it has become the de facto standard for metagenomic trimming due to its robust overlap-detection algorithm [27]. Its alignment-based approach offers fine control, making it popular in studies where library metadata are known or where partial adapters must be trimmed accurately without over-trimming the biological sequence.

Trimmomatic, one of the earliest widely adopted trimmers, remains in use for its sliding-window trimming strategy and its flexible modules for leading/trailing trimming, palindrome-mode adapter detection, and paired-end synchronization. Introduced in 2014, it became foundational to early shotgun metagenomics

Table 2
Commonly used tools for adapter removal and quality filtering

Tool	Primary functions	Strengths	Limitations
Cutadapt	Adapter trimming, overlap-based removal	Highly precise; fine parameter control	Requires known adapter sequences
Trimmomatic	Sliding-window trimming, adapter removal, paired-end sync	Stable; widely used; flexible modules	Slower; less automated
fastp	Auto adapter detection, quality trimming, QC report	Fast; all-in-one; strong defaults	Less tunable for edge cases
Atropos	Cutadapt-like trimming with additional filters	High accuracy; Python API	Less widely adopted
Seqtk	Simple trimming & quality filtering	Very fast	Limited adapter functionality

pipelines [28] and continues to be used in many legacy workflows due to its stability and reproducibility.

fastp represents a newer generation of QC tools that integrate adapter detection, base-quality trimming, per-read complexity assessment, poly-G tail detection for two-color chemistry, and QC reporting into a single executable. Its high speed, automatic adapter inference, and built-in HTML reporting make it especially suitable for large metagenomic datasets and for situations where metadata are incomplete [29, 9]. Comparative benchmarking studies have shown that fastp often matches or exceeds Trimmomatic and Cutadapt in trimming accuracy while being significantly faster [9].

Despite differences in design philosophy, all three tools share several operational principles grounded in best practices from large-scale metagenomic projects such as the Human Microbiome Project and the Earth Microbiome Project [30]. First, trimming parameters must be identical across samples to avoid introducing batch-specific artifacts, which can lead to spurious differences in diversity or composition [31]. Second, adapter removal should precede quality filtering so that synthetic sequences do not distort quality-score profiles or sliding-window decisions [32]. Third, paired-end synchronization is essential to avoid orphaned reads that cause mapping and assembly errors. Finally, the output of trimming should be inspected—typically with FastQC or MultiQC—to ensure that adapter contamination and low-quality tails are indeed removed without excessive loss of read length or coverage [33].

Below are concise, instructional examples illustrating typical usage in metagenomic workflows. These commands are intentionally minimal and emphasize clarity rather than optimization.

Cutadapt

```
cutadapt \
-a AGATCGGAAGAGC \
-A AGATCGGAAGAGC \
-o sample_R1.trim.fq.gz \
-p sample_R2.trim.fq.gz \
sample_R1.fq.gz sample_R2.fq.gz
```

Trimmomatic

```
trimmomatic PE \
sample_R1.fq.gz sample_R2.fq.gz \
R1.paired.fq.gz R1.unpaired.fq.gz \
R2.paired.fq.gz R2.unpaired.fq.gz \
SLIDINGWINDOW:4:20 MINLEN:50
```

fastp

```
fastp \
-i sample_R1.fq.gz \
-I sample_R2.fq.gz \
-o sample_R1.clean.fq.gz \
-O sample_R2.clean.fq.gz \
-h sample_fastp.html \
-j sample_fastp.json
```

4 Host and Nontarget Read Removal

4.1 When Host Removal Is Necessary (and the Risks of Not Doing It)

Host and nontarget read removal is not required for all metagenomic studies; its necessity depends on sample type and the balance between host background and microbial biomass. The primary goal is to enhance microbial signal, reduce false annotations, and control computational cost. In samples where host DNA dominates, failure to remove host reads can produce stronger artifacts than any downstream analytical step.

Situations Where Host Removal Is Essential

1. *Human and Animal-Associated Samples (High Host Background)*

   Oral, skin, nasal, lung, blood, and tissue biopsies often contain overwhelming human DNA. Multiple studies report *60–99% host reads* in such datasets [34].

   If not removed:

   - Loss of microbial coverage; low-abundance taxa become undetectable.
   - Assembly collapses due to host-dominated graphs.
   - Host fragments falsely classified as microbes [35].
   - Computational burden increases by an order of magnitude.

2. *Plant Host Samples (Chloroplast/mitochondria-Dominated)*

   Leaf and root tissues contain abundant organellar DNA. Chloroplast reads are frequently misclassified as bacteria due to homology [36].

   If not removed:

   - Artificial inflation of bacterial diversity.
   - Incorrect pathway annotations from plant organelles.

3. *Virome Studies Where Host DNA Exceeds 95%*

   Most virus-enriched preparations contain >95% host DNA [37].

If not removed:

- Viral genome recovery becomes impossible.

Situations Where Host Removal Is Generally Unnecessary

- Soil, sediments, wastewater, marine water, river water.
- Samples where no single organism is a "host" of the ecosystem.
- Removing host sequences would distort the real ecological structure [38].

4.2 General Workflow for Host and Nontarget Read Removal

Despite variation in software, the workflow across studies is highly standardized:

1. *Define the host reference.*
   - Human: GRCh38, CHM13.
   - Mouse: GRCm39.
   - Plant: nuclear + mitochondrial + chloroplast genomes.
   - Non-model species: closest phylogenetic reference, or de novo host assembly [39].
2. *Map raw or trimmed reads to the host reference.*
   Widely used high-speed aligners:
   - *Bowtie2* [40].
   - *BWA-MEM* [41].
   - *Minimap2* for long reads [42].
3. *Filter mapped reads.*
   Retain only the unmapped reads using SAMtools or BBMap:
   - SAMtools: samtools view -b -f 4.
   - BBMap: bbsplit.sh with host genome.
4. *Validate the depletion step.*
   Use FastQC/MultiQC to confirm:
   - Remaining host content is minimal.
   - Read length and quality are not inadvertently reduced.
   - Microbial coverage increases [43].

4.3 Practical Procedures and Typical Commands

Below are concise and commonly used procedures applicable to human, plant, and animal microbiome studies.

A. *Human or Animal Samples*

Step 1: Index host genome.

```
bowtie2-build GRCh38.fa human_index
```

Step 2: Align reads to the host reference.

```
bowtie2 -x human_index -1 sample_R1.fq -2 sample_R2.fq \
--very-sensitive -p 8 -S host.sam
```

Step 3: Extract unmapped reads only.

```
samtools view -b -f 12 -F 256 host.sam > host_removed.bam
samtools fastq host_removed.bam -1 clean_R1.fq -2
clean_R2.fq
```

Empirically, this removes *70–99%* of reads in oral/skin/nasal datasets [44]

B. *Plant Samples (Including Chloroplast and Mitochondria)*

Reference Setup

Concatenate nuclear, mitochondrial, and chloroplast genomes:

```
cat nuclear.fa mito.fa chloro.fa > plant_host.fa
```

Align and filter are identical to the human workflow.

Plant organelles often account for *50–80%* of reads [45].

C. *Non-Model Host Samples*

For insects, amphibians, fish, etc., where genome references are incomplete:

1. Use the *closest phylogenetic reference* (e.g., same genus).
2. Or generate a *pseudo-reference* using:
   - De novo host assembly from high-depth host reads.
   - Reference-guided scaffolding [46].

5 Final Evaluation of Clean Reads

After trimming, quality filtering, and host-removal, a brief final evaluation is required to confirm whether the processed reads are suitable for downstream metagenomic analysis. This step does not duplicate raw QC; instead, it ensures that preprocessing has *improved* data quality without introducing technical distortions. International pipelines, such as KneadData, Sunbeam, MGnify, and nf-core/mag, all include a similar "final check" before assembly [47].

The primary goal of this evaluation is to verify *data integrity*—that sufficient sequencing information has been retained and that the distribution of read lengths, GC content, and quality scores remains biologically reasonable. Clean reads are generally considered high-quality when they satisfy four practical and widely adopted criteria.

1. *Adequate read length.* Excessive trimming can remove biological signal and reduce mappability. For Illumina short reads, the modal read length should remain ≥*80–90%* of the original (e.g., ≥120–135 bp for 150 bp reads), and the mean length should generally be ≥*75–100 bp* [48].
2. *Stable per-base quality.* Clean reads should retain high Q-scores, typically ≥*70–85% Q30* and ≥*90–95% Q20*, matching Illumina performance benchmarks. Pronounced quality collapse at read ends or large shifts between samples often indicate library preparation inconsistencies [49].
3. *Consistent GC composition.* GC distributions should resemble those of raw reads or known environmental expectations. GC shifts >5% or bimodal peaks commonly reflect contamination, over-trimming, or disproportionate removal of specific sequence types [50].
4. *Sufficient sequencing depth.* Downstream assembly or profiling generally requires ≥*1–5 million paired-end reads per sample*, and many metagenomic assemblies perform best with ≥*10–20 million reads*. Samples retaining unusually low read counts (<20–30% of the raw reads) should be flagged for reprocessing or exclusion [51].

A second objective of the final evaluation is to confirm *between-sample consistency*, which is essential for comparative metagenomics. MultiQC summaries allow researchers to quickly identify samples with abnormal read retention, GC shifts, duplication rates, or host-removal percentages. Such outliers often arise from sample degradation, DNA extraction issues, or inconsistent filtering parameters and may generate spurious clustering or false differential abundance patterns [52]. Detecting these anomalies before assembly prevents extensive downstream artifacts.

Tools used in this stage are deliberately simple and lightweight. *FastQC* provides read-level metrics (length distribution, Q-scores, GC content), while *MultiQC* aggregates cross-sample patterns and highlights outliers [53]. For host-associated samples, residual host signals can be confirmed through *Bowtie2/BWA host-mapping statistics.* For microbial complexity and duplication assessment, researchers may optionally use *seqkit, BBTools*, or *Picard MarkDuplicates.*

Finally, researchers should document a compact summary of post-QC metrics—such as retained read count, mean read length, GC percentage, Q20/Q30, duplication rate, and host-removal fraction—in the Methods or Supplementary Information. This practice enhances transparency and ensures that the clean reads entering assembly and profiling are demonstrably high-quality, biologically representative, and suitable for robust metagenomic inference.

Conflict of Interest

The authors have no conflicts of interest to declare that are relevant to the content of this chapter.

References

1. Jeske JT, Gallert C (2022) Microbiome analysis via OTU and ASV-based pipelines—a comparative interpretation of ecological data in WWTP systems. Bioengineering 9(4):146
2. Meziti A, Rodriguez-R LM, Hatt JK, Peña-Gonzalez A, Levy K, Konstantinidis KT (2021) The reliability of metagenome-assembled genomes (MAGs) in representing natural populations: insights from comparing MAGs against isolate genomes derived from the same fecal sample. Appl Environ Microbiol 87(6):e02593–e02520
3. Shaffer JP, Nothias L-F, Thompson LR, Sanders JG, Salido RA, Couvillion SP, Brejnrod AD, Lejzerowicz F, Haiminen N, Huang S (2022) Standardized multi-omics of earth's microbiomes reveals microbial and metabolite diversity. Nat Microbiol 7(12):2128–2150
4. Eisenhofer R, Minich JJ, Marotz C, Cooper A, Knight R, Weyrich LS (2019) Contamination in low microbial biomass microbiome studies: issues and recommendations. Trends Microbiol 27(2):105–117
5. FastQC A (2015) Quality control tool for high throughput sequence data. BibSonomy. Accessed on 17 Mar 2022. Available online: https://www.bibsonomy.org/bibtex/f230a919c34360709aa298734d63dca31
6. Ewels P, Magnusson M, Lundin S, Käller M (2016) MultiQC: summarize analysis results for multiple tools and samples in a single report. Bioinformatics 32(19):3047–3048
7. Martin M (2011) Cutadapt removes adapter sequences from high-throughput sequencing reads. EMBnet J 17(1):10–12
8. Bolger AM, Lohse M, Usadel B (2014) Trimmomatic: a flexible trimmer for Illumina sequence data. Bioinformatics 30(15): 2114–2120
9. Chen S, Zhou Y, Chen Y, Gu J (2018) fastp: an ultra-fast all-in-one FASTQ preprocessor. Bioinformatics 34(17):i884–i890
10. Constantinides B, Hunt M, Crook DW (2023) Hostile: accurate decontamination of microbial host sequences. Bioinformatics 39(12): btad728
11. Hornung BV, Zwittink RD, Kuijper EJ (2019) Issues and current standards of controls in microbiome research. FEMS Microbiol Ecol 95(5):fiz045
12. Gohl DM, Vangay P, Garbe J, MacLean A, Hauge A, Becker A, Gould TJ, Clayton JB, Johnson TJ, Hunter R (2016) Systematic improvement of amplicon marker gene methods for increased accuracy in microbiome studies. Nat Biotechnol 34(9):942–949
13. Smith DD, Tyler AD (2025) Considerations of depth, coverage, and other read quality metrics. In: Manual of molecular microbiology: fundamentals and applications. Wiley, pp 244–261
14. Stepanauskas R, Fergusson EA, Brown J, Poulton NJ, Tupper B, Labonté JM, Becraft ED, Brown JM, Pachiadaki MG, Povilaitis TJ (2017) Improved genome recovery and integrated cell-size analyses of individual uncultured microbial cells and viral particles. Nat Commun 8(1):84
15. Weyrich LS, Farrer AG, Eisenhofer R, Arriola LA, Young J, Selway CA, Handsley-Davis M, Adler CJ, Breen J, Cooper AJ (2019) Laboratory contamination over time during low-biomass sample analysis. Mol Ecol Resour 19(4):982–996
16. De Barba M, Miquel C, Lobréaux S, Quenette P, Swenson J, Taberlet PJ (2017) High-throughput microsatellite genotyping in ecology: improved accuracy, efficiency, standardization and success with low-quantity and degraded DNA. Mol Ecol Resour 17(3): 492–507
17. Ferone M, Gowen A, Fanning S, Scannell AG (2020) Microbial detection and identification methods: bench top assays to omics approaches. Compr Rev Food Sci Food Saf 19(6):3106–3129
18. Pfeifer SJ (2017) From next-generation resequencing reads to a high-quality variant data set. Heredity (Edinb) 118(2):111–124
19. Schubert M, Lindgreen S, Orlando LJ (2016) AdapterRemoval v2: rapid adapter trimming, identification, and read merging. BMC Res Notes 9(1):88
20. Wingett SW, Andrews S (2018) FastQ screen: a tool for multi-genome mapping and quality control. F1000Res 7:1338

21. Wood DE, Lu J, Langmead B (2019) Improved metagenomic analysis with Kraken 2. Genome Biol 20(1):257
22. De Coster W, D'hert S, Schultz DT, Cruts M, Van Broeckhoven C (2018) NanoPack: visualizing and processing long-read sequencing data. Bioinformatics 34(15):2666–2669
23. Fukasawa Y, Ermini L, Wang H, Carty K, Cheung M-S (2020) LongQC: a quality control tool for third generation sequencing long read data. G3: Genes, Genomes, Genetics 10(4):1193–1196
24. Orlando L, Allaby R, Skoglund P, Der Sarkissian C, Stockhammer PW, Ávila-Arcos MC, Fu Q, Krause J, Willerslev E, Stone AC (2021) Ancient DNA analysis. Nat Rev Method Primers 1(1):14
25. Weirather JL, de Cesare M, Wang Y, Piazza P, Sebastiano V, Wang X-J, Buck D, Au KF (2017) Comprehensive comparison of Pacific Biosciences and Oxford Nanopore Technologies and their applications to transcriptome analysis. F1000Res 6:100
26. Wang W, Vilella F, Alama P, Moreno I, Mignardi M, Isakova A, Pan W, Simon C, Quake SR (2020) Single-cell transcriptomic atlas of the human endometrium during the menstrual cycle. Nat Med 26(10):1644–1653
27. Alic AS, Ruzafa D, Dopazo J, Blanquer I (2016) Objective review of de novo standalone error correction methods for NGS data. Wiley Interdiscipl Rev: Comput Mol Sci 6(2):111–146
28. Quince C, Walker AW, Simpson JT, Loman NJ, Segata N (2017) Shotgun metagenomics, from sampling to analysis. Nat Biotechnol 35(9):833–844
29. Estaki M, Jiang L, Bokulich NA, McDonald D, González A, Kosciolek T, Martino C, Zhu Q, Birmingham A, Vázquez-Baeza Y (2020) QIIME 2 enables comprehensive end-to-end analysis of diverse microbiome data and comparative studies with publicly available data. Curr Protoc Bioinformatics 70(1):e100
30. Knight R, Vrbanac A, Taylor BC, Aksenov A, Callewaert C, Debelius J, Gonzalez A, Kosciolek T, McCall L-I, McDonald D (2018) Best practices for analysing microbiomes. Nat Rev Microbiol 16(7):410–422
31. Yu Y, Mai Y, Zheng Y, Shi L (2024) Assessing and mitigating batch effects in large-scale omics studies. Genome Biol 25(1):254
32. Fages A, Hanghøj K, Khan N, Gaunitz C, Seguin-Orlando A, Leonardi M, Constantz CM, Gamba C, Al-Rasheid KA, Albizuri S (2019) Tracking five millennia of horse management with extensive ancient genome time series. Cell 177(6):1419–1435.e1431
33. Chen S, Huang T, Zhou Y, Han Y, Xu M, Gu J (2017) AfterQC: automatic filtering, trimming, error removing and quality control for fastq data. BMC Bioinform 18(Suppl 3):80
34. Muus C, Luecken MD, Eraslan G, Sikkema L, Waghray A, Heimberg G, Kobayashi Y, Vaishnav ED, Subramanian A, Smillie C (2021) Single-cell meta-analysis of SARS-CoV-2 entry genes across tissues and demographics. Nat Med 27(3):546–559
35. Guo J, Bolduc B, Zayed AA, Varsani A, Dominguez-Huerta G, Delmont TO, Pratama AA, Gazitúa MC, Vik D, Sullivan MB (2021) VirSorter2: a multi-classifier, expert-guided approach to detect diverse DNA and RNA viruses. Microbiome 9(1):37
36. Hancock JF (2012) Plant evolution and the origin of crop species. Cabi
37. Barba M, Czosnek H, Hadidi A (2014) Historical perspective, development and applications of next-generation sequencing in plant virology. Viruses 6(1):106–136
38. Hale RC, Seeley ME, La Guardia MJ, Mai L, Zeng EY (2020) A global perspective on microplastics. J Geophys Res Oceans 125(1):e2018JC014719
39. Garg R, Patel RK, Tyagi AK, Jain M (2011) De novo assembly of chickpea transcriptome using short reads for gene discovery and marker identification. DNA Res 18(1):53–63
40. Langmead B, Wilks C, Antonescu V, Charles R (2019) Scaling read aligners to hundreds of threads on general-purpose processors. Bioinformatics 35(3):421–432
41. Han Y, Gao S, Muegge K, Zhang W, Zhou B (2015) Advanced applications of RNA sequencing and challenges. Bioinform Biol Insights 9. https://doi.org/10.4137/BBI.S28991
42. Kovaka S, Fan Y, Ni B, Timp W, Schatz MC (2021) Targeted nanopore sequencing by real-time mapping of raw electrical signal with UNCALLED. Nat Biotechnol 39(4):431–441
43. Gyuris A, Navarrete-Perea J, Jo A, Cristea S, Zhou S, Fraser K, Wei Z, Krichevsky AM, Weissleder R, Lee H (2019) Physical and molecular landscapes of mouse glioma extracellular vesicles define heterogeneity. Cell Rep 27(13):3972–3987.e3976
44. Yan A (2025) The human intestinal virome at the mucosal-luminal interface. Université d'Ottawa| University of Ottawa
45. Jin J-J, Yu W-B, Yang J-B, Song Y, DePamphilis CW, Yi T-S, Li D-Z (2020) GetOrganelle: a

fast and versatile toolkit for accurate de novo assembly of organelle genomes. Genome Biol 21(1):241

46. Alonge M, Wang X, Benoit M, Soyk S, Pereira L, Zhang L, Suresh H, Ramakrishnan S, Maumus F, Ciren D (2020) Major impacts of widespread structural variation on gene expression and crop improvement in tomato. Cell 182(1):145–161.e123
47. Schumilo S, Aschermann C, Gawlik R, Schinzel S, Holz T (2017) {kAFL}:{Hardware-Assisted} feedback fuzzing for {OS} kernels. In: 26th USENIX security symposium (USENIX Security 17), pp 167–182
48. Lopes I, Altab G, Raina P, De Magalhães JP (2021) Gene size matters: an analysis of gene length in the human genome. Front Genet 12: 559998
49. Van Dijk EL, Jaszczyszyn Y, Thermes CJ (2014) Library preparation methods for next-generation sequencing: tone down the bias. Exp Cell Res 322(1):12–20
50. Kumar A, Sharma CJ (2022) Recent update of the various sources originating ghost peaks in gas chromatography: a review. J Chromatography A 1685:463625
51. McKnight DT, Huerlimann R, Bower DS, Schwarzkopf L, Alford RA, Zenger KR (2019) microDecon: a highly accurate read-subtraction tool for the post-sequencing removal of contamination in metabarcoding studies. Environ DNA 1(1):14–25
52. Reitmeier S, Hitch TC, Treichel N, Fikas N, Hausmann B, Ramer-Tait AE, Neuhaus K, Berry D, Haller D, Lagkouvardos IJ (2021) Handling of spurious sequences affects the outcome of high-throughput 16S rRNA gene amplicon profiling. ISME Commun 1(1):31
53. Yu Y, Hou W, Liu Y, Wang H, Dong L, Mai Y, Chen Q, Li Z, Sun S, Yang JJ (2024) Quartet RNA reference materials improve the quality of transcriptomic data through ratio-based profiling. Nat Biotechnol 42(7):1118–1132

Chapter 4

Metagenomic Assembly and Gene Prediction

Bo Li, Xue Yang, Tongyi Zhao, Jiying Xu, Qiyu Meng, Qidong Yin, and Yanting Zou

Abstract

Metagenomic assembly and gene prediction connect quality-controlled reads to downstream microbiome analyses. This chapter outlines core assembly strategies, including per-sample versus co-assembly and short-read versus hybrid approaches, and highlights key parameters and metrics for evaluating assembly quality. Gene prediction from contigs and the construction of nonredundant gene catalogs are introduced as fundamental steps for representing community coding potential. The resulting contigs and gene sets provide essential input for metagenome-assembled genome (MAG) reconstruction, as well as taxonomic and functional annotation in subsequent chapters.

Key words Metagenomic assembly, Contig, Hybrid sequencing, Gene prediction

1 Introduction

High-throughput metagenomic sequencing generates massive numbers of short reads from complex microbial communities, but these raw fragments are difficult to interpret directly. Assembly and gene prediction provide the critical bridge between quality-controlled reads and downstream ecological and functional analysis. By reconstructing longer contigs and identifying coding sequences, metagenomic workflows shift from base-level data to biologically meaningful units—putative genomes, genes, and gene catalogs—that can be annotated and compared across samples and environments [1, 2].

Metagenomic assembly is considerably more challenging than single-genome assembly because the reads come from many different organisms in the same sample, often with very different abundances and with many shared or exchanged DNA segments [3]. In such mixtures, sequences from dominant community members are assembled preferentially, whereas low-abundance genomes tend to remain highly fragmented or even completely unassembled

Tao Huang (ed.), *Microbial 'Omics: Methods and Protocols*, Methods in Molecular Biology, vol. 3033,
https://doi.org/10.1007/978-1-0716-5264-0_4,

[4]. Community complexity, sequencing depth, and library design all influence contiguity, accuracy, and computational cost of metagenomic assembly. Over the past decade, specialized assemblers, such as IDBA-UD, MEGAHIT, and metaSPAdes, have been developed to address these issues, introducing multi-k assembly strategies, more efficient graph representations (in particular, the succinct de Bruijn graph used by MEGAHIT), and metagenome-specific heuristics tailored to uneven coverage and complex communities [5, 6]. Study design choices, such as per-sample versus co-assembly and short-read versus hybrid (short + long read) approaches, also have a substantial impact on assembly outcomes and must be aligned with research goals [7, 8].

Once contigs have been generated, gene prediction converts anonymous DNA fragments into a catalog of putative coding sequences that captures the community's functional potential. In prokaryote-dominated metagenomes, tools such as Prodigal (metagenomic mode), MetaGeneMark, and FragGeneScan infer protein-coding genes from fragmentary sequences lacking complete genomic context [9–11]. Predicted genes are often clustered into non-redundant gene catalogs to reduce sequence redundancy, lower computational demands for annotation, and provide stable reference units for integrating large cohorts and multiple projects [12]. In special cases, metagenomes enriched in eukaryotic DNA may require prediction strategies that account for introns and more complex gene structures [13].

Assembly and gene prediction do not exist in isolation but form the substrate for genome-resolved and gene-centric analyses in later steps. Contigs and their coverage profiles are the basis for reconstructing metagenome-assembled genomes (MAGs) through binning [14, 15], while predicted genes underpin taxonomic assignment, functional annotation, and pathway reconstruction in downstream chapters. The quality of these products depends directly on assembly contiguity, error rates, and the completeness of predicted coding sequences. Building on the preprocessed reads from previous steps, this chapter therefore focuses on the concepts, tools, and evaluation metrics that govern metagenomic assembly and gene prediction and prepares the data structures required for MAG reconstruction and annotation in subsequent chapters.

2 Inputs and Design Considerations

2.1 Defining Suitable Input Reads

Metagenomic assembly should start with preprocessed, well-documented reads rather than raw sequencing files. In practice, suitable input consists of adapter-free, quality-trimmed reads that have been processed according to the workflow described previously, with paired-end relationships preserved (matching read counts and synchronized R1/R2 order). Any inconsistency between mate files will cause loss of pairing information and can fragment assemblies or make some assemblers fail.

For human- or animal-associated samples, assembling libraries that still contain a large proportion of host DNA is inefficient and typically produces assemblies dominated by host-derived contigs, which then complicate binning and annotation. In such cases, it is usually preferable to remove host reads by mapping to an appropriate reference genome (e.g., GRCh38 for human, or combined nuclear/mitochondrial/chloroplast genomes for plants) and to assemble only the unmapped fraction [16]. In contrast, for soils, sediments, wastewater, or seawater, where no single macroscopic host dominates, host depletion is generally not required and may even remove genuine eukaryotic signal.

In all cases, a minimal set of library metadata should be recorded alongside the input files: sequencing platform, read length, single-end versus paired-end layout, approximate insert-size distribution, and whether host-removal or duplicate-removal has been applied [17]. These variables do not enter the assembler directly, but they constrain tool choice and parameter ranges, and they are essential for interpreting assembly statistics and comparing results across samples and studies.

2.2 Per-Sample Assembly Versus Co-Assembly

A second key design decision is whether to perform per-sample assembly or co-assembly. Assembling each sample independently preserves genuine between-sample differences, keeps contigs naturally associated with individual samples, and reduces the risk of chimeric contigs that combine sequences from distinct environments or time points. This strategy is particularly appropriate for heterogeneous clinical cohorts, case-control designs, or studies where strong sample-to-sample variation is expected [18]. Its main limitation is that low-abundance genomes may never achieve sufficient coverage in any single sample to assemble well.

Co-assembly, in which reads from biologically related samples (e.g., time series, spatial gradients, replicate reactors, or individuals from the same niche) are pooled before assembly, can increase effective coverage for rare genomes and improve recovery of metagenome-assembled genomes (MAGs) [19]. The trade-off is higher computational demand and a greater risk of strain-level chimerism when closely related populations coexist in different samples.

In practice, many studies adopt a compromise: grouping samples by environment, host, or treatment, performing co-assembly within each group, and then mapping sample-specific reads back to the joint assembly. This design leverages shared depth while still allowing downstream, sample-resolved coverage estimation and binning, and it connects naturally to the MAG reconstruction workflow introduced in later chapters.

3 Principles and Strategies of Metagenomic Assembly

3.1 De Bruijn Graph Assembly in Metagenomics

Most short-read metagenomic assemblers are built on the de Bruijn graph framework. Instead of directly overlapping full reads, the assembler breaks each read into a series of short words of length *k* (k-mers). In a common implementation, each (k–1)-mer is represented as a node, and each k-mer becomes a directed edge connecting its prefix (k–1)-mer to its suffix (k–1)-mer. A contiguous path through this graph corresponds to a plausible reconstruction of an underlying DNA sequence: by walking along well-supported paths and concatenating the last base of each edge, the assembler recovers contigs. Sequencing errors typically generate low-frequency k-mers that form short dead-end branches, which can be pruned away using coverage and topology criteria [20, 21].

In a single-genome dataset, this graph is often dominated by a few long, linear paths interrupted by branches where repeats or errors occur. In metagenomics, however, the de Bruijn graph encodes all genomes in the community simultaneously, making it intrinsically more complex. Closely related species and strains share large stretches of sequence; their reads produce identical k-mers that collapse into shared nodes and edges, so what would be two separate genome paths in isolation becomes a braided subgraph. Mobile genetic elements—plasmids, phages, transposons—and horizontally transferred genes further complicate the picture by acting as hubs that connect otherwise distinct genomic backgrounds. As a result, the assembler must distinguish between true branching caused by biological diversity and spurious branching caused by sequencing errors and low coverage.

A second complication is uneven coverage. In real communities, some genomes are extremely abundant while others are rare. Abundant genomes generate high-coverage paths that resemble single-genome graphs, whereas low-abundance genomes contribute only a few k-mers, producing weakly supported paths that are difficult to separate from error-induced branches [22]. Aggressive graph simplification may therefore remove genuine low-abundance genomes, while permissive settings retain many erroneous branches and inflate the complexity of the graph. Metagenomic assemblers address this by combining coverage-based heuristics (removing very low-frequency k-mers) with strategies that take local graph structure into account when trimming or merging paths.

The choice of *k* controls how tangled the graph becomes. With a *small k*, many different genomic regions share the same k-mers, leading to a highly connected graph that tends to merge similar sequences and collapse closely related strains. With a *large k*, k-mers are more specific, which helps separate closely related genomes but requires higher coverage and makes low-abundance paths prone to fragmentation. Modern metagenomic assemblers therefore adopt *multi-k or iterative-k strategies*, starting from smaller *k* values to ensure connectivity and gradually increasing *k* to refine paths and resolve ambiguities [23, 24]. Conceptually, the assembly problem in metagenomics can be viewed as finding a compromise between graph simplification and preservation of biological diversity in this multi-genome de Bruijn graph.

3.2 Short-Read Versus Hybrid Assembly

From a practical perspective, the main question is not whether short reads or long reads are "better," but which strategy is appropriate for a given study design. Short-read Illumina data remain the default choice for most metagenomic projects because they provide high base accuracy, stable quality profiles, and cost-effective depth across many samples [25]. For large cohorts, time-series surveys, or studies whose primary goals are community composition, functional potential, and gene catalog coverage, short-read-only assembly is usually sufficient: contig-level resolution supports taxonomic profiling, pathway reconstruction, and comparative analyses, while keeping sequencing and computational costs manageable [26].

Hybrid assembly, which combines accurate short reads with long reads from platforms such as Oxford Nanopore or PacBio, becomes attractive when the focus shifts to genome-resolved questions. Long reads can bridge repeats, separate closely related strains, and recover complete chromosomes, plasmids, and phage genomes that remain fragmented in short-read assemblies [27, 28]. Typical scenarios include reconstruction of key pathogens or antibiotic-resistance plasmids, detailed characterization of low-diversity but genomically complex systems (e.g., bioreactors), and studies where circularization of MAGs is an explicit objective. In these settings, fewer samples are sequenced, but each sample is characterized more deeply and with multiple technologies.

A simple rule of thumb is therefore:

- Short-read-only assembly for many samples, broad ecological or functional comparisons, and when contig-level resolution is adequate.
- Hybrid assembly for fewer samples, genome-centric questions, and when resolving repeats, plasmids, or strain structure justifies the additional cost and complexity of long-read data and error-correction steps.

The rest of this chapter assumes short-read input as the baseline, while noting where hybrid workflows require extra steps (e.g., long-read correction or separate long-read assembly) before integrating into the same downstream processes of binning, annotation, and quantification.

4 Practical Metagenomic Assembly Workflow

4.1 Choice of Assemblers

A relatively small set of assemblers underpins most contemporary metagenomic workflows. For short-read Illumina data, metagenome-aware de Bruijn graph assemblers remain the standard; for long-read or hybrid datasets, specialized tools exploit long-range information to improve contiguity and resolve complex regions. Tool choice is therefore primarily driven by the sequencing technology used, community complexity, and whether the study is gene-centric or genome-resolved.

Short-Read Metagenomic Assemblers

For Illumina shotgun data, three assemblers are particularly widely used:

- *MEGAHIT* is a scalable de Bruijn graph assembler optimized for large metagenomic datasets. It uses succinct graph representations and multi-*k* strategies to balance contiguity, speed, and memory usage, making it suitable for projects with many samples or large co-assemblies. It is often the default choice when computational efficiency is a major constraint, and the primary goals are contig-level analyses, gene catalog construction, and routine MAG recovery.
- *metaSPAdes* extends the SPAdes assembler with metagenome-specific heuristics, including careful error correction and iterative graph simplification. In many benchmarks, it produces highly contiguous assemblies for abundant genomes, which benefits downstream binning and functional annotation. The trade-off is higher runtime and memory requirements than MEGAHIT, so it is typically preferred for smaller projects, subsets of samples, or studies where maximizing contig continuity and MAG quality is more important than throughput.
- *IDBA-UD* is an earlier metagenomic assembler that introduced iterative *k* values while explicitly modeling uneven sequencing depth. Although it has been partially superseded by MEGAHIT and metaSPAdes in many pipelines, it remains present in legacy workflows and provides a useful conceptual reference for multi-*k* metagenomic assembly.

Most of these assemblers can be used either per sample or for co-assembly simply by pooling reads from multiple, biologically related samples (e.g., time series, spatial gradients, replicate bioreactors) before running the assembler. Co-assembly can improve coverage for low-abundance genomes but increases computational cost and the risk of chimeric contigs when closely related strains vary in abundance between samples. In practice, many genome-resolved pipelines group samples by environment or treatment, co-assemble within groups using MEGAHIT or metaSPAdes, and then map each sample's reads back to the joint contigs for binning and downstream analyses.

Long-Read and Hybrid Metagenomic Assemblers

When long reads (Oxford Nanopore or PacBio) are available, or when hybrid assemblies are planned, other assemblers come into play:

- *metaFlye* is a long-read assembler adapted for metagenomes. It constructs repeat-aware assembly graphs directly from long reads and is capable of generating highly contiguous assemblies, including complete chromosomes, plasmids, and phage genomes in moderate-complexity communities. Short reads are often used afterward to polish base-level errors.
- *OPERA-MS* is an explicit hybrid metagenomic assembler. It typically uses a short-read assembler to build initial contigs, then integrates long reads to scaffold and refine them, leveraging coverage and graph structure to improve contiguity and strain separation. This approach is especially useful in host-associated microbiomes where resolving plasmids and mobile elements is a key objective.
- *hybridSPAdes* and related SPAdes modules support hybrid assembly of short and long reads. Although originally developed for microbial isolates, they can be applied to low-complexity metagenomes or enrichment cultures, where a limited number of genomes dominate and long-read coverage is relatively high.
- For high-fidelity PacBio datasets, tools such as hifiasm-meta use HiFi reads to produce very contiguous, strain-aware metagenomic assemblies. These methods are computationally demanding and are currently used mainly in focused genome-resolved projects, rather than large cohort surveys.

In most *hybrid workflows*, the division of labor is straightforward: Illumina reads provide high-accuracy base information and QC, while long-read or hybrid assemblers (metaFlye, OPERA-MS, hybridSPAdes, hifiasm-meta) supply the backbone contigs and scaffolds that bridge repeats and structural variants. Short-read assemblers such as MEGAHIT or metaSPAdes remain valuable in these projects—as baseline short-read assemblies, for samples without long-read data, or as references for comparing contiguity and completeness.

A practical summary is:

- Use *short-read assemblers* (MEGAHIT, metaSPAdes) as the default for Illumina-only metagenomes, large sample numbers, and gene-centric or survey-style studies.

- Consider *long-read or hybrid assemblers* (metaFlye, OPERA-MS, hybridSPAdes, hifiasm-meta) when the study is genome-resolved, sample numbers are moderate, and resolving repeats, plasmids, or strain structure justifies the extra sequencing and computational cost.

In Table 1, these assemblers can be compared side by side in terms of input type, key features, strengths, limitations, and typical use cases, providing a quick reference for selecting an appropriate tool for a given metagenomic project.

4.2 Key Parameters and Recommended Settings

Although modern metagenomic assemblers expose many options, a relatively small set of parameters has the greatest impact on assembly outcome: *k-mer size (or multi-k scheme), minimum contig length thresholds, and resource settings (threads and memory)*. Choosing and documenting these parameters consistently is often more important than fine-tuning rarely used options [40–42] (Lindgreen et al. 2016; Sczyrba et al. 2017). Example parameter choices for short-read metagenomic assembly, including typical k-mer schemes, minimum contig length thresholds, and recommended compute resources, are summarized in Table 2.

1. k-mer Size and Multi-*k* Strategies

   In de Bruijn graph assemblers, the k-mer size determines the trade-off between sensitivity and specificity. Smaller *k* values increase graph connectivity and help assemble low-coverage or highly variable regions, but they also merge repeats and closely related genomes more easily, leading to chimeric contigs. Larger *k* values reduce graph tangling and improve specificity but require higher coverage to support nodes and edges; rare taxa or low-coverage regions may fragment or disappear when *k* is too large [30, 31, 50].

   Most metagenomic assemblers (e.g., MEGAHIT, metaSPAdes, IDBA-UD) therefore use *multi-k* schemes: they start with small *k* to build a "backbone" graph and then gradually increase *k* to refine paths and break spurious connections [30, 31, 33, 51]. In practice, default multi-*k* presets are usually adequate. For 150 bp paired-end reads from moderately complex communities, multi-*k* ranges such as 21–99 (MEGAHIT) or 21–127 (metaSPAdes) are widely used in benchmarking studies and large-scale projects [6, 30, 31]. Users should avoid manually restricting *k* to a single very small value, which tends to over-merge related genomes, or to a single very large value, which can severely fragment low-coverage organisms [41, 42, 52].

Table 1
Commonly used assemblers for metagenomic short-read, long-read, and hybrid data

Assembler	Input type	Key features	Strengths	Limitations	Typical use case	Representative reference
MEGAHIT	Short reads (Illumina)	Succinct de Bruijn graph; multi-*k*; metagenome-aware	Very fast; low memory; scales to large/co-assemblies	Contiguity sometimes lower than metaSPAdes for abundant genomes	Large surveys; complex communities; per-sample or co-assembly on many samples	[29, 30]
metaSPAdes	Short reads (Illumina)	SPAdes core + metagenome heuristics; built-in error correction	High contiguity; good for MAG recovery	Higher runtime and RAM than MEGAHIT	Medium-scale projects; genome-resolved studies; maximizing contig continuity	[3, 31]
IDBA-UD	Short reads (Illumina)	Iterative *k*; uneven-depth modeling	Robust on uneven coverage; historically influential	Less actively maintained; slower and less scalable	Legacy pipelines; conceptual reference for multi-*k* metagenomic assembly	[32, 33]
metaFlye	Long reads (Nanopore/ PacBio)	Long-read, repeat-aware assembly; metagenome mode	Highly contiguous assemblies; resolves repeats/ plasmids	Higher raw error; usually needs short-read polishing	Long-read metagenomes; recovery of complete chromosomes, plasmids, and phages	[34, 35]
OPERA-MS	Hybrid (short + long reads)	Hybrid scaffolding; integrates coverage and graph structure	Strong contiguity gains; improved plasmid/strain resolution	More complex workflow; requires both data types	Host-associated microbiomes; plasmid and mobile-element–focused studies	[36, 37]
hybridSPAdes	Hybrid (short + long reads)	SPAdes-based hybrid modes	Good for low-complexity or enrichment metagenomes	Less suited for very complex communities; higher resource use	Enrichment cultures; low-diversity systems with moderate long-read coverage	[38]
hifiasm-meta	Long reads (PacBio HiFi)	HiFi-specific, strain-aware metagenomic assembly	Very high contiguity and accuracy; good strain separation	High computational cost; mainly for focused projects	Deep, genome-resolved studies using HiFi data; high-quality MAG and genome closure	[39]

Table 2
Example parameter choices for short-read metagenomic assembly

Parameter	Typical setting (150 bp PE reads)	Notes
k-mer scheme	Multi-*k* (e.g., 21–99 or 21–127)	Use assembler defaults; avoid single very small or very large *k* [30, 31, 43]
Minimum contig length	500–1000 bp for general analyses	Raise to 1–2 kb for binning; lower for viral/mobile element studies [44, 45]
CPU threads	8–32	Use as many as available without oversubscribing shared nodes
RAM	32–128 GB for typical per-sample assemblies	Large co-assemblies may require 256 GB or more [41, 46, 47]
Input reads	Cleaned, paired-end reads after Chapter 1 QC	Keep QC consistent across samples and report all settings [48, 49]

2. Minimum Contig Length Thresholds

Assemblers typically allow users to specify a *minimum contig length* for reporting results. Very short contigs (e.g., <500 bp) contribute little to taxonomic or functional interpretation and are disproportionately enriched in assembly errors, chimeras, and low-complexity sequence [41, 53]. However, setting the threshold too high will discard genuine short genes, plasmid fragments, and viral contigs, which can be particularly problematic in gene- and mobile element-centric studies [54, 55] (Roux et al. 2015).

For most short-read metagenomes, *500–1000 bp* is a reasonable threshold for general-purpose analyses and is commonly used in comparative evaluations of assemblers and pipelines [41, 42, 56]. When the primary goal is *genome-resolved binning*, many workflows filter contigs shorter than 1–2 kb before binning, because such fragments do not carry enough single-copy marker genes or tetranucleotide signal to support robust clustering [45, 57]. Conversely, when focusing on *viral metagenomics* or mobile genetic elements, thresholds may be relaxed to capture shorter, biologically meaningful contigs [55, 58]. The key is to choose a length cutoff consistent with the intended downstream analysis and to apply it uniformly across all samples and assemblies.

3. Resource Settings: Threads and Memory

Metagenomic assembly is computationally demanding. Although assemblers such as MEGAHIT and metaSPAdes automatically adapt to available memory and thread counts, performance and reproducibility still depend on sensible resource settings. As a rule of thumb, users should allocate *as many CPU threads as available without oversubscription* (e.g., 8–32 threads on a typical server) and ensure that physical memory is sufficient for the expected graph size. Under-provisioned memory leads to crashes or aggressive internal simplification, while overoptimistic thread counts on shared systems can cause severe slowdowns.

Benchmarking studies of metagenomic assemblers report that, for typical Illumina metagenomes(5–20 million paired-end reads)from moderate-complexity communities, *32–128 GB RAM* is usually adequate for MEGAHIT or metaSPAdes, whereas very deep datasets or large co-assemblies may require 256 GB or more [31, 41, 42, 47]. When resources are limited, it is preferable to reduce the number of samples per assembly—for example, by splitting a large co-assembly into smaller groups—rather than drastically changing assembler default parameters, which can markedly degrade assembly quality [41, 47, 59].

4. Reproducibility and Reporting

   Beyond individual parameter values, *consistent and transparent reporting* is critical. Differences in *k* ranges, minimum contig length, or read-filtering thresholds can produce assemblies that differ as much as those generated by different tools [41, 60]. To help standardize practice, comparative evaluations and best-practice guidelines for metagenomics emphasize explicit documentation of assembler name and version, k-mer schemes, length thresholds, and resource settings as part of the Methods or Supplementary Information [61]. Including these details greatly improves comparability between studies and facilitates reuse of assembled contigs and MAGs [62].

4.3 Assembly Quality Assessment

Once an assembly has been generated, its suitability for downstream analysis must be evaluated using a combination of graph-agnostic summary statistics and read-mapping-based diagnostics. For metagenomes, quality assessment focuses less on achieving an "optimal" single metric and more on identifying assemblies that are clearly under-assembled, over-fragmented, or strongly biased by a subset of genomes [63].

At the most basic level, assemblers report global statistics such as the total length of assembled contigs, N50/L50, maximum contig length, and contig count. These metrics provide a coarse overview of continuity and yield. For example, higher N50 values and longer maximum contigs generally indicate more contiguous assemblies, while an excessive number of very short contigs suggests over-fragmentation or insufficient coverage. However, global statistics must be interpreted cautiously in metagenomics: a few very long contigs from dominant taxa can inflate N50 even when low-abundance genomes remain highly fragmented

[64]. Therefore, continuity metrics are most informative when compared across assemblers or parameter sets on the same dataset, rather than as absolute performance targets.

Read-mapping-based evaluation is essential to assess how well the assembly captures the information present in the reads. Mapping the quality-controlled reads back to the assembled contigs (e.g., with Bowtie2, BWA-MEM, or BBMap) yields the overall mapping rate, coverage depth distribution, and identification of regions with unusually low or high coverage. High mapping rates (for example, >80–90% for many shotgun datasets) indicate that the assembly recapitulates most of the sequenced content, whereas low mapping rates suggest overaggressive filtering, assembler failure, or large fractions of unassembled reads [65]. Coverage profiles also highlight highly repetitive or strain-variable regions (very high coverage) and underrepresented genomes (low, patchy coverage), which can inform subsequent binning and re-assembly strategies.

Specialized tools such as QUAST and metaQUAST provide standardized assembly reports for metagenomic data, including length distributions, GC content, N50/L50, and, when reference genomes or mock communities are available, estimates of misassemblies, mismatches, and genome fraction recovered [66]. In real environmental datasets without complete references, metaQUAST can still be used in a reference-free mode to summarize contig statistics and read-mapping-based metrics. These reports are particularly valuable when comparing multiple assemblers (e.g., MEGAHIT vs. metaSPAdes) or parameter settings on the same dataset, as they help identify systematic gains in contiguity or losses in coverage.

For genome-resolved metagenomics, assembly assessment is closely tied to the quality of metagenome-assembled genomes (MAGs). Tools such as CheckM evaluate completeness and contamination of genome bins based on single-copy marker genes [45]. Although detailed MAG evaluation is the focus of the following chapter, it is useful to note that poor global assembly quality—extreme fragmentation, low mapping rates, or strong GC bias—almost always translates into incomplete or highly contaminated bins. Conversely, assemblies that show balanced contig lengths, high read-mapping rates, and plausible GC distributions typically support the recovery of higher-quality MAGs [67].

In practice, metagenomic assembly quality assessment usually combines a small set of metrics into a compact summary table for each assembly: total assembled base pairs, number of contigs above a chosen threshold (e.g., ≥1 kb), N50/L50, maximum contig length, GC content range, read-mapping rate, and basic coverage statistics. An example is shown in Table 3. Reporting these metrics alongside the parameter settings described later (k-mer scheme, minimum contig length, threads, and memory) provides sufficient transparency for most comparative studies and enables other

Table 3
Commonly used metrics for metagenomic assembly evaluation

Metric	Interpretation in metagenomes	Typical use
Total assembled length	Overall yield of assembled sequence	Compare yield between assemblers/parameter sets
Number of contigs ≥1 kb	Fragmentation vs. continuity (excluding very short contigs)	Diagnose over-fragmentation
N50/L50	Contiguity of assembly; dominated by abundant genomes	Compare continuity across assemblies on same dataset
Maximum contig length	Upper bound of contig size; reflects best-assembled genomes	Identify highly contiguous dominant genomes
GC content distribution	Biological plausibility; detection of GC-biased loss or contamination	Screen for bias or unexpected GC peaks
Read-mapping rate	Fraction of reads represented in assembly	Detect under-assembly or overly aggressive filtering
Coverage depth distribution	Evenness across contigs and genomes	Identify over- or underrepresented regions and taxa

researchers to judge whether the assembly is appropriate for taxonomic profiling, gene catalog construction, or genome-resolved analyses.

5 Gene Prediction from Metagenomic Assemblies

Once contigs have been assembled and assessed, the next step in most metagenomic workflows is to identify protein-coding genes and, in some cases, noncoding features. Gene prediction converts anonymous nucleotide contigs into a set of open reading frames (ORFs) that can be used for taxonomic profiling, functional annotation, pathway reconstruction, and comparative analyses. In metagenomes, however, gene prediction is complicated by incomplete genomes, fragmented genes at contig ends, high sequence divergence, and the simultaneous presence of bacteria, archaea, viruses, and eukaryotes [68] (Fig. 1).

5.1 Rationale and Challenges in Metagenomic Gene Calling

In single, well-annotated genomes, gene prediction can exploit curated training sets, clear codon usage patterns, and known gene structures. In contrast, metagenomic contigs are typically:

- Fragmented—many contigs represent partial chromosomes or plasmids; genes may be truncated at both ends.
- Taxonomically diverse—different organisms have different GC content, codon usage, and gene length distributions.

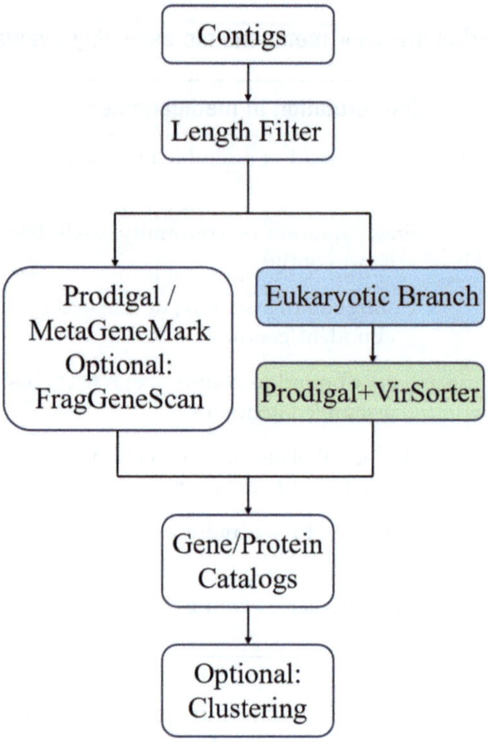

Fig. 1 Gene prediction workflow for metagenomic assemblies

- Unevenly covered—low-abundance genomes contribute few contigs; high-abundance genomes may dominate.
- Structurally heterogeneous—the same assembly may contain bacterial, archaeal, viral, and eukaryotic sequences.

These characteristics make *model-based genome-specific gene predictors* less suitable, because a single parameter set cannot capture the diversity of the community. For this reason, most metagenomic pipelines adopt *"meta-mode" gene finders*, which are explicitly designed to work on mixed, fragmentary contigs without prior taxonomic labels [69].

A second challenge is the *trade-off between sensitivity and specificity*. Aggressive gene calling (many short ORFs, permissive start codons) increases sensitivity to novel small proteins but also generates large numbers of spurious ORFs, inflating apparent gene richness and complicating downstream annotation. Conversely, overly conservative models miss true genes from atypical genomes (e.g., low-GC bacteria, divergent lineages, or highly reduced genomes). Metagenomic gene prediction must therefore strike a balance: reasonably sensitive across diverse taxa, but not so permissive that false positives dominate functional profiles [70].

Finally, input contig filtering strongly affects gene calling. Most workflows restrict gene prediction to contigs above a certain length (often ≥500–1000 bp), both to reduce computation and to avoid calling ORFs on very short, error-prone fragments [65, 71]. For genome-resolved analyses, only contigs that pass assembly quality criteria (Subheading 4.3) are typically used for gene prediction and subsequent binning.

5.2 Gene Prediction Tools for Metagenomic Assemblies

Most metagenomic workflows rely on a small set of gene prediction tools to identify protein-coding genes on assembled contigs. In contrast to single-genome projects, metagenomic contigs are fragmentary, taxonomically mixed, and span a broad range of GC contents and codon usage, so gene callers must perform reasonably well without genome-specific training [72]. In practice, this has led to a de facto standard: metagenome-aware prokaryotic gene callers run on assembled contigs ≥500–1000 bp, supplemented by specialized tools for eukaryotic and viral sequences.

5.2.1 Prokaryotic Gene Prediction

For bacteria and archaea, two tools dominate current metagenomic pipelines:

Prodigal (metagenomic mode).

Prodigal is one of the most widely used prokaryotic gene callers and includes a dedicated metagenomic mode (-p meta) designed for anonymous, fragmentary contigs [73]. Unlike its genome-trained mode, which learns parameters from a single genome, metagenomic mode uses generalized models that tolerate variable GC content and partial genes. It explicitly handles ORFs truncated at contig ends and outputs both nucleotide (.fna) and amino acid (.faa) sequences, which can be passed directly to downstream annotation tools (e.g., Prokka, eggNOG-mapper, KEGG-based pipelines). Because it is very fast, easy to integrate, and license-friendly, Prodigal in metagenomic mode has become the *default gene caller* in many shotgun metagenomics workflows (e.g., MGnify, MetaWRAP, ATLAS, nf-core/mag) [68, 74].

A typical command on assembled contigs is:

```
prodigal \
-i contigs.fa \
-a proteins.faa \
-d genes.fna \
-p meta
```

Most studies restrict input to contigs above a length threshold (often ≥500–1000 bp; see Subheading 4.2), which reduces spurious ORFs on very short fragments and focuses computation on sequences that are informative for functional and genome-resolved analyses [75].

MetaGeneMark and FragGeneScan as Complements

MetaGeneMark extends the GeneMark family to metagenomes by using pretrained models covering diverse GC contents and applying GC-stratified calling to contigs [76]. Benchmarking studies show that MetaGeneMark and Prodigal (meta) are broadly comparable in accuracy on bacterial and archaeal metagenomes, with MetaGeneMark sometimes performing slightly better on GC-extreme genomes, and Prodigal offering a better balance of speed and ease of integration [41, 42, 77, 78]. MetaGeneMark is therefore a common alternative in large gene-catalog projects and marine or soil metagenomes with broad taxonomic diversity.

FragGeneScan and its faster re-implementation FragGeneScan-RS were originally developed for short, error-prone reads, explicitly modeling sequencing errors within a hidden Markov model [71]. While they can also be applied to contigs, their main strength lies in scenarios where assembly is shallow or not performed (e.g., quick functional screening of reads, extremely low coverage data). In assembly-centric workflows based on high-quality Illumina data, FragGeneScan is now used mainly as a niche complement rather than a primary gene caller.

A concise comparison of commonly used prokaryotic gene prediction tools is given in Table 4.

In summary, Prodigal (meta) on contigs ≥500–1000 bp represents a sensible default for prokaryotic gene prediction, with MetaGeneMark as a well-established alternative and FragGeneScan reserved for special cases involving short or error-prone sequences.

5.2.2 Eukaryotic and Viral Gene Prediction

Many metagenomic datasets also contain eukaryotic microbes (protists, fungi) and viruses, whose gene structure differs substantially from prokaryotes. Simply applying prokaryotic gene callers to all contigs can mis-annotate intron-containing genes, ignore untranslated regions, and distort functional profiles in eukaryote-rich habitats.

For eukaryotic contigs, standard genome projects rely on tools such as AUGUSTUS or GeneMark-ES/ET, often trained with species-specific or transcriptomic data. In metagenomes, where such training data are rarely available, MetaEuk provides a practical alternative: it performs protein-profile-guided gene discovery on contigs, using sensitive similarity searches against large protein databases to reconstruct exon–intron gene models [13]. This approach is particularly useful for eukaryote-enriched samples, such as plankton metagenomes or host-associated microbiomes in which protists and fungi contribute substantially to biomass.

Table 4
Commonly used prokaryotic gene prediction tools

Tool	Input type	Metagenome-specific features	Strengths	Limitations	Representative use case
Prodigal (meta mode)	Assembled contigs	"-p meta" mode; handles partial genes; no single-genome training	Very fast; easy to run; widely adopted in pipelines	Not suitable for eukaryotic intron-rich genes	Default caller in most short-read metagenomes
MetaGeneMark	Assembled contigs	Pretrained models; GC-aware stratification	Good performance across diverse GC contents	License constraints for some variants; slower than Prodigal	Large gene catalogs; diverse environmental datasets
FragGeneScan/ FGS-RS	Reads or short/ error-prone contigs	Explicit error model for frame shifts and Ns	Robust on noisy or very short sequences	Computationally heavier; less needed on high-quality contigs	Read-level or shallow metagenomic surveys

For viral contigs, most dsDNA phages and many other viral genomes have compact, intron-free gene structures but short coding sequences and high coding density. Many viral metagenomics pipelines still employ Prodigal (sometimes with tailored parameters) to predict viral ORFs, in combination with viral identification tools such as VirSorter, VirFinder, or VIBRANT [79]. Several virus-oriented gene callers (e.g., MetaGeneAnnotator, VIGA) have been proposed, but their use is less standardized than Prodigal-based workflows and often tied to specific virome pipelines.

In practice, most general-purpose metagenomic analyses adopt a two-step strategy:

1. Treat the majority of contigs as prokaryotic by default, running Prodigal (meta) or MetaGeneMark on contigs above a length threshold to generate a core gene catalog.
2. For contigs taxonomically assigned to eukaryotes or viruses (e.g., after preliminary classification), optionally run eukaryote-aware tools such as MetaEuk, and apply viral gene-calling workflows on virus-enriched subsets when viromics is a primary objective.

This separation avoids forcing all contigs into a single prokaryotic gene model while keeping the overall pipeline simple enough to be applied routinely to large metagenomic datasets.

5.3 Practical Workflow and Recommended Settings for Gene Prediction

In practice, gene prediction on metagenomic assemblies follows a relatively simple and reproducible workflow. The goal is to generate a clean, well-documented gene and protein catalog that can be used consistently across taxonomic, functional, and genome-resolved analyses.

5.3.1 Standard Workflow on Prokaryotic Contigs

A minimal, robust prokaryotic gene prediction pipeline can be summarized as:

1. *Select contigs for gene prediction.*
   - Start from assemblies that have passed the quality criteria in Subheading 4.3.
   - Restrict to contigs above a length threshold (commonly ≥500–1000 bp) to avoid calling ORFs on very short, error-prone fragments.
   - Optionally exclude contigs flagged as likely eukaryotic or viral by preliminary classification to avoid mis-annotation.
2. Example: keep contigs ≥1000 bp

```
seqkit seq -m 1000 contigs.fa > contigs_1kb.fa
```

3. *Run a metagenome-aware gene caller (Prodigal or MetaGeneMark).*
   - Use *Prodigal in metagenomic mode (-p meta)* or MetaGeneMark on the filtered contigs.
   - Ensure that the same tool and parameters are applied to all samples in a study.
4. Prodigal meta-mode on ≥1 kb contigs

```
prodigal \
-i contigs_1kb.fa \
-a proteins.faa \
-d genes.fna \
-p meta \
-o prodigal.gff
```

5. *Store nucleotide, amino acid, and coordinate outputs.*
   - Nucleotide sequences (genes.fna) are useful for read-mapping–based quantification.
   - Protein sequences (proteins.faa) are used for functional and taxonomic annotation.
   - GFF or similar coordinate files preserve gene boundaries on contigs and support later genome binning and visualization.
6. Optionally post-process and cluster genes.

   - For large projects, predicted proteins are often clustered at a chosen identity threshold (e.g., 90–95%) to build a nonredundant gene catalog before annotation.
   - Tools such as CD-HIT or MMseqs2 are commonly used for fast clustering.
7. Simple protein clustering with MMseqs2 (example)

```
mmseqs createdb proteins.faa proteins_db
   mmseqs linclust proteins_db clusters tmp --min-seq-id 0.9
        mmseqs createtsv proteins_db proteins_db clusters nr_clusters.tsv
```

8. Document all settings.
   - Record contig length thresholds, gene caller name and version, and any non-default parameters.
   - Use consistent file naming (e.g. <sample>.contigs.fa, <sample>.genes.fna, <sample>.proteins.faa) to simplify downstream scripting and reproducibility.

5.3.2 Handling Eukaryotic- and Viral-Enriched Subsets

When preliminary analysis indicates substantial eukaryotic or viral contributions, a branch of the workflow can target those contigs with specialized tools while keeping the core prokaryotic pipeline unchanged:

1. Identify eukaryotic and viral contigs.
   - Use taxonomic classifiers (e.g., Kraken2, Kaiju, or marker-gene-based approaches) to flag contigs associated with eukaryotes or viruses.
   - Split the assembly into subsets: prokaryotic majority, eukaryote-enriched, viral-enriched.
2. Run eukaryote-aware gene prediction where appropriate:
   - On contigs classified as eukaryotic, apply tools such as MetaEuk to derive exon–intron structures and protein-coding genes.
   - Keep eukaryotic predictions in separate files (e.g., proteins_euk.faa) to avoid mixing annotation models.
3. Apply viral gene-calling workflows for virome-focused projects.
   - For viral contigs, either reuse Prodigal (with documented parameters) or employ virus-oriented pipelines (e.g., VirSorter/VIBRANT plus their recommended gene callers).
   - Again, maintain separate protein sets (e.g., proteins_viral.faa) to facilitate dedicated virome analyses.

5.3.3 Minimal Reporting for Methods

For most studies, a short, standardized description of gene prediction in the Methods is sufficient, for example:

Protein-coding genes were predicted on quality-checked metagenomic assemblies using Prodigal (version) in metagenomic mode (-p meta) on contigs ≥1000 bp. Predicted nucleotide sequences and amino acid translations were retained for downstream functional and taxonomic annotation. For samples enriched in eukaryotic sequences, contigs classified as eukaryotic were additionally processed with MetaEuk (version) using default parameters, and predicted proteins were analyzed separately.

This level of detail makes it clear which contigs were used, which tools and parameters were applied, and how non-prokaryotic sequences were handled, while staying consistent with the pipeline structure laid out in the chapter.

6 Linking Assembly and Gene Prediction to Downstream Analyses

Metagenomic assembly and gene prediction are not endpoints; they provide the structural backbone for all subsequent taxonomic, functional, and ecological inferences. How contigs are filtered, how genes are called, and how identifiers are propagated will directly shape the reliability of metagenome-assembled genomes (MAGs), gene catalogs, and abundance profiles. This section briefly summarizes how the outputs of this chapter connect to downstream analyses that are developed in later chapters.

6.1 From Contigs to MAGs and Community Structure

Quality-checked contigs are the primary substrate for binning and MAG reconstruction, which aim to group contigs originating from the same genome based on sequence composition and coverage patterns. Binning methods (e.g., MetaBAT, CONCOCT, MaxBin) use tetranucleotide frequencies and per-sample coverage profiles derived from mapping clean reads back to contigs; thus, the mapping step depends critically on the trimmed, host-depleted reads and the assembly quality described in Chapters 1 and 2 [80, 81].

Once bins are constructed, their completeness, contamination, and strain heterogeneity are typically assessed with tools such as CheckM or GUNC [45, 82]. These metrics, in turn, rely on the presence and integrity of single-copy marker genes—features that may be fragmented or missed if assembly and gene prediction are suboptimal. Incomplete or chimeric assemblies can inflate apparent species richness, create artificial MAGs, or obscure real population structure. Conversely, well-assembled contigs with accurately predicted genes support high-quality MAGs that form the basis for genome-resolved analyses of metabolism, phylogeny, and population genetics [83].

At the community level, contig-level information also feeds into taxonomy-aware abundance estimates. Contigs or MAGs can be taxonomically annotated using marker-gene approaches (e.g., GTDB-Tk) or whole-genome similarity, and the coverage of each taxon is then quantified by read mapping. Consistent assembly and gene prediction across samples therefore enable robust comparisons of community structure along environmental gradients or host phenotypes.

6.2 From Genes to Functional and Pathway Profiles

Predicted genes and proteins constitute the *functional layer* of metagenomic analysis. The protein catalog generated in Subheading 5 is typically aligned against reference databases, such as KEGG, eggNOG, COG, CAZy, or CARD, to infer orthologous groups, enzyme commission (EC) numbers, and higher-level pathway assignments [84]. The sensitivity and specificity of these annotations depend on several upstream decisions:

- Length and quality of contigs. Short or error-prone contigs tend to produce truncated ORFs that are difficult to annotate and may inflate the number of "hypothetical proteins."
- Gene calling thresholds. Very permissive gene prediction produces many short ORFs with weak homology, biasing functional profiles toward poorly characterized categories.
- Redundancy handling. Large projects often cluster proteins at 90–95% identity to form nonredundant gene catalogs; abundance is then propagated back to cluster representatives. The way this clustering is done affects apparent gene richness and the resolution of strain-level variation [85].

Gene-level abundance is typically estimated by mapping clean reads back to predicted genes or proteins and computing counts, coverage, or transcripts-per-million–like measures. These abundances are then aggregated into functional categories (e.g. KEGG modules, eggNOG orthologous groups) and used for multivariate ordination, differential abundance testing, and pathway enrichment analyses. Any systematic bias introduced during assembly or gene prediction—such as preferential loss of low-GC genes or aggressive trimming of particular taxa—will propagate into these functional profiles and may be misinterpreted as biological differences if not carefully controlled.

6.3 Practical Recommendations for Downstream Compatibility

To ensure that assembly and gene prediction outputs are maximally useful for downstream analyses developed in later chapters, several practical recommendations are worth emphasizing:

1. *Stable identifiers and clear file structure.*

   Contig IDs should be unique and consistent across all files (assembly FASTA, gene coordinates, MAG bin files). Gene IDs should encode their parent contig where possible (e.g., contig00001_00001). This makes it straightforward to aggregate information from binning, annotation, and quantification.

2. *Explicit length and quality thresholds.*

   The contig length cutoff for assembly evaluation and gene prediction (e.g., ≥500 bp or ≥ 1000 bp) should be stated clearly and applied uniformly. Downstream tools for MAG reconstruction and functional profiling can then be configured with matching thresholds, avoiding silent exclusion of short contigs at later stages.

3. *Separation of prokaryotic, eukaryotic, and viral subsets.*

   Where eukaryotic or viral contigs are expected, keeping their predicted genes in separate FASTA/GFF files allows tailored annotation (e.g., MetaEuk for eukaryotes, viral pipelines for phages) without contaminating prokaryotic gene catalogs. This separation also simplifies comparative analyses that target specific biological compartments (bacterial metabolism vs. viral auxiliary genes, etc.).

4. *Propagation of coverage information.*

   Read-mapping statistics obtained during assembly assessment (coverage depth per contig) should be retained and linked to both MAGs and genes. Coverage patterns support bin quality evaluation, detection of strain mixtures, and more accurate abundance estimates for metabolic pathways [86, 87].

5. *Minimal reporting for reproducibility.*

   Methods sections should state the assembler and version, major parameters (k-mer strategy, minimum contig length), gene caller and mode (e.g., Prodigal -p meta on contigs ≥1 kb), and any clustering or filtering applied to gene catalogs. This level of detail allows other researchers to interpret downstream taxonomic, functional, and statistical analyses in a transparent way and to reproduce the pipeline when necessary.

By viewing assembly and gene prediction as the *central bridge* between raw sequencing reads and downstream analyses—MAG reconstruction, taxonomic profiling, functional annotation, and ecological statistics—researchers can design pipelines that are not only technically sound at each individual step but also coherent as an integrated whole. The following chapters build directly on these outputs, detailing how contigs, genes, and coverage profiles are converted into biologically meaningful insights about the structure and function of microbial communities.

Conflict of Interest

The authors have no conflicts of interest to declare that are relevant to the content of this chapter.

References

1. Qin JJ, Li RQ, Raes J, Arumugam M, Burgdorf KS, Manichanh C, Nielsen T, Pons N, Levenez F, Yamada T, Mende DR, Li JH, Xu JM, Li SC, Li DF, Cao JJ, Wang B, Liang HQ, Zheng HS, Xie YL, Tap J, Lepage P, Bertalan M, Batto JM, Hansen T, Le Paslier D, Linneberg A, Nielsen HB, Pelletier E, Renault P, Sicheritz-Ponten T, Turner K, Zhu HM, Yu C, Li ST, Jian M, Zhou Y, Li YR, Zhang XQ, Li SG, Qin N, Yang HM, Wang J, Brunak S, Dore J, Guarner F, Kristiansen K, Pedersen O, Parkhill J, Weissenbach J, Bork P, Ehrlich SD, Wang J, Consortium M (2010) A human gut microbial gene catalogue established by metagenomic sequencing. Nature 464(7285):59–U70. https://doi.org/10.1038/nature08821
2. Methé BA, Nelson KE, Pop M, Creasy HH, Giglio MG, Huttenhower C, Gevers D, Petrosino JF, Abubucker S, Badger JH, Chinwalla AT, Earl AM, FitzGerald MG, Fulton RS, Hallsworth-Pepin K, Lobos EA, Madupu R, Magrini V, Martin JC, Mitreva M, Muzny DM, Sodergren EJ, Versalovic J, Wollam AM, Worley KC, Wortman JR, Young SK, Zeng Q, Aagaard KM, Abolude OO, Allen-Vercoe E, Alm EJ, Alvarado L, Andersen GL, Anderson S, Appelbaum E, Arachchi HM, Armitage G, Arze CA, Ayvaz T, Baker CC, Begg L, Belachew T, Bhonagiri V, Bihan M, Blaser MJ, Bloom T, Bonazzi VR, Brooks P, Buck G, Buhay CJ, Busam DA, Campbell JL, Canon SR, Cantarel BL, Chain PS, Chen IMA, Chen L, Chhibba S, Chu K, Ciulla DM, Clemente JC, Clifton SW, Conlan S, Crabtree J, Cutting MA, Davidovics NJ, Davis CC, DeSantis TZ, Deal C, Delehaunty KD, Dewhisrst FE, Deych E, Ding Y, Dooling DJ, Dugan SP, Dunne W, Durkin AS, Edgar RC, Erlich RL, Farmer CN, Farrell RM, Faust K, Feldgarden M, Felix VM, Fisher S, Fodor AA, Forney L, Foster L, Di Francesco V, Friedman J, Friedrich DC, Fronick CC, Fulton LL, Gao H, Garcia N, Giannoukos G, Giblin C, Giovanni MY, Goldberg JM, Goll J, Gonzalez A, Griggs A, Gujja S, Haas BJ, Hamilton HA, Harris EL, Hepburn TA, Herter B, Hoffmann DE, Holder ME, Howarth C, Huang KH, Huse SM, Izard J, Jansson JK, Jiang HY, Jordan C, Joshi V, Katancik J, Keitel W, Kelley ST, Kells C, Kinder-Haake S, King NB, Knight R, Knights D, Kong HH, Koren O, Koren S, Kota KC, Kovar CL, Kyrpides NC, La Rosa PS, Lee SL, Lemon KP, Lennon N, Lewis CM, Lewis L, Ley RE, Li K, Liolios K, Liu B, Liu Y, Lo CC, Lozupone CA, Lunsford RD, Madden T, Mahurkar AA, Mannon PJ, Mardis ER, Markowitz VM, Mavrommatis K, McCorrison JM, McDonald D, McEwen J, McGuire AL, McInnes P, Mehta T, Mihindukulasuriya KA, Miller JR, Minx PJ, Newsham I, Nusbaum C, O'Laughlin M, Orvis J, Pagani I, Palaniappan K, Patel SM, Pearson M, Peterson J, Podar M, Pohl C, Pollard KS, Priest ME, Proctor LM, Qin X, Raes J, Ravel J, Reid JG, Rho M, Rhodes R, Riehle KP, Rivera MC, Rodriguez-Mueller B, Rogers YH, Ross MC, Russ C, Sanka RK, Sankar P, Sathirapongsasuti JF, Schloss JA, Schloss PD, Schmidt TM, Scholz M, Schriml L, Schubert AM, Segata N, Segre JA, Shannon WD, Sharp RR, Sharpton TJ, Shenoy N, Sheth NU, Simone GA, Singh I, Smillie CS, Sobel JD, Sommer DD, Spicer P, Sutton GG, Sykes SM, Tabbaa DG, Thiagarajan M, Tomlinson CM, Torralba M, Treangen TJ, Truty RM, Vishnivetskaya TA, Walker J, Wang L, Wang Z, Ward DV, Warren W, Watson MA, Wellington C, Wetterstrand KA, White JR, Wilczek-Boney K, Wu YQ, Wylie KM, Wylie T, Yandava C, Ye L, Ye Y, Yooseph S, Youmans BP, Zhang L, Zhou YJ, Zhu YM, Zoloth L, Zucker JD, Birren BW, Gibbs RA, Highlander SK, Weinstock GM, Wilson RK, White O, Consortiu HMP (2012) A framework for human microbiome research. Nature 486(7402):215–221. https://doi.org/10.1038/nature11209
3. Vollmers J, Wiegand S, Kaster A-K (2017) Comparing and evaluating metagenome assembly tools from a microbiologist's perspective-not only size matters! PLoS One 12(1): e0169662
4. Bálint M, Bahram M, Eren AM, Faust K, Fuhrman JA, Lindahl B, O'Hara RB, Öpik M, Sogin ML, Unterseher M (2016) Millions of reads, thousands of taxa: microbial community structure and associations analyzed via marker genes. FEMS Microbiol Rev 40(5):686–700
5. Kunin V, Copeland A, Lapidus A, Mavromatis K, Hugenholtz P (2008) A bioinformatician's guide to metagenomics. Microbiol Mol Biol Rev 72(4):557–578
6. Pérez-Cobas AE, Gomez-Valero L, Buchrieser C (2020) Metagenomic approaches in microbial ecology: an update on whole-genome and marker gene sequencing analyses. Microb Genom 6(8):e000409
7. Cohen Y, Naseraldin H, Chaudhuri A, Pilati F (2019) Assembly systems in Industry 40 era: a road map to understand Assembly 40. Int J Adv Manufact Technol 105(9):4037–4054
8. Frank JA, Pan Y, Tooming-Klunderud A, Eijsink VG, McHardy AC, Nederbragt AJ, Pope PB (2016) Improved metagenome assemblies and taxonomic binning using long-read

circular consensus sequence data. Sci Rep 6 (1):25373

9. Castellana N, Bafna VJ (2010) Proteogenomics to discover the full coding content of genomes: a computational perspective. J Proteomics 73(11):2124–2135
10. Clamp M, Fry B, Kamal M, Xie X, Cuff J, Lin MF, Kellis M, Lindblad-Toh K, Lander ES (2007) Distinguishing protein-coding and noncoding genes in the human genome. Proc Natl Acad Sci U S A 104(49):19428–19433
11. Ezkurdia I, Juan D, Rodriguez JM, Frankish A, Diekhans M, Harrow J, Vazquez J, Valencia A, Tress ML (2014) Multiple evidence strands suggest that there may be as few as 19 000 human protein-coding genes. 23(22):5866–5878
12. Ejigu GF, Jung J (2020) Review on the computational genome annotation of sequences obtained by next-generation sequencing. Biology 9(9):295
13. Levy Karin E, Mirdita M, Söding J (2020) MetaEuk—sensitive, high-throughput gene discovery, and annotation for large-scale eukaryotic metagenomics. Microbiome 8 (1):48
14. Maguire F, Jia B, Gray KL, Lau WYV, Beiko RG, Brinkman FS (2020) Metagenome-assembled genome binning methods with short reads disproportionately fail for plasmids and genomic Islands. Microb Genom 6(10):e000436
15. Meziti A, Rodriguez-R LM, Hatt JK, Peña-Gonzalez A, Levy K, Konstantinidis KT (2021) The reliability of metagenome-assembled genomes (MAGs) in representing natural populations: insights from comparing MAGs against isolate genomes derived from the same fecal sample. Appl Environ Microbiol 87(6): e02593–e02520
16. Nielsen HB, Almeida M, Juncker AS, Rasmussen S, Li J, Sunagawa S, Plichta DR, Gautier L, Pedersen AG, Le Chatelier E (2014) Identification and assembly of genomes and genetic elements in complex metagenomic samples without using reference genomes. Nat Biotechnol 32(8):822–828
17. Straube N, Lyra ML, Paijmans JL, Preick M, Basler N, Penner J, Rödel MO, Westbury MV, Haddad CF, Barlow A (2021) Successful application of ancient DNA extraction and library construction protocols to museum wet collection specimens. Mol Ecol Resour 21(7):2299–2315
18. Willett W (2012) Nutritional epidemiology. Oxford university press
19. Michalak L, Gaby JC, Lagos L, La Rosa SL, Hvidsten TR, Tétard-Jones C, Willats WG, Terrapon N, Lombard V, Henrissat B (2020) Microbiota-directed fibre activates both targeted and secondary metabolic shifts in the distal gut. Nat Commun 11(1):5773
20. Miller JR, Koren S, Sutton G (2010) Assembly algorithms for next-generation sequencing data. Genomics 95(6):315–327
21. Georganas E, Egan R, Hofmeyr S, Goltsman E, Arndt B, Tritt A, Buluç A, Oliker L, Yelick K (2018) Extreme scale de novo metagenome assembly. In: SC18: International Conference for High Performance Computing, Networking, Storage and Analysis. IEEE, pp 122–134
22. Howe AC, Jansson JK, Malfatti SA, Tringe SG, Tiedje JM, Brown CT (2014) Tackling soil diversity with the assembly of large, complex metagenomes. Proc Natl Acad Sci U S A 111 (13):4904–4909
23. Dai Y, Wu Y, Zhou F, Barnard K (2021) Attentional local contrast networks for infrared small target detection. IEEE Trans Geosci Remote Sens 59(11):9813–9824
24. Hershberger J, Maxel M, Suri S (2007) Finding the k shortest simple paths: a new algorithm and its implementation. ACM Trans Algorithms 3(4):45
25. Tedersoo L, Albertsen M, Anslan S, Callahan B (2021) Perspectives and benefits of high-throughput long-read sequencing in microbial ecology. Appl Environ Microbiol 87(17): e00626–e00621
26. Chen X, Yin X, Shi X, Yan W, Yang Y, Liu L, Zhang T (2024) Melon: metagenomic long-read-based taxonomic identification and quantification using marker genes. Genome Biol 25 (1):226
27. Pollard MO, Gurdasani D, Mentzer AJ, Porter T, Sandhu MS (2018) Long reads: their purpose and place. Hum Mol Genet 27(R2): R234–R241
28. Warburton PE, Sebra RP (2023) Long-read DNA sequencing: recent advances and remaining challenges. Annu Rev Genomics Hum Genet 24(1):109–132
29. Pan S, Zhu C, Zhao X-M, Coelho LP (2022) A deep siamese neural network improves metagenome-assembled genomes in microbiome datasets across different environments. Nat Commun 13(1):2326
30. Li D, Liu CM, Luo Sadakane K, Lam TW (2015) MEGAHIT: ultra-fast single-node solution for large complex metagenomics assembly via de Bruijn graph. Bioinformatics 31:1676. https://doi.org/10.1093/bioinformatics
31. Nurk S, Meleshko D, Korobeynikov A, Pevzner PA (2017) metaSPAdes: a new versatile

metagenomic assembler. Genome Res 27 (5):824–834

32. Yue H, Yue W, Jiao S, Kim H, Lee Y-H, Wei G, Song W, Shu D (2023) Plant domestication shapes rhizosphere microbiome assembly and metabolic functions. Microbiome 11(1):70
33. Peng Y, Leung Yiu SM, Chin FYL (2012) IDBA-UD: assembler for single-cell and metagenomic sequencing data with highly uneven. Bioinformatics 28(11):1420–1428. https://doi.org/10.1093/bioinformatics
34. Li H, Durbin R (2024) Genome assembly in the telomere-to-telomere era. Nat Rev Genet 25(9):658–670
35. Bickhart DM, Behsaz B et al (2020) metaFlye: scalable long-read metagenome assembly using repeat graphs. Nat Methods 17:1103–1110. https://doi.org/10.1038/s41592-020
36. Meyer F, Fritz A, Deng Z-L, Koslicki D, Lesker TR, Gurevich A, Robertson G, Alser M, Antipov D, Beghini F (2022) Critical assessment of metagenome interpretation: the second round of challenges. Nat Methods 19(4):429–440
37. Bertrand SJ, Kalathiyappan M, Ng A, Nagarajan N (2019) Hybrid metagenomic assembly enables high-resolution analysis resistance determinants and mobile in human microbiomes. Nat Biotechnol 37(8). https://doi.org/10.1038/s41587-019
38. Jin H, You L, Zhao F, Li S, Ma T, Kwok L-Y, Xu H, Sun Z (2022) Hybrid, ultra-deep metagenomic sequencing enables genomic and functional characterization of low-abundance species in the human gut microbiome. Gut Microbes 14(1):2021790
39. Feng X, Cheng H, Portik D, Li H (2022) Metagenome assembly of high-fidelity long reads with hifiasm-meta. Nat Methods 19 (6):671–674
40. Han Z, Gao C, Liu J, Zhang J, Zhang SQ (2024) Parameter-efficient fine-tuning for large models: a comprehensive survey. arXiv preprint arXiv:240314608
41. Lindgreen S, Adair KL, Gardner PP (2016) An evaluation of the accuracy and speed of metagenome analysis tools. Sci Rep 6:19233. https://doi.org/10.1038/srep19233
42. Sczyrba A, Hofmann P, Belmann P et al (2017) Critical assessment of Metagenome Interpretation—a benchmark of metagenomics software. Nat Methods 14:1063–1071. https://doi.org/10.1038/nmeth.4458
43. Koren S, Walenz BP, Berlin K, Miller JR, Bergman NH, Phillippy AM (2017) Canu: scalable and accurate long-read assembly via adaptive k-mer weighting and repeat separation. Genome Res 27(5):722–736
44. Gregory AC, Zablocki O, Zayed AA, Howell A, Bolduc B, Sullivan MB (2020) The gut virome database reveals age-dependent patterns of virome diversity in the human gut. Cell Host Microbe 28(5):724–740.e728
45. Parks DH, Imelfort M, Skennerton CT, Hugenholtz P, Tyson GW (2015) CheckM: assessing the quality of microbial genomes recovered from isolates, single cells, and metagenomes. Genome Res 25(7):1043–1055
46. Gröschel AH, Walther A, Löbling TI, Schacher FH, Schmalz H, Müller AH (2013) Guided hierarchical co-assembly of soft patchy nanoparticles. Nature 503(7475):247–251
47. van der Walt S, Colbert SC, Varoquaux G (2011) The NumPy array: a structure for efficient numerical computation. Comput Sci Eng 13(2):22–30. https://doi.org/10.1109/MCSE.2011.37
48. Ewels P, Magnusson M, Lundin S, Käller M (2016) MultiQC: summarize analysis results for multiple tools and samples in a single report. Bioinformatics 32(19):3047–3048
49. Knight R, Vrbanac A, Taylor BC et al (2018) Best practices for analysing microbiomes. Nat Rev Microbiol 16:410–422. https://doi.org/10.1038/s41579-018-0029-9
50. Porter TM, Hajibabaei M (2018) Scaling up: a guide to high-throughput genomic approaches for biodiversity analysis. Mol Ecol 27(2):313–338
51. Piro VC (2018) Reference and taxonomy based methods for classification and abundance estimation of organisms in metagenomic samples
52. Shagam L, Gerasimov A, Kuznetsova T, Aksenova M, Akopova T, Dlin V (2017) Novel genotype-phenotype correlations in X-linked Alport syndrome: serum albumin level, age at onset of hematuria and hypertension. American Society of Human Genetics Annual meeting
53. Tørresen OK, Star B, Mier P, Andrade-Navarro MA, Bateman A, Jarnot P, Gruca A, Grynberg M, Kajava AV, Promponas VJ (2019) Tandem repeats lead to sequence assembly errors and impose multi-level challenges for genome and protein databases. Nucleic Acids Res 47 (21):10994–11006
54. Camargo AP, Roux S, Schulz F, Babinski M, Xu Y, Hu B, Chain PS, Nayfach S, Kyrpides NC (2024) Identification of mobile genetic elements with geNomad. Nat Biotechnol 42 (8):1303–1312
55. Roux S, Enault F, Hurwitz BL, Sullivan MB (2015) VirSorter: mining viral signal from

microbial genomic data. PeerJ 3:e985. https://doi.org/10.7717/peerj.985

56. Birol I, Raymond A, Jackman SD, Pleasance S, Coope R, Taylor GA, Yuen MMS, Keeling CI, Brand D, Vandervalk BP (2013) Assembling the 20 Gb white spruce (Picea glauca) genome from whole-genome shotgun sequencing data. Bioinformatics 29(12):1492–1497
57. Ayling M, Clark MD, Leggett RM (2020) New approaches for metagenome assembly with short reads. Brief Bioinform 21(2):584–594
58. Roux S, Camargo AP, Coutinho FH, Dabdoub SM, Dutilh BE, Nayfach S, Tritt A (2023) iPHoP: an integrated machine learning framework to maximize host prediction for metagenome-derived viruses of archaea and bacteria. PLoS Biol 21(4):e3002083
59. Cheng S, Xian W, Fu Y, Marin B, Keller J, Wu T, Sun W, Li X, Xu Y, Zhang Y (2019) Genomes of subaerial Zygnematophyceae provide insights into land plant evolution. Cell 179 (5):1057–1067.e1014
60. Levy S, Sutton G, Ng PC, Feuk L, Halpern AL, Walenz BP, Axelrod N, Huang J, Kirkness EF, Denisov G (2007) The diploid genome sequence of an individual human. PLoS Biol 5 (10):e254
61. Bharti R, Grimm DG (2021) Current challenges and best-practice protocols for microbiome analysis. Brief Bioinform 22(1):178–193
62. Rodriguez-R LM, Gunturu S, Harvey WT, Rosselló-Mora R, Tiedje JM, Cole JR, Konstantinidis KT (2018) The Microbial Genomes Atlas (MiGA) webserver: taxonomic and gene diversity analysis of archaea and bacteria at the whole genome level. Nucleic Acids Res 46 (W1):W282–W288
63. Zamora GOM (2013) De novo sequencing, assembly and analysis of the genome and transcriptome of the nematode Panagrolaimus superbus. National University of Ireland, Maynooth
64. Jiang C, Wang X, Li X, Inlora J, Wang T, Liu Q, Snyder M (2018) Dynamic human environmental exposome revealed by longitudinal personal monitoring. Cell 175(1):277–291.e231
65. Ekblom R, Wolf JB (2014) A field guide to whole-genome sequencing, assembly and annotation. Evol Appl 7(9):1026–1042
66. Blanot M (2020) Long-read, error prone metagenomics: systematic evaluation of assembly tools for nanopore sequencing. Universitat Politècnica de València
67. Chen L-X, Anantharaman K, Shaiber A, Eren AM, Banfield JF (2020) Accurate and complete genomes from metagenomes. Genome Res 30 (3):315–333
68. Aplakidou E, Vergoulidis N, Chasapi M, Venetsianou NK, Kokoli M, Panagiotopoulou E, Iliopoulos I, Karatzas E, Pafilis E, Georgakopoulos-Soares I (2024) Visualizing metagenomic and metatranscriptomic data: a comprehensive review. Comput Struct Biotechnol J 23:2011
69. Gao S-M, Schippers A, Chen N, Yuan Y, Zhang M-M, Li Q, Liao B, Shu W-S, Huang L-N (2020) Depth-related variability in viral communities in highly stratified sulfidic mine tailings. Microbiome 8(1):89
70. Matchado MS, Rühlemann M, Reitmeier S, Kacprowski T, Frost F, Haller D, Baumbach J, List M (2024) On the limits of 16S rRNA gene-based metagenome prediction and functional profiling. Microb Genom 10(2):001203
71. Rho M, Tang H, Ye Y (2010) FragGeneScan: predicting genes in short and error-prone reads. 38(20):e191
72. Pandey AK (2018) Microevolution of Neisseria lactamica during prolonged colonisation of the nasopharynx. University of Southampton
73. Pronk LJ, Medema MH (2022) Whokaryote: distinguishing eukaryotic and prokaryotic contigs in metagenomes based on gene structure. Microb Genom 8(5):000823
74. Yepes-García J, Falquet L (2024) Metagenome quality metrics and taxonomical annotation visualization through the integration of MAGFlow and BIgMAG. F1000Res 13:640
75. Chen Y, Wang Y, Paez-Espino D, Polz MF, Zhang T (2021) Prokaryotic viruses impact functional microorganisms in nutrient removal and carbon cycle in wastewater treatment plants. Nat Commun 12(1):5398
76. Li Q, Wu W, Zhu Y, Feng F, Ye J, Wang Z (2025) GENERanno: a genomic foundation model for metagenomic annotation. https://doi.org/10.1101/2025.06.04.656517
77. Sun Z, Huang S, Zhang M, Zhu Q, Haiminen N, Carrieri AP, Vázquez-Baeza Y, Parida L, Kim H-C, Knight R (2021) Challenges in benchmarking metagenomic profilers. Nat Methods 18(6):618–626
78. McIntyre AB, Ounit R, Afshinnekoo E, Prill RJ, Hénaff E, Alexander N, Minot SS, Danko D, Foox J, Ahsanuddin S (2017) Comprehensive benchmarking and ensemble approaches for metagenomic classifiers. Genome Biol 18 (1):182
79. Kieft K, Zhou Z, Anantharaman K (2020) VIBRANT: automated recovery, annotation and curation of microbial viruses, and evaluation of viral community function from genomic sequences. Microbiome 8(1):90
80. Sieber CM, Probst AJ, Sharrar A, Thomas BC, Hess M, Tringe SG, Banfield JF (2018)

Recovery of genomes from metagenomes via a dereplication, aggregation and scoring strategy. Nat Microbiol 3(7):836–843

81. Kieft K, Adams A, Salamzade R, Kalan L, Anantharaman K (2022) vRhyme enables binning of viral genomes from metagenomes. Nucleic Acids Res 50(14):e83
82. Orakov A, Fullam A, Coelho LP, Khedkar S, Szklarczyk D, Mende DR, Schmidt TS, Bork P (2021) GUNC: detection of chimerism and contamination in prokaryotic genomes. Genome Biol 22(1):178
83. Pinto Y, Bhatt AS (2024) Sequencing-based analysis of microbiomes. Nat Rev Genet 25 (12):829–845
84. Sequeira JC, Rocha M, Alves MM, Salvador AFJC (2022) UPIMAPI, reCOGnizer and KEGGCharter: Bioinformatics tools for functional annotation and visualization of (meta)-omics datasets. Comput Struct Biotechnol J 20:1798–1810
85. Truong DT, Tett A, Pasolli E, Huttenhower C, Segata N (2017) Microbial strain-level population structure and genetic diversity from metagenomes. Genome Res 27(4):626–638
86. Anyansi C, Straub TJ, Manson AL, Earl AM, Abeel T (2020) Computational methods for strain-level microbial detection in colony and metagenome sequencing data. Front Microbiol 11:1925
87. Albertsen M, Hugenholtz P, Skarshewski A, Nielsen KL, Tyson GW, Nielsen PH (2013) Genome sequences of rare, uncultured bacteria obtained by differential coverage binning of multiple metagenomes. Nat Biotechnol 31 (6):533–538

Chapter 5

Absolute Quantification of Bacteria in the Microbiome and Its Application

Jia-Xin Guo and Yi-Zhou Gao

Abstract

The advent of genomics and deep sequencing technologies has facilitated the development of absolute quantification techniques, which offer researchers more objective and precise sequencing outcomes. Unlike traditional relative quantification methods, which provide comparative data, absolute quantification delivers definitive measurements of genes or taxa. This analytical approach mitigates the potential for extraneous influences when comparing disparate samples, thereby reducing analytical errors. The implementation of absolute quantification techniques enhances our comprehension of microbial community structures, ecological dynamics, and their associations with host health or disease conditions. This chapter emphasizes a straightforward and broadly applicable method for genomic quantification, which necessitates the incorporation of a specified amount of internal standard DNA into the samples, eliminating the need for subsequent adjustments during library construction and sequencing. By assessing the proportion of internal standard DNA across various samples, sequencing data can be transformed into absolute quantification metrics. The internal standard method for absolute quantification is versatile and can be effectively utilized across multiple domains, including disease diagnosis, microbial ecology research, the fermentation industry, and environmental monitoring. Overall, absolute quantification methods furnish a more accurate and holistic perspective for microbiome research.

Key words Quantitative microbiome, Internal standard method, Next-generation sequencing analysis, Spike DNA, Multiple internal standards

1 Introduction

Next-generation sequencing (NGS) has significantly advanced our understanding of the composition and functionality of the microbiome, revealing numerous novel physiological mechanisms involved in disease progression [1]. Nonetheless, a primary challenge in taxonomic analysis using NGS lies in the establishment of a universal standard for characterizing microbial composition. Traditional profiling methods have relied on the percentage of target genes relative to the total microbial load [2, 3]; however, this approach complicates objective comparisons of changes in target

Tao Huang (ed.), *Microbial 'Omics: Methods and Protocols*, Methods in Molecular Biology, vol. 3033,
https://doi.org/10.1007/978-1-0716-5264-0_5,

gene expression across samples with varying total microbial loads [4]. For example, a reduction in the percentage of a target strain may arise from either a decrease in the abundance of that specific microorganism or an increase in the abundance of nontarget strains.

An absolute quantification profiling of the microbiome is essential for understanding microbial dynamics. Vandeputte et al. [5] were one of the pioneers in introducing this concept, employing flow cytometry as a method for quantification. In their approach, the total bacterial load was assessed via flow cytometry prior to 16S rRNA sequencing, allowing for the normalization of depth sequencing data in an absolute manner. This methodology effectively elucidated the variations in microbial abundance across different samples; however, its complexity and high cost have limited its widespread application. To address these challenges, Jian et al. [6] proposed the use of quantitative PCR (qPCR) for determining total bacterial load. In this technique, the bacterial load is quantified by amplifying a 450 bp fragment that encompasses the V3-V4 hypervariable regions of the 16S rRNA gene. This method has also been adapted for samples containing non-bacterial organisms, such as fungi; however, the accuracy of the absolute quantification might be influenced by two aspects: a. the copy numbers of the 16S rRNA gene in a specific strain cannot be estimated [7]; b. when the Ct value is large, the small error in the amplification efficiency of the PCR reaction might be amplified exponentially, resulting in a large error [8].

Both cell-counting [9] and qPCR methods present distinct advantages and limitations. Flow cytometry is capable of counting only viable cells; however, the presence of substantial amounts of extracellular DNA in samples may introduce bias in the results. Additionally, variations in environmental conditions or cell lysis can affect the consistency of extracellular DNA, potentially leading to significant discrepancies when compared to cellular DNA. The sensitivity of qPCR can be influenced by the choice of primers, which may limit its ability to provide absolute quantification [10]. To mitigate the issue of low sensitivity associated with qPCR, researchers such as Hindson et al. and Kim et al. [11, 12] have employed digital droplet PCR (ddPCR) [13] or other molecular quantification techniques. Furthermore, Gianluca et al. [14] conducted a comparative analysis of relative and absolute quantification methods for microbial profiles in stool samples, examining the biases inherent in the results obtained from three absolute quantification techniques: flow cytometry, qPCR, and ddPCR. Their findings indicated that absolute quantification methods yielded more accurate and strongly correlated results compared to relative quantification; however, the use of different absolute quantification methods resulted in varying quantitative microbial profiles.

A viable approach to mitigate significant bias in sequencing data is the incorporation of internal standards derived from the sequencing source [15]. The internal standard methodology standardizes the size and proportion of the genome across samples by introducing spike DNA, which facilitates homogenization through the added internal standard. Typically, the internal standard manifests in two forms: one involves the addition of a foreign molecule or cell that is entirely distinct from the sample, while the other utilizes an unchanged characteristic of the sample itself; the former is more commonly employed. A critical consideration in the design of spike DNA is to ensure that the added internal standard genes are absent from the original sample, as their presence could result in measurement inaccuracies. Consequently, it is imperative to possess a preliminary understanding of the sample and to judiciously select the appropriate internal standard gene when conducting quantitative experiments utilizing internal standards.

Jones et al. [16] conducted a pioneering study that assessed variations in microbial profiles by incorporating *Shewanella oneidensis* into human fecal samples. Their findings indicated that the internal standard method for analyzing precious, low-abundance samples yielded more precise quantitative data compared to the quantitative polymerase chain reaction (qPCR) method. This approach effectively mitigated the risk of overinterpretation of microbial profiles and the emergence of conflicting results. Nonetheless, it is important to note that intracellular labeling may introduce measurement inaccuracies stemming from challenges associated with the cultivation of the selected microorganisms, DNA extraction processes, and anomalies in spike gene copy numbers. These issues can be addressed by adjusting gene measurements through preliminary experimentation or by consulting published literature for estimates of variations in spike gene copy numbers [17].

Spike DNA molecules are increasingly utilized in internal standard methodologies as opposed to cellular components. These spike molecules are typically chosen for genes that are absent from the sample, with a preference for synthetic genes. The use of synthetic genes facilitates their exclusion from the sample genome and allows for tailored design to fulfill specific criteria regarding length, GC content, and mass density. It is recommended that synthetic genes be constructed to exceed 800 base pairs in length, avoid internal complementation, maintain a balanced GC content, and possess the capability to be linked to a barcode during subsequent library preparation. To mitigate potential quantitative inaccuracies associated with relying on a single internal standard, it is advisable to incorporate a mixture of multiple internal standards into the samples. This approach enables the selection of the most suitable internal standard for inter-sample quantification. For instance, if a sample was added the same amount of three internal standards, but the measured relative abundance is 1:3:1, it means the second internal standard cannot be used as the reference internal standard due to its existence in the original sample. The implementation of multiple mixed internal standards can significantly enhance the precision of absolute abundance estimations [18].

The quantity of internal standards incorporated into the sample should be determined by the size of the sample. It is essential to add internal standards in a manner that facilitates their effective detection within the sample, while also ensuring that the sequencing capacity is not compromised by an excessive number of internal standards. According to Lin et al. [19], the optimal proportion of internal standards to be added to the genome ranges from 1% to 3%. Furthermore, careful consideration must be given to the thorough mixing of the internal standards with the sample.

The internal standard method has been extensively employed across various research domains. Guo et al. [20] developed an internal standard plasmid incorporating the *RID1* gene from plants, which was utilized to assess the microbial composition within the root systems of plants. Similarly, Lin et al. [19] created an internal standard plasmid containing the 18S rRNA of *Schizosaccharomyces pombe*, enabling precise quantification of microorganisms per unit of seawater. Beyond environmental microbiome studies, quantitative microbiomics has also been applied to investigate human gut microbiota, where enhanced comprehension of gut microbiota has facilitated improved prediction and treatment of diseases linked to these microbial communities. Rao et al. [21] employed an internal plasmid that included sequences from *Salinibacter ruber*, *Trichoderma reesei*, and *Haloarcula hispida* to quantify microbial abundance in seawater samples. This multiplexed internal standard plasmid effectively captured the intricate microbial composition and dynamics of bacteria, fungi, and archaea present in infant feces. Tkacz et al. [22] synthesized chimeric DNA spikes using multiple representative strains from *Rhizobia* (soil), *Firmicutes* (human gut), *Bifidobacteria* (human gut), and *Enterobacteriaceae* (food). These DNA spikes were capable of analyzing soil samples while maintaining adequate sequencing breadth, even for less prevalent species. Zhou et al. [23] demonstrated that the incorporation of spike DNA enhanced the accuracy of pneumonia detection in patients. Furthermore, Jin et al. [4] investigated the microbial content in various regions of the intestinal flora and the intestinal mucus layer of mice by introducing three bacterial strains—*Bacillus subtilis* 168, *Pseudomonas putida* KT2440 [24], and *Pandoraea* sp. MCB032 [25]—as internal standards in their analyses.

The extensive implementation of absolute quantification of the microbiome utilizing the internal standard method is poised to significantly advance scientific knowledge and medical progress. This approach will facilitate a more precise understanding of the

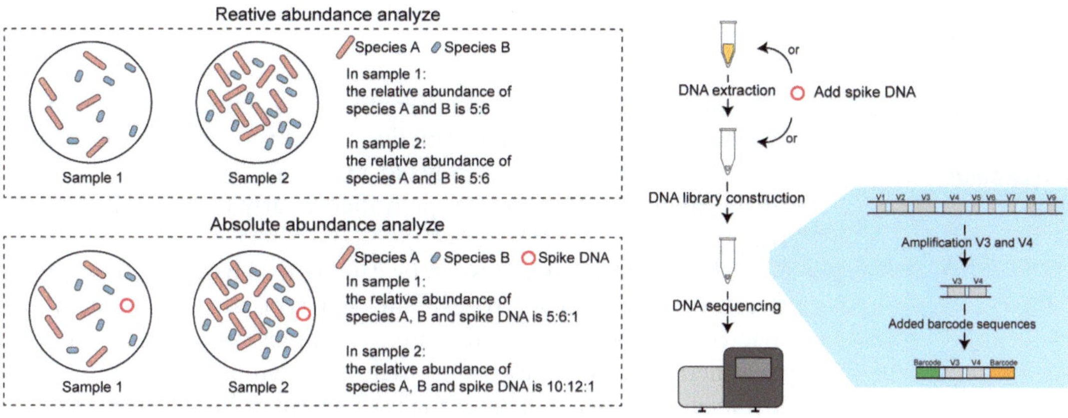

Fig. 1 The technical roadmap of the absolute quantity method

composition and dynamics of microbial communities, elucidate the functions of various microbial populations, and clarify their specific interactions with the environment. Furthermore, the high reproducibility associated with this method will enhance the identification of rare and critical microorganisms that are not amenable to culture. The application of absolute quantification of the microbiome extends to personalized medicine, disease diagnosis, and the development of probiotics. These applications are centered on the formulation of diagnostic therapies informed by gut microbiome analysis, wherein the identification and characterization of the gut microbiome associated with particular diseases or healthy populations will enhance the safety and efficacy of therapeutic interventions. The technical roadmap of the absolute quantity method is shown in Fig. 1. Quantitative information about the bacteria in the sample will be determined by sequencing the V3 and V4 regions of 16S.

2 Materials

Kits and regents: Genomic DNA extraction kit, Gel recovery kit, DNA Library Prep Kits for Illumina Systems, agar, PBS buffer (phosphate buffer saline), TAE (Tris-Acetate-EDTA) Electrophoresis Buffer 1X (Trimethylol aminomethane 0.04 M, Acetic acid 0.02 M, Ethylene Diamine Tetraacetic Acid 1 mM).

DNA: a plasmid for internal standard (spike DNA).

Equipment: microcentrifuge, Nanodrop, Gel electrophoresis apparatus, Illumina HiSeq.

3 Methods

3.1 Internal Standard and Samples Preparation

The internal standard must adhere to specific criteria: (1) the spiked DNA should possess a distinctive characteristic fraction that is entirely absent from the analyzed samples; (2) the concentration of the spiked DNA must be quantified clearly before added to the real samples.

Macrogenomic studies can be categorized based on their objectives into two primary types. The first type involves comparative macrogenomic studies that examine samples subjected to different treatments across various spatial and temporal contexts. For instance, this may include analyses of microbial composition in soybean inter-root systems under varying fertilizer applications, or comparisons of microbial communities in the intestines of diseased subjects versus control groups, with the aim of identifying potential biomarkers. Such studies necessitate the inclusion of at least two subgroups, each comprising a minimum of three biological replicates. In cases where individual variability is significant, such as in gut microbiota studies, it is advisable to utilize more than five biological replicates. A comprehensive sampling plan should be established prior to the experiment, ensuring that samples are collected according to predetermined sampling points. If the study involves correlation analyses with environmental factors, the number of replicates for these factors should align with the number of sequenced samples.

The second type of study focuses on the exploration of microbial composition in rare samples, such as those obtained from deep-sea environments. This approach may also extend to the investigation of microbial communities in unique substrates, including deep-sea sediments, crater soils, and lunar soils [26]. Macrogenomics, when integrated with other genomic methodologies, can facilitate an in-depth examination of microbial community functions at the genomic, transcriptional, and metabolic levels. It is imperative that samples are collected in an aseptic environment and that appropriate transportation methods to the laboratory are selected based on the sample type, while maintaining the lowest possible temperatures to mitigate DNA/RNA degradation. Prompt extraction of sample DNA/RNA is also essential.

3.2 Internal Standard Addition and DNA Extraction

Spike DNA should be incorporated into the samples prior to extraction, with the weight of the internal standard constituting 5–8% of the total gene pool by weight. Care should be taken to minimize pipetting errors when adding the spike DNA to the samples. The extraction of DNA from the mixed samples should be conducted in accordance with the guidelines provided by the Genomic DNA extraction kit. In instances where the extracted DNA cannot be processed immediately, it is essential to store it at −80 °C.

3.3 DNA Amplification

The target regions for sequencing in the samples were amplified using polymerase chain reaction (PCR) with a PCR apparatus. The primers employed in the PCR process must consist of two distinct regions: one segment that is complementary to the sequences at both termini of the target gene, and another segment that incorporates a barcode. To ensure the accuracy and reliability of subsequent data analyses, two critical criteria must be met: (1) the amplification should utilize a minimal number of cycles; and (2) the number of cycles for amplifying each sample must be consistent. A representative sample should be randomly selected for preliminary experiments to confirm that the majority of samples can yield amplification products at appropriate concentrations with low cycle numbers. The number of cycles in this PCR reaction system typically ranges from 5 to 25. Following the reaction, the target gene fragments within the PCR system can be isolated using gel electrophoresis or magnetic beads.

3.4 DNA Library Construction

The amplification of DNA sequences is conducted in accordance with the operational guidelines provided by DNA Library Preparation Kits to construct a sequencing library. This process encompasses four primary steps: (1) The addition of adapter tsequences to the termini of the target DNA region via PCR; (2) the utilization of a gel recovery kit to excise and recover the PCR products; (3) the elution of the products using Tris-HCl buffer, followed by detection through 2% agarose gel electrophoresis; and (4) the denaturation of the DNA using sodium hydroxide, resulting in the formation of single-stranded DNA fragments.

3.5 Computer Sequencing

The sequencing of genes within the DNA library was conducted utilizing the Illumina sequencing system. In this process, the splice sequences of DNA fragments are designed to complement the base sequences that are affixed to the sequencing chip, resulting in their immobilization on the chip surface. Following this, PCR is employed to synthesize the DNA fragments based on the immobilized base sequences, thereby facilitating the synthesis of the target DNA fragments directly on the chip.

After the processes of denaturation and annealing, one end of the DNA fragment on the chip randomly pairs with an adjacent primer, leading to further immobilization and the formation of a "bridge." PCR amplification continues, resulting in the generation of DNA clusters. These clusters are subsequently linearized into single-stranded DNA, which allows for the reading of sequence information through fluorescent signals in subsequent reactions.

In the next phase, a single strand of DNA is combined with a modified DNA polymerase and deoxynucleotide triphosphates (dNTPs) that are labeled with four distinct fluorescent markers, enabling the incorporation of a single base per cycle. After each reaction cycle, the reaction plate's surface is scanned with a laser to identify the nucleotide types of each template sequence that has been polymerized during that cycle. Prior to the next round of amplification and reading, the "fluorophores" and "terminating groups" are chemically cleaved to restore the 3′ end viscosity, allowing for the continuation of nucleotide polymerization. The fluorescence signals collected during each cycle are quantified to derive the sequence of the template DNA fragments.

3.6 Raw Data Processing

The workflow for processing raw data is outlined as follows. Initially, paired-end (PE) reads obtained from sequencing undergo sample splitting. Subsequently, the double-ended reads are subjected to quality control and filtering based on sequencing quality. Concurrently, these reads are spliced together based on their overlapping relationships, resulting in optimized data that has undergone quality-controlled splicing. This optimized data is then processed using sequence noise reduction techniques, such as DADA2 or Deblur, to derive amplicon sequence variant (ASV) representative sequences and their corresponding abundance information.

During the construction of the sequencing library, barcode sequences are incorporated into DNA fragments to differentiate between various libraries. Consequently, the raw downstream data typically contains barcode information, which is extraneous to the intended analysis and must be removed. Additionally, the raw data may include low-quality bases, ambiguous bases, and short sequences, all of which could compromise the reliability of subsequent analyses. Therefore, it is imperative to perform quality control filtering on the raw sequencing data prior to analysis to yield high-quality clean data, thereby ensuring the accuracy of the ensuing analytical results. The specific criteria for raw data filtering are as follows: (1) removal of reads containing 0.1% or more of ambiguous bases (N bases); (2) elimination of reads that include sequencing junction sequences; (3) exclusion of reads with over 50% low-quality bases ($Q < 20$); and (4) for samples derived from host environments, it is essential to compare the sequences against host genomes to filter out host sequences, thereby minimizing their interference in subsequent analyses. In this context, host genome sequences will be filtered out based on the sample source; for instance, if the sample originates from human, mouse, or rat, filtering will occur directly according to the source, while for other sources, host sequences must be provided for comparison.

The methodologies and software utilized in the genomic data processing are elaborated upon in the following sections. Quality control (QC) of the raw genomic reads and de-hosting procedures were conducted using SOAPnuke software, which facilitated the filtration of raw data for QC purposes. Subsequently, Bowtie2 was employed to align the host sequences, enabling the removal of these sequences to yield clean data. The processed reads were then assembled and functionally annotated. The assembly of the QC and de-hosted sequences was executed using MEGAHIT [27], resulting in the generation of contigs. Gene sequences within these contigs were predicted using the MetaGeneMark software. To eliminate redundancy among the identified genes, CD-HIT software [28] was applied, followed by the use of Salmon software to compile a relative abundance table for each gene. Functional annotation of the non-redundant genes was achieved through comparisons with the eggNOG, KEGG, BacMet, CARD, COG, CAZy, and Swiss-Prot databases, utilizing either DIAMOND or RGI for this purpose. For species annotation of the genomes, Kraken2 [29] was employed alongside self-constructed databases, which included screenings of the NCBI NT database or the UHGG database, to quantify the number of species sequences present in the samples. Bracken2 was then utilized to estimate the actual abundance of these species within the samples, thereby completing the species annotation process. It is noteworthy that alternative software options exist, providing flexibility in the methodologies employed.

3.7 Internal Standard Quantification

The quantification of the internal standard could be done after sequencing, and this is processed in two steps. First is the internal standard DNA searching in the genome sequence using BLASTn homology search with a bit score cutoff of 50. This step picks up all the potential internal standard reads, and the reads verify process using BLASTx with a bit score cutoff of 40. The verity step avoids counting the false positive reads by homology search of BLASTx between different translated proteins derived from the reads. Following these two steps, the quantification of the internal standard DNA is determined.

3.8 Metagenome Normalization

The gene pool of an individual sample size, along with transcriptional abundance, can be determined using the subsequent formula:

$$A_{\text{total}} = {A_{\text{reads}} \times S_{\text{added}}}/{S_{\text{reads}}}$$

A_{total} = total gene numbers in one sample.

A_{reads} = the numbers of total protein-encoding reads.

S_{added} = the number of internal standard DNA added in the sample.

S_{reads} = the numbers of internal standard DNA read.

In order to assess the variations in the abundances of specific genes across different samples, the adjusted abundances of individual genes may be determined utilizing the subsequent formula:

$$G_{\text{target}} = {G_{\text{reads}} \times S_{\text{added}}}/{S_{\text{reads}}}$$

G_{added} = the number of proteins encoding reads of the target gene.

G_{target} = the number of target gene in the sample.

G_{target} can be transferred into the gene abundance per volume or weight by dividing the mass or volume of the sample.

4 Notes

The internal standard method for absolute quantification offers notable advantages, including a streamlined procedure and enhanced reproducibility. However, the introduction of an internal standard is a primary source of bias in this methodology. Specifically, when an internal standard is incorporated into a sample for quantification, it is imperative that the internal standard closely resembles the target species and is absent from the sample itself. Variations in the copy number of the internal standard gene can result in discrepancies and inaccuracies in the quantification results. The molecular internal standard approach is demonstrably more effective than the cellular internal standard method, as the specificity and copy stability of synthetic gene sequences facilitate more convenient research applications. Nonetheless, the successful implementation of the internal standard method necessitates a preliminary understanding of the sample, ensuring that the added internal standard sequence is not present in the sample. Furthermore, the assessment of sample quality can introduce additional challenges to the internal standardization process. For instance, when samples of identical weight but differing hydration levels are treated with the same quantity of internal standard, the analytical results for samples with higher hydration may be underestimated. Therefore, it is advisable to adequately dry samples prior to the addition of the internal standard and subsequent weighing, as the inherent characteristics of samples with varying densities, such as soils, or those with significant hydration discrepancies, can lead to inherent bias. Enhanced isolation and screening of such samples may mitigate some of this bias. Additionally, variations in nucleic acid extraction efficiency among samples can contribute to errors in the absolute quantification using internal standards. For example, differences in the effectiveness of cell lysis during nucleic acid extraction and selective access to specific genes can introduce variability. The issue of gene preference during PCR reactions, which can occur during library construction or other procedural steps, is a common challenge not only in internal standard methods but also in other absolute quantification techniques. Although it is theoretically assumed that all molecules in a PCR reaction exhibit uniform amplification efficiency, practical considerations such as differences in GC content or primer specificity can lead to discrepancies between the true abundance of certain genes and their measured abundance.

For different samples, the amplified genes selected from the samples are different. When the selected sample is a bacterium or archaeon, 16S is usually chosen; when the selected sample is eukaryotic cell, 18S is selected; and when the selected sample is a fungus, internally transcribed spacer (ITS) should be selected.

Various types of errors can arise from internal standard absolute quantification methods. To mitigate sample-to-sample variability associated with this approach, it is advisable to incorporate multiple quality control measures into the procedure. For instance, the inclusion of internal standards in technical duplicates is recommended, along with ensuring that the ratio of internal standard reads obtained in these duplicates falls within an acceptable margin of error. Additionally, employing multiple internal standards and calibrating to the appropriate concentration through gradient dilution can enhance accuracy. To minimize the impact of PCR processes on sequencing outcomes, it is preferable to utilize modified enzymes whenever feasible or to limit the number of PCR cycles to the minimum necessary.

Conflict of Interest

This work was funded by the Natural Science Foundation of Heilongjiang Province (YQ2025C033). The authors have no conflicts of interest to declare that are relevant to the content of this chapter.

References

1. Luo M, Zhang D, Tian C, Sun B, Huang T, Zhang H, Shi Y, Gao Y-Z, Luo R, Zhu J-K (2023) The innovation life: an international journal of life sciences. Innov Life 1(1): 100001. https://doi.org/10.59717/j.xinn-life.2023.100001
2. Sun DL, Gao YZ, Ge XY, Shi ZL, Zhou NY (2020) Special features of bat microbiota differ from those of terrestrial mammals. Front Microbiol 11:1040. https://doi.org/10.3389/fmicb.2020.01040
3. Xu C, Kadosh D, Sun D, Zeng G, Gao Y-Z (2023) Editorial: omics-originated exploration of pathogenic patterns and molecular mechanisms in human and animal fungal pathogens. Front Microbiol 14. https://doi.org/10.3389/fmicb.2023.1243709
4. Jin W-Y, Guo J-X, Zhang M, Teng L-Z, Chao Y, Sansonetti PJ, Gao Y-Z (2023) Absolute quantification of the microbiota spatial distribution in the murine large intestine. Innov Life 1(2):100030. https://doi.org/10.59717/j.xinn-life.2023.100030
5. Vandeputte D, Kathagen G, D'Hoe K, Vieira-Silva S, Valles-Colomer M, Sabino J, Wang J, Tito RY, De Commer L, Darzi Y, Vermeire S, Falony G, Raes J (2017) Quantitative microbiome profiling links gut community variation to microbial load. Nature 551(7681): 507–511. https://doi.org/10.1038/nature24460
6. Jian C, Luukkonen P, Yki-Jarvinen H, Salonen A, Korpela K (2020) Quantitative PCR provides a simple and accessible method for quantitative microbiota profiling. PLoS One 15(1):e0227285. https://doi.org/10.1371/journal.pone.0227285
7. Jin W-Y, Peng J, Dai J, Tang R, Guo J-X, Zhao H, Wang J, Zhang S, Gao Y-Z (2024) Bacterial load in meconium. iMeta 3(1):e173. https://doi.org/10.1002/imt2.173
8. Gao YZ, Liu H, Chao HJ, Zhou NY (2016) Constitutive expression of a nag-like dioxygenase gene through an internal promoter in the 2-chloronitrobenzene catabolism gene cluster of *pseudomonas stutzeri* ZWLR2-1. Appl

Environ Microbiol 82(12):3461–3470. https://doi.org/10.1128/AEM.00197-16

9. Zhang M, Tang R, Li F-X, Jin W-Y, Guo J-X, Teng L-Z, Meng G, Sansonetti PJ, Gao Y-Z (2024) Optimization of ultrasound-mediated DNA transfer for bacteria and preservation of frozen competent cells. Microbiol Spectrum 12(12):e00978–e00924. https://doi.org/10.1128/spectrum.00978-24
10. Smith CJ, Osborn AM (2009) Advantages and limitations of quantitative PCR (Q-PCR)-based approaches in microbial ecology. FEMS Microbiol Ecol 67(1):6–20. https://doi.org/10.1111/j.1574-6941.2008.00629.x
11. Hindson CM, Chevillet JR, Briggs HA, Gallichotte EN, Ruf IK, Hindson BJ, Vessella RL, Tewari M (2013) Absolute quantification by droplet digital PCR versus analog real-time PCR. Nat Methods 10(10):1003–1005. https://doi.org/10.1038/nmeth.2633
12. Kim TG, Jeong SY, Cho KS (2014) Comparison of droplet digital PCR and quantitative real-time PCR for examining population dynamics of bacteria in soil. Appl Microbiol Biotechnol 98(13):6105–6113. https://doi.org/10.1007/s00253-014-5794-4
13. Manzari C, Oranger A, Fosso B, Piancone E, Pesole G, D'Erchia AM (2020) Accurate quantification of bacterial abundance in metagenomic DNAs accounting for variable DNA integrity levels. Microb Genom 6(10). https://doi.org/10.1099/mgen.0.000417
14. Galazzo G, van Best N, Benedikter BJ, Janssen K, Bervoets L, Driessen C, Oomen M, Lucchesi M, van Eijck PH, Becker HEF, Hornef MW, Savelkoul PH, Stassen FRM, Wolffs PF, Penders J (2020) How to count our microbes? The effect of different quantitative microbiome profiling approaches. Front Cell Infect Microbiol 10:403. https://doi.org/10.3389/fcimb.2020.00403
15. Chen K, Hu Z, Xia Z, Zhao D, Li W, Tyler JK (2015) The overlooked fact: fundamental need for spike-in control for virtually all genome-wide analyses. Mol Cell Biol 36(5):662–667. https://doi.org/10.1128/MCB.00970-14
16. Jones MB, Highlander SK, Anderson EL, Li W, Dayrit M, Klitgord N, Fabani MM, Seguritan V, Green J, Pride DT, Yooseph S, Biggs W, Nelson KE, Venter JC (2015) Library preparation methodology can influence genomic and functional predictions in human microbiome research. Proc Natl Acad Sci USA 112(45):14024–14029. https://doi.org/10.1073/pnas.1519288112
17. Stoddard SF, Smith BJ, Hein R, Roller BR, Schmidt TM (2015) rrnDB: improved tools for interpreting rRNA gene abundance in bacteria and archaea and a new foundation for future development. Nucleic Acids Res 43 (Database issue):D593–D598. https://doi.org/10.1093/nar/gku1201
18. Stammler F, Glasner J, Hiergeist A, Holler E, Weber D, Oefner PJ, Gessner A, Spang R (2016) Adjusting microbiome profiles for differences in microbial load by spike-in bacteria. Microbiome 4(1):28. https://doi.org/10.1186/s40168-016-0175-0
19. Lin Y, Gifford S, Ducklow H, Schofield O, Cassar N (2019) Towards quantitative microbiome community profiling using internal standards. Appl Environ Microbiol 85(5). https://doi.org/10.1128/AEM.02634-18
20. Guo X, Zhang X, Qin Y, Liu YX, Zhang J, Zhang N, Wu K, Qu B, He Z, Wang X, Zhang X, Hacquard S, Fu X, Bai Y (2020) Host-associated quantitative abundance profiling reveals the microbial load variation of root microbiome. Plant Commun 1(1): 100003. https://doi.org/10.1016/j.xplc.2019.100003
21. Rao C, Coyte KZ, Bainter W, Geha RS, Martin CR, Rakoff-Nahoum S (2021) Multi-kingdom ecological drivers of microbiota assembly in preterm infants. Nature 591(7851):633–638. https://doi.org/10.1038/s41586-021-03241-8
22. Tkacz A, Hortala M, Poole PS (2018) Absolute quantitation of microbiota abundance in environmental samples. Microbiome 6(1):110. https://doi.org/10.1186/s40168-018-0491-7
23. Zhou H, Ouyang C, Han X, Shen L, Ye J, Fang Z, Chen W, Chen A, Ma Q, Hua M, Zhu J, Wu X, Lin X, Shui Y, Zhou C, Fang K, Du J, Huang Z, Wang G, Lv Q, Huang W, Wang J, Hua X, Zhou J, Liu C, Yu Y (2022) Metagenomic sequencing with spiked-in internal control to monitor cellularity and diagnosis of pneumonia. J Infect 84(1):e13–e17. https://doi.org/10.1016/j.jinf.2021.09.018
24. Gao Y-Z, Wang Y, Ji M, Zhou N-Y, Huang WE (2023) A whole-cell hydrogen peroxide biosensor and its application in visual food analysis. The Innovation Life 1(1):100011. https://doi.org/10.59717/j.xinn-life.2023.100011
25. Jiang XW, Liu H, Xu Y, Wang SJ, Leak DJ, Zhou NY (2009) Genetic and biochemical analyses of chlorobenzene degradation gene clusters in *Pandoraea* sp. strain MCB032. Arch Microbiol 191(6):485–492. https://doi.org/10.1007/s00203-009-0476-9
26. Yue Z, Gou S, Sun S, Yang W, Chen Y, Wang Y, Lin H, Di K, Lin Y, Li X, Wu F (2024) Geological context of the Chang'e-6 landing area and implications for sample analysis.

Innovation 5(5). https://doi.org/10.1016/j.xinn.2024.100663

27. Li D, Luo R, Liu CM, Leung CM, Ting HF, Sadakane K, Yamashita H, Lam TW (2016) MEGAHIT v1.0: a fast and scalable metagenome assembler driven by advanced methodologies and community practices. Methods 102: 3–11. https://doi.org/10.1016/j.ymeth.2016.02.020
28. Kanehisa M, Sato Y, Morishima K (2016) BlastKOALA and GhostKOALA: KEGG tools for gunctional characterization of genome and metagenome sequences. J Mol Biol 428(4): 726–731. https://doi.org/10.1016/j.jmb.2015.11.006
29. Liu Y, Ghaffari MH, Ma T, Tu Y (2024) Impact of database choice and confidence score on the performance of taxonomic classification using Kraken2. aBIOTECH 5(4): 465–475. https://doi.org/10.1007/s42994-024-00178-0

Chapter 6

Network Analysis in Microbiome Research: Methods, Tools, and Applications

Jiajun Liu, Yixue Li, and Tao Huang

Abstract

The human microbiome operates as a complex, interconnected ecosystem where microbial interactions dictate community stability, host health, and disease progression. Understanding these dynamics requires moving beyond simple taxonomic catalogs to systems-level network analyses. This chapter reviews network methodologies in microbiome research, progressing from traditional correlation-based approaches to advanced artificial intelligence techniques. We systematically cover co-occurrence, protein-protein interaction, metabolic, multi-omics integrated, and evolutionary transmission networks. Computational tools—spanning general platforms and specialized pipelines—are compared alongside topology metrics and community detection algorithms. Furthermore, we highlight the integration of graph neural networks and protein language models, discussing current challenges in data standardization, model interpretability, and the merging of mechanistic and data-driven paradigms.

Key words Microbiome network analysis, Viral evolution networks, Graph neural networks, Multi-omics integration, Community detection, Network topology

1 Introduction

Bacteria, archaea, fungi, and viruses exist not in isolation, but within complex ecosystems, interacting with hosts through metabolic exchange, immune modulation, and competitive exclusion. While early microbiome research heavily relied on taxonomic cataloging and differential abundance analysis, these linear approaches often fall short of capturing the functional dynamics and resilience of microbial communities. Consequently, the field is increasingly shifting toward systems-level network frameworks to interrogate community assembly, stability, and perturbation responses, moving toward a truly mechanistic understanding of microbiome dynamics [1, 2]. Integrative multi-omics approaches have further enriched this paradigm by linking taxonomic, transcriptomic, and metabolomic layers to reveal holistic host-microbiome interactions. Concurrently, the advent of whole-genome sequencing enables the

Tao Huang (ed.), *Microbial 'Omics: Methods and Protocols*, Methods in Molecular Biology, vol. 3033,
https://doi.org/10.1007/978-1-0716-5264-0_6,

reconstruction of evolutionary and transmission networks at single-nucleotide resolution, transforming epidemiological surveillance and lineage tracing. As the scale, heterogeneity, and dimensionality of these biological networks expand, traditional statistical models increasingly struggle to capture non-linear, high-order dependencies. To process this complex topological data, the field is undergoing a critical transition toward deep learning architectures. Graph neural networks (GNNs) and protein language models (PLMs) are being deployed to bridge the gap between raw sequence data and systems-level interactions, automating functional annotation directly from sequence and structural modalities [3–6]. By learning latent representations of network nodes and edges, these deep learning frameworks offer unprecedented predictive power for structural-functional relationships, bypassing the limitations of homology-based pipelines.

Despite these methodological advances, distinguishing true ecological interactions from compositional artifacts remains a significant computational hurdle. The following sections present the theoretical foundations of network inference, evaluate various network types and their biological applications, and systematically compare topology metrics and community detection algorithms. Finally, we explore state-of-the-art computational tools and AI-enhanced graph representation learning, concluding with a discussion on the ongoing challenges of model interpretability and data standardization.

2 Theoretical Foundations and Network Inference Methods

Co-occurrence networks derive from pairwise taxon correlations computed across samples, where edges connect taxa that increase or decrease together, or inversely. Pearson and Spearman correlations provide intuitive measures but generate spurious associations in relative abundance data due to compositional constraints. SparCC addresses this through iterative log-ratio transformations, assuming data sparsity [7]. Method selection for microbiome co-occurrence network construction should consider dataset sparsity, sample size, and analytical goals.

Gaussian graphical models estimate conditional dependencies in microbial association networks through sparse inverse covariance selection. SPIEC-EASI implements this for compositional microbiome data by combining centered log-ratio (CLR) transformation with graphical lasso regularization, enabling identification of direct ecological associations while excluding spurious correlation [8]. Bayesian network approaches provide alternative frameworks for directed associations in regulatory or causal networks from longitudinal or interventional data.

Table 1
Network topology metrics and biological interpretation

Scope	Metric	Definition	Biological significance
Global	Average path length	Average of shortest path lengths between all reachable node pairs	Measures metabolite or information transfer efficiency. Shorter paths indicate rapid cross-feeding or influence propagation
	Network diameter	Maximum shortest path distance between any pair of nodes	Defines the maximum propagation distance; limits how fast a perturbation spreads across the community
	Average clustering coefficient	Mean of local clustering coefficients for all nodes	Quantifies overall community cohesiveness. High values indicate "small-world" properties, common in microbial guilds where taxa form tight functional clusters
	Network density	Ratio of actual edges to possible edges	Measures community complexity and connectedness. High density often correlates with increased stability and resistance to invasion, though extremely high density can imply fragility to systemic collapse
Local	Node degree	Number of connections (in/out-degree for directed graphs)	Identifies hub taxa or keystone species. High-degree taxa often dominate biomass, drive metabolic flux, or serve as super-spreaders in transmission networks
	Betweenness centrality	Proportion of shortest paths passing through a node	Identifies connectors or bottlenecks. These taxa bridge distinct functional modules (e.g., linking carbohydrate degraders to methanogens); their removal can fragment the community
	Closeness centrality	Reciprocal of average shortest path distance to all other nodes	Indicates taxa positioned to rapidly affect or sense changes in the entire community; often metabolically versatile organisms central to cross-feeding networks
	Eigenvector centrality	Measure of influence based on neighbors' centrality	Identifies taxa connected to other influential taxa. Useful for finding "hidden influencers" that may not have the highest degree but interact with key players
	Local clustering coefficient	Density of connections among a node's neighbors	Measures the cohesiveness of a taxon's immediate functional niche. High local clustering suggests the taxon is embedded in a tight-knit, cooperative guild

Network topology metrics quantify structural properties with biological significance (Table 1). Degree centrality identifies hub taxa in gut microbiome networks, key enzymes in pathway networks, or dominant variants in viral evolution networks. Betweenness centrality identifies nodes lying on many shortest paths—metabolic bottlenecks, transmission bridges, or connectors between functional modules. Closeness centrality measures average distance to all other nodes, indicating how quickly a taxon can influence the community. Eigenvector centrality assigns higher scores to nodes connected to other high-scoring nodes, distinguishing hubs linked to influential neighbors from those linked to peripheral taxa. At the structural level, clustering coefficients and modularity quantify local cohesiveness and functional compartmentalization, while stability analysis evaluates interaction fluctuations within the microecological network [8]. Beyond structural characterization, these topological features significantly enhance machine learning predictive power. For instance, the HiRisk-Detector model, by integrating node degree and connectivity metrics from SARS-CoV-2 haplotype networks, successfully discriminated variants of concern 27 days prior to WHO announcements [9].

Community detection algorithms partition networks into modules (Table 2). Louvain optimizes modularity for large undirected networks, widely applied to microbiome co-occurrence networks [10]. Leiden improves upon Louvain by guaranteeing well-connected communities [11]. Infomap minimizes information flow description length, handling directed networks by optimizing random walk traversal [12]. Walktrap employs random walks to identify communities [13]. Girvan-Newman iteratively removes high-betweenness edges revealing hierarchical structure [14]. Algorithm choice should match network characteristics and biological interpretation needs.

3 Network Types in Microbiome Research

Co-occurrence networks represent statistical associations inferred from abundance data, where nodes denote taxa (operational taxonomic units, OTUs; amplicon sequence variants, ASVs; or species) or functional groups, and edges capture positive associations (mutualistic interactions, shared niches) or negative associations (competitive exclusion, antagonism). Statistical methods range from simple Pearson correlation to sophisticated approaches accounting for compositional constraints through graphical models. Recent advances employ consensus network inference that integrates multiple statistical approaches to distinguish true ecological associations from compositional artifacts [15]. Applications include identifying co-abundance modules, comparing topology between health and disease states, and detecting shifts in association patterns across conditions [16].

Protein-protein interaction (PPI) networks represent proteins as nodes and physical binding, functional coupling, or pathway co-membership as edges, derived from experiments, predictions, or literature mining. The Search Tool for the Retrieval of Interacting Genes/Proteins (STRING) provides confidence-scored networks for thousands of organisms, including microbial taxa [17]. In microbiome analysis, PPI networks facilitate the functional

Table 2
Community detection algorithms: features and applications

Algorithm	Network type	Mechanism	Typical microbiome applications and interpretation
Louvain	Undirected	Greedy modularity optimization. Rapidly maximizes modularity by iteratively merging nodes.	Standard for co-occurrence networks. Identifies Co-Abundant Groups (CAGs) or broad ecological guilds. Used to detect large-scale dysbiosis patterns (e.g., enterotypes)
Leiden	Undirected	Refined modularity. Improves on Louvain by guaranteeing well-connected communities and fixing resolution limit issues	High-resolution module detection. Preferable for finding tightly coupled functional consortia where Louvain might produce disconnected artifacts
Infomap	Directed/ undirected	Information flow/random walk. Minimizes map equation description length; follows the flow of probability	Metabolic and transmission flow. Ideal for directed networks (e.g., metabolic flux, virus transmission). Identifies modules based on flow constraints (e.g., source-sink dynamics)
Walktrap	Undirected	Random walk similarity. Nodes are grouped if random walks are likely to remain within the cluster	Local niche discovery. Effective for finding dense subgraphs representing local micro-niches or strictly mutualistic pairs, often robust to noise
Girvan-Newman	Undirected	Edge betweenness removal. Iteratively removes "bridge" edges to reveal hierarchical structure	Hierarchical structure analysis. Useful for small networks to demonstrate how communities split (e.g., defining core vs. peripheral microbiome members) but computationally slow

annotation of metagenomic gene catalogs by propagating information across protein association graphs. Building on this, graph convolutional networks (GCNs) approaches like DeepFRI integrate protein contact maps with sequence embeddings to learn structure–function relationships, enabling large-scale prediction of Gene Ontology and enzyme commission annotations. When applied to infant gut microbiome cohorts, such workflows have demonstrated near-complete functional coverage of nonredundant gene catalogs, significantly outperforming traditional homology-based pipelines [5, 6].

Metabolic and functional networks map biochemical capabilities where nodes represent metabolites (glucose, short-chain fatty acids), enzymatic reactions, genes encoding metabolic functions, or functional modules (carbon fixation, nitrogen cycling), while edges denote biochemical transformations, substrate-product relationships, or functional profile co-occurrence. Construction approaches include constraint-based stoichiometric modeling (implemented via COnstraint-Based Reconstruction and Analysis for Python, COBRApy; and Microbial Community Modeling, MICOM), pathway-centric inference based on the MetaCyc Metabolic Pathway Database (MetaCyc) and the Kyoto Encyclopedia of Genes and Genomes (KEGG), and correlation networks based on functional gene abundance. Applications identify core metabolic bottlenecks, infer syntrophic cross-feeding, and link topology changes with health-disease transitions. Recent multi-omics integration reveals regulatory rewiring in inflammatory conditions and metabolic vulnerabilities [18].

Multi-omics networks link nodes across molecular layers—taxa, transcripts, proteins, metabolites, host features—with edges capturing cross-layer correlations or regulatory relationships. Methods include canonical correlation analysis, graph-based community detection, and ensemble learning. Recent implementations integrate consensus strategies to delineate disease modules and regulatory circuits, supporting identification of metabolic bottlenecks, biomarker panels, and host-microbiome interactions [18, 19].

Evolutionary and transmission networks reconstruct pathogen relationships, with nodes for variants/strains and edges for mutational or transmission events. HIV-TRACE identifies transmission clusters via genetic distance thresholding without tree inference [20]. VENAS combines allele-adjusted distances and Louvain detection for variant tracing [21]. McAN builds directed haplotype networks as minimum-cost arborescences for lineage identification [22]. LNA models mobility networks from spatiotemporal data to forecast epidemic risks [23]. HiRisk-Detector applies machine learning to McAN-derived topological features (e.g., betweenness, out-degree, depth) for early warning of high-risk variants, preceding official classification by approximately one month [9]. TempSnap-Trace extends McAN with temporal snapshots and Infomap detection for automated lineage tracing [24]. These methods support outbreak source tracing, super-spreader detection, and trans-border transmission mapping.

4 Computational Tools, Databases, and Analysis Platforms

General network platforms provide foundational graph computation. igraph and Networkx are widely used libraries offering comprehensive graph algorithms, including shortest path calculation, centrality metrics, and community detection [25, 26]. Gephi offers interactive force-directed visualization [27]; Cytoscape supports plugin-based workflows, including EnrichmentMap for pathway analysis [28]. PyTorch Geometric accelerates graph neural network training on GPUs [29].

Table 3
Common tools for microbiome network inference and analysis

Category	Tool	Key algorithm/features	Applicability/best for
Correlation-based	SparCC	Iterative log-ratio transformation; assumes sparse correlations	Compositional data (16S/metagenomics). Best for datasets with high sparsity and compositionality bias. Gold standard for robust correlation recovery
	FastSpar	C++ implementation of SparCC with threading	Large-Scale Compositional Data. Provides the robustness of SparCC but scales to thousands of samples (orders of magnitude faster)
	FlashWeave	Probabilistic graphical model with heuristics; ultrafast	Massive metagenomics (>1 k samples). Handles heterogeneous data and confounding factors; extremely efficient for cross-study meta-analysis
	CoNet	Ensemble approach (Pearson, Spearman, Bray-Curtis, etc.) + permutation.	Consensus Network Inference. Best when robustness is prioritized over speed; reduces false positives by requiring agreement across multiple metrics
Graphical models	SPIEC-EASI	CLR transform + Graphical Lasso (glasso) or Inverse Covariance	Direct Interaction Recovery. Distinguishing direct ecological interactions from indirect associations. Ideal for smaller, well-sampled datasets (n > 100)
	gCoda	Variational approximation for logistic normal distribution	Compositional conditional dependence. Similar to SPIEC-EASI but performs better in certain low-sample regimes
Specialized/ multi-omics	WGCNA	Weighted correlation network analysis; soft thresholding	Functional/transcriptional modules. Best for metatranscriptomics or functional gene profiles where compositionality is less critical than module detection
Pipelines and analysis	NetCoMi	Wrapper for multiple methods (SparCC, Cclasso, etc.) + differential analysis	Comparative Network Analysis. Excellent for comparing networks between groups (e.g., healthy vs. disease) with statistical rigor
AI & databases	STRING	Database of known/predicted PPIs with confidence scores	Functional annotation. Essential for projecting metagenomic gene lists onto known protein interaction networks
	WSGMB	GNNs & PLMs/Signed Graph Neural Networks	Biomarker discovery. Advanced tasks: Identifying disease-predictive network motifs (WSGMB)

Microbiome-specific software focuses on compositional constraints in abundance data (Table 3). SparCC and its C++ implementation FastSpar estimate latent correlations by iterating log-ratio transforms under a sparsity assumption, thereby reducing compositional artefacts [7, 30]. FlashWeave scales association inference to large metagenomic cohorts with heuristics tailored to heterogeneous, high-dimensional data [31]. CoNet combines several similarity measures with permutation-based null models to down-weight metric-specific false positives [32]. SPIEC-EASI and gCoda apply sparse inverse covariance estimation to centered log-ratio data to recover conditional dependencies rather than marginal correlations [8, 33]. WGCNA builds weighted correlation networks to detect co-expression modules in functional or transcriptomic profiles [34]. NetCoMi wraps multiple inference backends and adds tools for differential network comparison between groups [35]. STRING provides confidence-scored PPI networks with functional enrichment [17]; deep learning approaches enable functional prediction from multimodal protein data [5].

5 AI-Enhanced Graph Representation Learning

Standard deep learning workflow integrates raw sequencing data into latent representations for network inference. GNNs and PLMs enable end-to-end learning by modeling topological relationships through message-passing, where node representations are iteratively updated by aggregating neighbor information. GCNs extend convolution to irregular graphs through spectral or spatial formulations. Graph attention networks (GAT) learn attention weights prioritizing informative neighbors, enabling focus on relevant connections. These architectures can process metagenomic sequences or OTU abundance tables: protein language models such as Evolutionary Scale Modeling (ESM) first generate sequence embeddings capturing evolutionary patterns [36], which are then fused with structural information through graph neural networks that model proteins as hierarchical 3D graphs. For instance, DeepFRI demonstrates multimodal protein analysis by combining GCN-based processing of protein contact graphs with sequence representations from a pretrained PLM, enabling scalable prediction of Gene Ontology and enzyme commission annotations [5].

Applications demonstrate effectiveness across microbiome analysis tasks. WSGMB employs signed weighted graph neural architectures for the discovery of microbial biomarkers via health-disease network comparison [37]. Temporal graph neural network models predict microbial community dynamics by learning interaction relationships from historical abundance time series, achieving accurate forecasts up to several months ahead in wastewater treatment and gut microbiome datasets [38]. For metagenomic protein function modeling, GCN-based architectures typified by DeepFRI couple graph convolutions on protein structure graphs with sequence representations from a pretrained PLM, outperforming sequence-only CNN baselines especially for proteins sharing limited sequence similarity with annotated homologs [5]. Extended to large infant gut metagenome cohorts, these DeepFRI-derived functional profiles provide almost complete annotation of microbial gene catalogs and enable downstream network-based analyses of microbiome functional potential and host–microbe interactions [6]. This integration of sequence-based PLMs with structure-aware GNNs alleviates limitations of sequence-only models in capturing functional properties governed by three-dimensional organization [5].

6 Conclusion

Network topologies identify modular compartments, high-degree hub taxa, and shifts in assembly over time; these patterns link to differences in community stability between healthy and diseased states. In epidemiology, transmission networks track infection routes, classify emerging variants, and connect multi-omics layers to map host-microbe regulatory circuits.

Correlation-based edges cannot establish causation; separating true interactions from shared environmental drivers requires longitudinal cohorts paired with perturbation experiments. Cross-study comparison remains limited by inconsistent sampling and sequencing protocols, so standardized benchmark datasets are needed. Current AI models lack interpretability—coupling graph neural networks with constraint-based metabolic models could produce testable hypotheses for targeted therapies and outbreak surveillance rather than black-box predictions.

Acknowledgments

This work was supported by Major Project of Guangzhou National Laboratory (GZNL2024A01003), National Key R&D Program of China (2022YFF1203202), and Strategic Priority Research Program of Chinese Academy of Sciences (XDB38050200, XDB38040202, XDA26040304).

Conflicts of Interest The authors have no conflicts of interest to declare that are relevant to the content of this chapter.

References

1. Kajihara KT, Hynson NA (2024) Networks as tools for defining emergent properties of microbiomes and their stability. Microbiome 12:184. https://doi.org/10.1186/s40168-024-01868-z
2. Tomasulo A, Simionati B, Facchin S (2024) Microbiome one health model for a healthy ecosystem. Sci One Health 3:100065. https://doi.org/10.1016/j.soh.2024.100065
3. Zhou T, Zhao F (2025) AI-empowered human microbiome research. Gut. https://doi.org/10.1136/gutjnl-2025-335946
4. Przymus P, Rykaczewski K, Martín-Segura A, Truu J, Carrillo De Santa Pau E, Kolev M,

Naskinova I, Gruca A, Sampri A, Frohme M, Nechyporenko A (2025) Deep learning in microbiome analysis: a comprehensive review of neural network models. Front Microbiol 15:2024. https://doi.org/10.3389/fmicb.2024.1516667

5. Gligorijević V, Renfrew PD, Kosciolek T, Leman JK, Berenberg D, Vatanen T, Chandler C, Taylor BC, Fisk IM, Vlamakis H, Xavier RJ, Knight R, Cho K, Bonneau R (2021) Structure-based protein function prediction using graph convolutional networks. Nat Commun 12:3168. https://doi.org/10.1038/s41467-021-23303-9
6. Maranga M, Szczerbiak P, Bezshapkin V, Gligorijevic V, Chandler C, Bonneau R, Xavier RJ, Vatanen T, Kosciolek T (2023) Comprehensive functional annotation of metagenomes and microbial genomes using a deep learning-based method. mSystems 8:e01178–22. https://doi.org/10.1128/msystems.01178-22
7. Friedman J, Alm EJ (2012) Inferring correlation networks from genomic survey data. PLoS Comput Biol 8:e1002687. https://doi.org/10.1371/journal.pcbi.1002687
8. Kurtz ZD, Müller CL, Miraldi ER, Littman DR, Blaser MJ, Bonneau RA (2015) Sparse and compositionally robust inference of microbial ecological networks. PLoS Comput Biol 11:e1004226. https://doi.org/10.1371/journal.pcbi.1004226
9. Li L, Li C, Li N, Zou D, Wenming Zhao, Luo H, Xue Y, Zhang Z, Bao Y, Song S (2024) Machine learning early detection of SARS-CoV-2 high-risk variants. Adv Sci:2405058. https://doi.org/10.1002/advs.202405058
10. Blondel V, Guillaume J-L, Lambiotte R, Lefebvre E (2008) Fast unfolding of communities in large networks. J Stat Mech Theory Exp 2008. https://doi.org/10.1088/1742-5468/2008/10/P10008
11. Traag VA, Waltman L, van Eck NJ (2019) From Louvain to Leiden: guaranteeing well-connected communities. Sci Rep 9:5233. https://doi.org/10.1038/s41598-019-41695-z
12. Rosvall M, Bergstrom CT (2008) Maps of random walks on complex networks reveal community structure. Proc Natl Acad Sci USA 105:1118–1123. https://doi.org/10.1073/pnas.0706851105
13. Pons P, Latapy M (2005) Computing communities in large networks using random walks. In: Yolum pInar, Güngör T, Gürgen F, Özturan C (eds) Computer and information sciences—ISCIS 2005. Springer, Berlin Heidelberg, Berlin, Heidelberg, pp 284–293
14. Girvan M, Newman MEJ (2002) Community structure in social and biological networks. Proc Natl Acad Sci 99:7821–7826. https://doi.org/10.1073/pnas.122653799
15. Champion C, Momal R, Le Chatelier E, Sola M, Mariadassou M, Berland M (2024) OneNet—one network to rule them all: consensus network inference from microbiome data. PLoS Comput Biol 20:1–18. https://doi.org/10.1371/journal.pcbi.1012627
16. Luo M, Zhu J, Jia J, Zhang H, Zhao J (2024) Progress on network modeling and analysis of gut microecology: a review. Appl Environ Microbiol 90:e00092–e00024. https://doi.org/10.1128/aem.00092-24
17. Szklarczyk D, Kirsch R, Koutrouli M, Nastou K, Mehryary F, Hachilif R, Gable AL, Fang T, Doncheva NT, Pyysalo S, Bork P, Jensen LJ, von Mering C (2023) The STRING database in 2023: protein–protein association networks and functional enrichment analyses for any sequenced genome of interest. Nucleic Acids Res 51:D638–D646. https://doi.org/10.1093/nar/gkac1000
18. Chetty A, Blekhman R (2024) Multi-omic approaches for host-microbiome data integration. Gut Microbes 16:2297860. https://doi.org/10.1080/19490976.2023.2297860
19. Muller E, Shiryan I, Borenstein E (2024) Multi-omic integration of microbiome data for identifying disease-associated modules. Nat Commun 15:2621. https://doi.org/10.1038/s41467-024-46888-3
20. Kosakovsky Pond SL, Weaver S, Leigh Brown AJ, Wertheim JO (2018) HIV-TRACE (TRAnsmission cluster engine): a tool for large scale molecular epidemiology of HIV-1 and other rapidly evolving pathogens. Mol Biol Evol 35:1812–1819. https://doi.org/10.1093/molbev/msy016
21. Ling Y, Cao R, Qian J, Li J, Zhou H, Yuan L, Wang Z, Ma L, Zheng G, Zhao G, Wang Z, Zhang G, Li Y (2022) An interactive viral genome evolution network analysis system enabling rapid large-scale molecular tracing of SARS-CoV-2. Sci Bullet 67:665–669. https://doi.org/10.1016/j.scib.2022.01.001
22. Li L, Xu B, Tian D, Wang A, Zhu J, Li C, Li N, Zhao W, Shi L, Xue Y, Zhang Z, Bao Y, Zhao W, Song S (2023) McAN: a novel computational algorithm and platform for constructing and visualizing haplotype networks. Brief Bioinform 24:bbad174. https://doi.org/10.1093/bib/bbad174

23. Reis Pinheiro CA, Galati M, Summerville N, Lambrecht M (2021) Using network analysis and machine learning to identify virus spread trends in COVID-19. Big Data Res 25: 100242. https://doi.org/10.1016/j.bdr.2021.100242
24. Liu J, Li D, Li Y, Huang T (2025) TempSnaptrace: a temporal snapshot-based framework for haplotype network tracing. Biosaf Health. https://doi.org/10.1016/j.bsheal.2025.11.003
25. Hagberg A, Schult DA, Swart PJ (2008) Exploring network structure, dynamics, and function using NetworkX. Proceedings of the 7th Python in Science Conference (SciPy 2008):11–15.
26. Csárdi G, Nepusz T (2006) The igraph software package for complex network research. InterJournal Complex Systems:1695.
27. Bastian M, Heymann S, Jacomy M (2009) Gephi: an open source software for exploring and manipulating networks. Proceedings of the International AAAI Conference on Web and Social Media 3(1):361–362.
28. Shannon P, Markiel A, Ozier O, Baliga NS, Wang JT, Ramage D, Amin N, Schwikowski B, Ideker T (2003) Cytoscape: a software environment for integrated models of biomolecular interaction networks. Genome Res 13:2498–2504. https://doi.org/10.1101/gr.1239303
29. Fey M, Lenssen JE (2019) Fast Graph Representation Learning with PyTorch Geometric. arXiv e-prints arXiv:1903.02428. https://doi.org/10.48550/arXiv.1903.02428
30. Watts SC, Ritchie SC, Inouye M, Holt KE (2019) FastSpar: rapid and scalable correlation estimation for compositional data. Bioinformatics 35:1064–1066. https://doi.org/10.1093/bioinformatics/bty734
31. Tackmann J, Rodrigues JFM, von Mering C (2019) Rapid inference of direct interactions in large-scale ecological networks from heterogeneous microbial sequencing data. Cell Syst 9: 286–296.e8. https://doi.org/10.1016/j.cels.2019.08.002
32. Faust K, Raes J (2016) CoNet app: inference of biological association networks using Cytoscape. F1000Res 5:1519. https://doi.org/10.12688/f1000research.9050.2
33. Fang H, Huang C, Zhao H, Deng M (2017) gCoda: conditional dependence network inference for compositional data. J Comput Biol 24: 699–708. https://doi.org/10.1089/cmb.2017.0054
34. Langfelder P, Horvath S (2008) WGCNA: an R package for weighted correlation network analysis. BMC Bioinform 9:559. https://doi.org/10.1186/1471-2105-9-559
35. Peschel S, Müller CL, von Mutius E, Boulesteix AL, Depner M (2021) NetCoMi: network construction and comparison for microbiome data in R. Brief Bioinform 22(4):bbaa290. https://doi.org/10.1093/bib/bbaa290
36. Lin Z, Akin H, Rao R, Hie B, Zhu Z, Lu W, Smetanin N, Verkuil R, Kabeli O, Shmueli Y, Costa A dos S, Fazel-Zarandi M, Sercu T, Candido S, Rives A (2023) Evolutionary-scale prediction of atomic-level protein structure with a language model. Science 379:1123–1130. https://doi.org/10.1126/science.ade2574
37. Pan S, Jiang X, Zhang K (2024) WSGMB: weight signed graph neural network for microbial biomarker identification. Brief Bioinform 25(1):bbad448. https://doi.org/10.1093/bib/bbad448
38. Andersen KS, Zhao K, Agerskov A de L, Sørensen CB, Holmager TJ, Nierychlo M, Peces M, Guo C, Nielsen PH (2025) Predicting microbial community structure and temporal dynamics by using graph neural network models. Nat Commun 16:9124. https://doi.org/10.1038/s41467-025-64175-7

Chapter 7

Unlocking Enzyme Discovery: Leveraging Multi-Omics, Machine Learning, and De Novo Design

Hongming Xia, Chunxiu Zhou, Baotong Fu, and Huawen Han

Abstract

Enzymes are fundamental protein catalysts essential to life processes and widely applied in industrial and healthcare sectors. However, the broader application of natural enzymes is constrained by their inherent catalytic limitations, and traditional discovery methods such as microbial enrichment are often slow and low-throughput. Driven by advances in multi-omics and artificial intelligence, a range of novel screening strategies has been developed, enabling significant enhancements in both catalytic efficiency and stability of enzymes. This chapter assesses high-throughput approaches, such as metagenomics, metaproteomics, machine learning, and de novo design, comparing their respective advantages and limitations for enzyme discovery. Furthermore, we discuss the application potential of lignocellulose-degrading and plastic-degrading enzymes in biomass conversion and plastic waste recycling.

Key words Enzyme discovery, Metagenomics, Metaproteomics, Machine learning, De novo enzyme

1 Introduction

Enzymes, a class of protein macromolecules [1], serve as highly efficient biological catalysts that are involved in nearly universal life processes of living organisms [2], including cellular metabolism [3], energy conversion [4], and signal transduction [5]. A typical example is the central dogma of molecular biology, where enzymes such as DNA polymerase and RNA polymerase are responsible for DNA replication, transcription, and translation [6, 7]. Enzymes are historically and broadly used in daily life and healthcare [8], in food industries such as baking and brewing, where amylases and proteases contribute to optimizing product quality [9]. As for the production of major industrial products, enzymes as biocatalysts provide sustainability and efficiency benefits over chemicals [10], such as cutinase in laundry detergents for the removal of lipid stains [11]. Since James B. Sumner first discovered crystallized urease, numerous studies have discovered a variety of tool enzymes that

Tao Huang (ed.), *Microbial 'Omics: Methods and Protocols*, Methods in Molecular Biology, vol. 3033,
https://doi.org/10.1007/978-1-0716-5264-0_7,

have laid the groundwork for molecular genetic manipulation, such as DNA polymerases, restriction endonucleases, green fluorescent proteins, and Cas9. The corresponding CRISPR-Cas9 gene-editing system acts as a pair of molecular scissors that precisely cut DNA, facilitating accurate genome engineering [12]. Given the catalytic limitations of natural enzymes in industrial applications and environmental remediation fields, enzyme engineering has shifted from random mutagenesis to rational design, fueled by genetic and protein engineering breakthroughs [13]. Directed evolution emulates natural selection by subjecting enzyme genes to rounds of mutation and screening, markedly enhancing catalytic efficiency, stability, or specificity [14–16]. With the revolutionary advancement in protein structure prediction, AlphaFold2 has not only dramatically boosted our ability to discover new enzymes from genomic data [17] but also provides reliable structural models for enzymes without experimental data, essential for understanding catalytic mechanisms and guiding design [17, 18].

Enzymes belong to a trade-off between catalytic specificity and catalytic promiscuity. Mounting evidence shows that catalytic promiscuity is an equally fundamental enzymatic trait [19]. In general, the evolution of new functions in enzymes spans tens of thousands of years [20]. During the adaptation to specific physiological and environmental stresses, enzymes evolve adaptive changes that improve their efficiency and specificity toward native substrates [21]. Conversely, the weak activity of enzymes with nonnative substrates, a result of ancient promiscuity, offers a diverse genetic pool for new functions [22]. These evidences showed that specificity and promiscuity are complementary aspects of an enzyme's functional versatility and evolutionary potential. Enzymatic promiscuity is primarily categorized into substrate promiscuity, condition promiscuity, and catalytic (or reaction) promiscuity [19]. Although ancestral enzymes are often characterized by catalytic promiscuity and low catalytic efficiency [23], the pressure of natural selection fosters functional diversification and optimization, prompting descendant enzymes to gradually acquire higher catalytic efficiency and more refined specificity for specific substrates [24–26].

The enduring need for enzymes in industrial applications accelerates the discovery of novel and high-efficient enzymes [27]. Initial efforts relied predominantly on traditional microbial cultivation, a process involving the isolation of single colonies and subsequent screening of their metabolites for target activities [28]. The vast enzymatic repertoire of the biosphere remains largely untapped, as most microorganisms from extreme environments are uncultivable [29]. Metagenomics circumvents this cultivation bottleneck by directly extracting environmental DNA and cloning it into surrogate hosts for functional screening, enabling high-throughput exploration of genetic resources from uncultured microbes

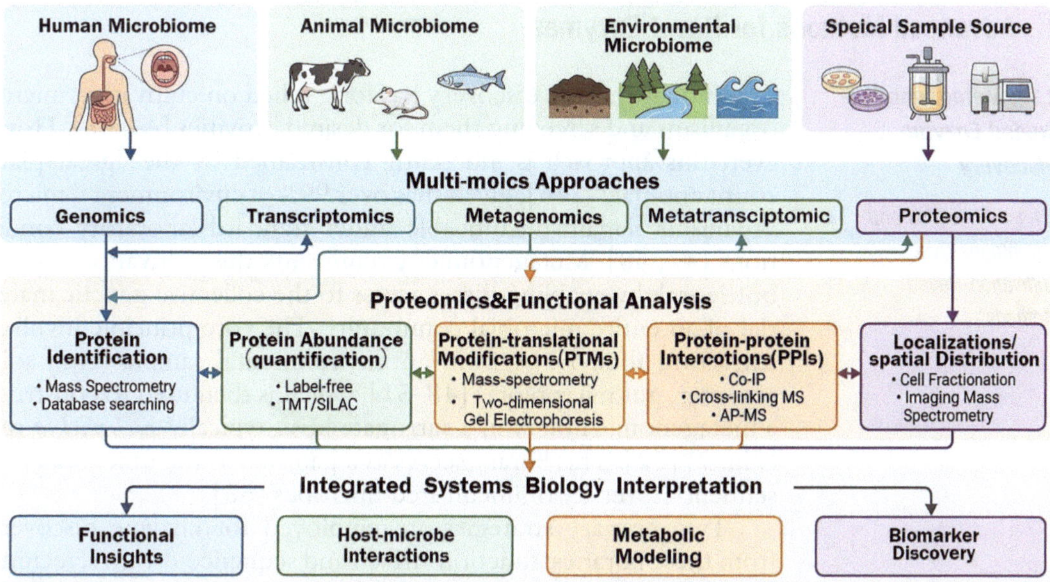

Fig. 1 Application of multi-omics across various fields of microbiome research

[30, 31]. This approach has successfully identified novel enzymes, such as a salt-tolerant cellulase from soil [32] and a thermostable xylanase from a hot spring [33]. As a complementary strategy, metaproteomics provides a direct snapshot of ecosystem function by identifying all proteins expressed in situ (Fig. 1) [34]. Concurrently, data-driven approaches leveraging machine learning and protein language models (LLMs) are accelerating enzyme discovery and engineering [35]. By learning the complex mapping between sequence, structure, and function from massive datasets, these models can accurately predict enzymatic properties, like substrate specificity, optimal pH, and binding sites [36, 37]. However, the utility of many natural enzymes is constrained by their limited stability and efficiency under industrial process conditions, such as extreme pH or high temperatures [38]. De novo design addresses this by computationally creating novel enzymes from first principles [39], independent of any known protein fold or sequence. This strategy has yielded highly active and thermostable catalysts and even designed RNA enzymes and metalloenzymes for cerium photoredox catalysis from scratch, demonstrating a powerful route to generating bespoke biocatalysts for demanding applications [40–43].

In this chapter, we systematically review state-of-the-art technologies for enzyme discovery and screening, encompassing innovative strategies, including metagenomics, metaproteomics, and machine learning-driven approaches. We further discuss the application potential and future perspectives of these methodologies in enzyme mining and related fields.

2 Advanced Methods for Novel Enzymes

2.1 Metagenomics-Guided Enzyme Discovery

2.1.1 Principles and Advantages Over Cultivation-Based Methods*

Traditional enzyme discovery has long relied on cultivating microorganisms and screening them for desired activities [44, 45]. However, this approach is inherently constrained by the "great plate count anomaly" [46], given that over 99% of environmental microorganisms remain uncultivable under standard laboratory conditions [47, 48]. Metagenomics circumvents this cultivation-related bottleneck by enabling direct access to the collective genetic material of an entire microbial community. The core principle involves extracting total DNA from an environmental sample (e.g., soil, seawater, gut microbiota) [48–51], which is then used to construct a metagenomic library in a surrogate host, typically *Escherichia coli* (Fig. 2) [52]. This library serves as a gateway to the functional and sequence diversity of uncultured microbes [53].

Two primary strategies are employed for enzyme discovery from these libraries: function-driven and sequence-driven screening [30, 47, 54, 55]. Function-driven screening is a direct method that relies solely on detectable enzyme activity, requiring no prior

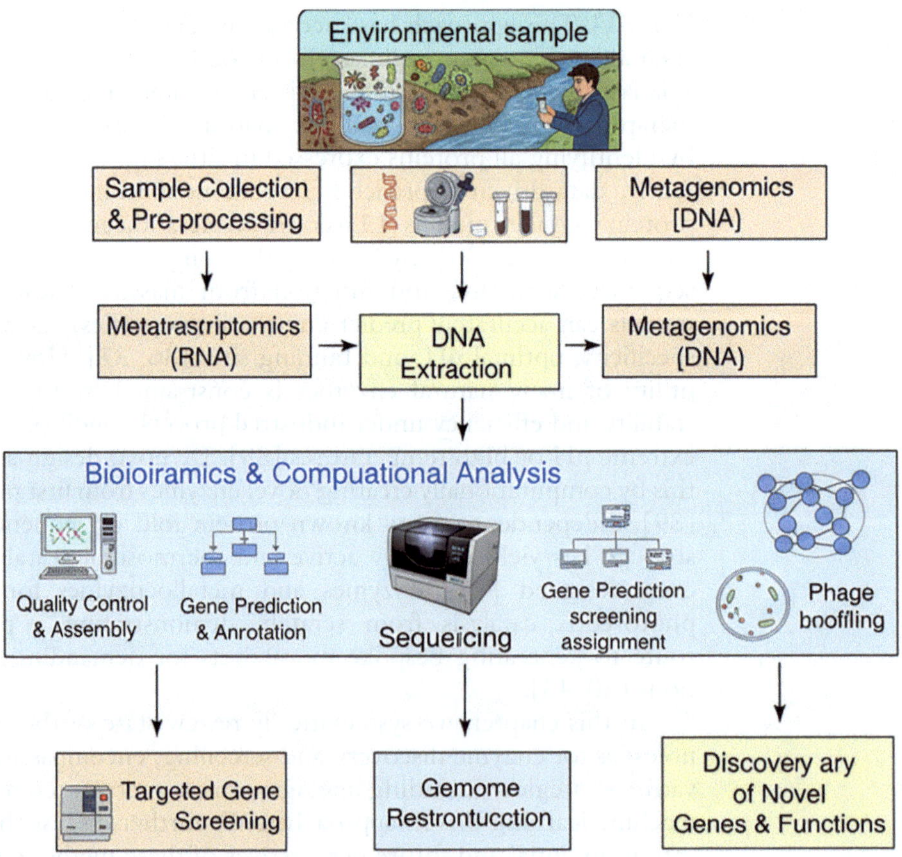

Fig. 2 Workflow for enzyme screening via metagenomics

sequence information [56]. Metagenomic libraries are screened on selective media containing specific substrates. Positive clones are identified through indicators, such as color changes from chromogenic substrates or exclusive growth when the substrate is the sole carbon source [51, 57]. The genes responsible for the activity are then identified by sequencing the active clones [58]. In contrast, a sequence-driven screening method uses conserved DNA sequences or protein domains from known enzyme families as probes or primers to retrieve homologous genes from metagenomic datasets [59]. While this method is highly efficient for identifying members of known enzyme families, its primary limitation is the inability to discover truly novel enzymes that lack sequence similarity to existing references [60].

2.1.2 Enzyme Screening Via Metagenomics

Metagenomics has proven to be a key tool for unlocking the reservoir of enzymatic resources encoded by uncultured microorganisms in diverse environments, achieving remarkable successes across various ecosystems. A landmark analysis of the cow rumen microbiome, which involved 268 gigabases of sequence data, identified 27,755 putative carbohydrate-active genes. Subsequent characterization of 90 candidate proteins revealed that 57% were cellulolytic enzymes, with enzymes MH-9 and MH-10 harboring two novel catalytic domains [61]. Screening of a deep-sea sediment library yielded six novel lipolytic enzymes, among which EM3L4 exhibited optimal catalytic efficiency at low temperatures (30–35 °C) and high salt concentrations (1 M NaCl). Furthermore, the enzyme showed tolerance to metal ions such as Mn^{2+} and Mg^{2+}, and detergents such as Tween-20, −40, highlighting its potential for industrial applications [62]. Recent studies revealed that gut microbiomes also act as a hotspot of mining novel enzymes. For example, two novel glycoside hydrolases were identified from the obligate anaerobe *Flavonifractor plautii* via functional gene screening encoded in our library of 19,500 expressed harboring gut bacterial DNA. These enzymes can efficiently convert type A and B blood to universal type O by removing terminal antigens, offering a solution to blood supply shortages [63]. By integrating both sequence- and function-driven approaches, researchers have mined diverse microbial consortia for high-efficiency pollutant-degrading enzymes, including oxygenases, laccases, and nitrilases [53]. A recent study combined metagenomics with multi-omics to identify 40 potential selenite reductases from 13,933 oxidoreductase candidates, one of which (*Ca*SrpA) was optimized via site-directed mutagenesis for effective bioremediation of selenium-contaminated soil [64]. In addition, a novel PET-degrading esterase, *Tm*Fae-PETase, was discovered from the gut microbiome of plastic-feeding mealworms (*Tenebrio molitorlarvae*) possessing broad substrate hydrolytic activity toward polyethylene terephthalate (PET), methyl ferulate (MFA), methyl cinnamate (MCA), and

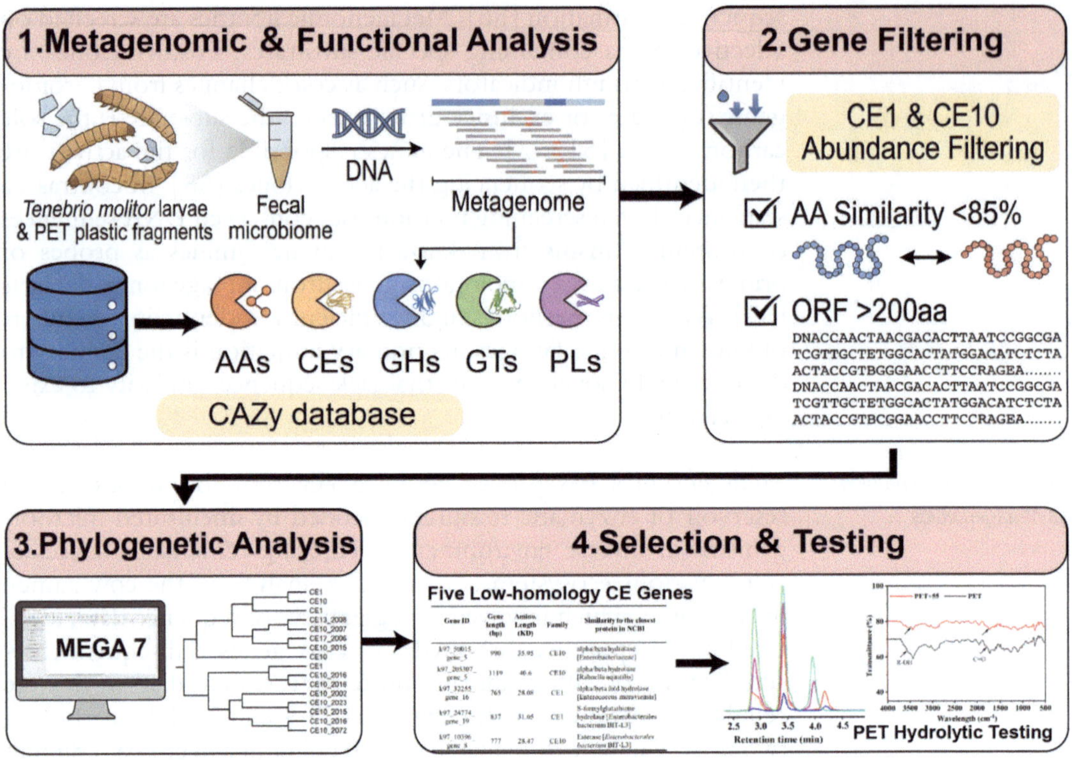

Fig. 3 Identification of a novel PET hydrolase from the lignocellulose degradation pathway in the gut microbiome of plastic-degrading mealworms (*Tenebrio molitor* larvae)

lignocellulose (Fig. 3) [65]. Notably, its hydrolytic activity against PET was significantly higher than that of known homologous hydrolases, indicating its potential as an efficient and promising catalyst for PET degradation [65].

2.1.3 Challenges in Enzyme Discovery from Metagenomes

Despite these advances, metagenomics-guided enzyme discovery still faces several limitations [66]. Obtaining high-quality DNA from oligotrophic habitats can be a major bottleneck, as samples are often contaminated with inhibitors (e.g., humic acids) that interfere with downstream molecular biology [67]. Function-driven screening is limited by the fact that many genes from diverse microorganisms may not be correctly transcribed, translated, or folded in the standard *E. coli* host, leading to a high rate of false negatives. Additionally, a significant portion of sequences in metagenomic datasets lack homology to known genes in reference databases. This "annotation gap" leaves the function of numerous potential novel enzymes uncharacterized, a problem especially acute for viral genomes [68]. To fully unlock the enzymatic potential concealed within the planet's microbial dark matter, addressing these limitations will be critical; this requires optimizing DNA extraction protocols, developing diverse heterologous expression systems, and advancing bioinformatic tools.

2.2 Metaproteomics-Guided Enzyme Discovery

2.2.1 Principles of Metaproteomics

Metaproteomics is regarded as one of the most direct approaches for elucidating microbial ecological functions, as it involves the analysis of the complete set of proteins expressed by microbial communities within environmental samples to obtain functional information [69]. Total proteins are extracted, digested, and separated before being identified through high-throughput mass spectrometry [70]. The resulting peptide spectra are subsequently matched against metagenomic or protein databases, which can determine the functional proteins and their taxonomic origins under defined environmental conditions [71]. This workflow enables the direct revelation of the actual metabolic status and functional activities of microbial communities, as opposed to inferring them from genetic potential alone [72]. Consequently, metaproteomics provides a snapshot of the actual metabolic state of microbial communities in situ.

Metaproteomics has been widely applied in environmental microbiology to clarify microbially mediated biogeochemical processes and explore host-microbe relationships, including those within the human gut microbiome. Moreover, its capacity to identify expressed enzymes makes it a valuable tool for enzyme discovery and functional screening (Fig. 4) [73–75].

2.2.2 Advantages of Metaproteomics

As an essential branch of functional omics, metaproteomics is employed to characterize the complete protein extracted directly

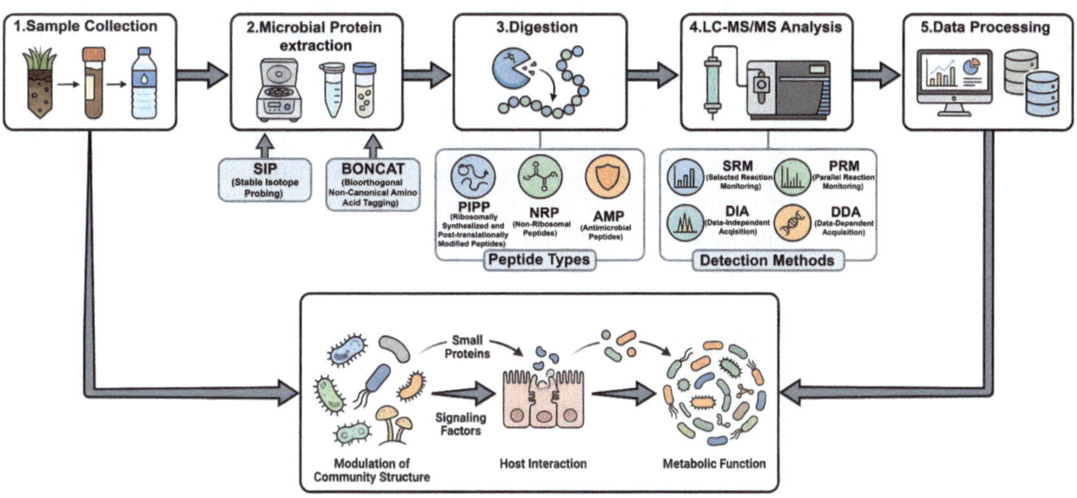

Fig. 4 Overview of metaproteomic workflow

from environmental samples [76]. This approach directly reveals the functional proteome actively expressed by a microbial community within its specific habitat [77]. Unlike metagenomics, which only reveals the genetic potential of a community, metaproteomics identifies the actual functional executors. Thus, it provides direct insights into the operational gene expression profiles of microorganisms [73]. In the field of enzyme resource mining, metaproteomics offers several unique advantages. First, it allows the direct identification of enzymes that are actively expressed in specific environmental conditions, effectively reducing the risk of missing functional genes caused by unsuccessful heterologous expression [78, 79]. Second, when the identified enzyme proteins are correlated with metagenomic data, a direct "gene-to-function" verification can be achieved. Such integration provides reliable experimental evidence for annotating open reading frames (ORFs) in metagenomic datasets [72]. Consequently, metaproteomics serves not only as a crucial complement to metagenomic functional predictions but also as a powerful tool for discovering novel enzymes with specific environmental adaptations.

2.2.3 Enzyme Discovery Via Metagenomics

As a powerful functional screening tool, metaproteomics enables the direct discovery of novel enzymes by identifying proteins expressed actively from diverse environmental sources. In one study, researchers analyzed waste cooking oil-contaminated soil by using metaproteomics. Through two-dimensional gel zymography, they analyzed the entire metaproteome for proteins showing activity toward fluorescent substrates and identified a novel lipase, ML-005. Functional validation demonstrated its strong capability for short-chain and medium-chain ester degradation [80]. This study highlighted a universal strategy for targeted mining of high-performance biocatalysts for short-chain and medium-chain esters. Metaproteomic analysis has also revealed the enzymatic potential of plastic-degrading microbiomes. In another study, a polyvinyl chloride (PVC)-degrading microbial consortium from Qinghai-Tibet Plateau meadow soil was found to express 123 lignin-associated and 298 PVC-related proteins (Fig. 5). Among them, an overexpressed dehalogenase SerB from *Rhodococcus*, belonging to the phosphoserine phosphatase family, demonstrates its ability in the bioremediation of recalcitrant plastics [81]. In another example, an activity-based metaproteomics (ABMP) strategy was applied to the mouse gut microbiome. Using activity-based probes (ABPs), researchers identified four novel α-galactosidases and characterized their catalytic mechanisms (Fig. 6). The enzyme AGAL5 not only efficiently hydrolyzed various natural sugars but also exhibited transglycosylation activity [79]. This ABMP approach provides an effective approach to explore the enzymatic potential of gut microbiota and shows promise for applications in human health and

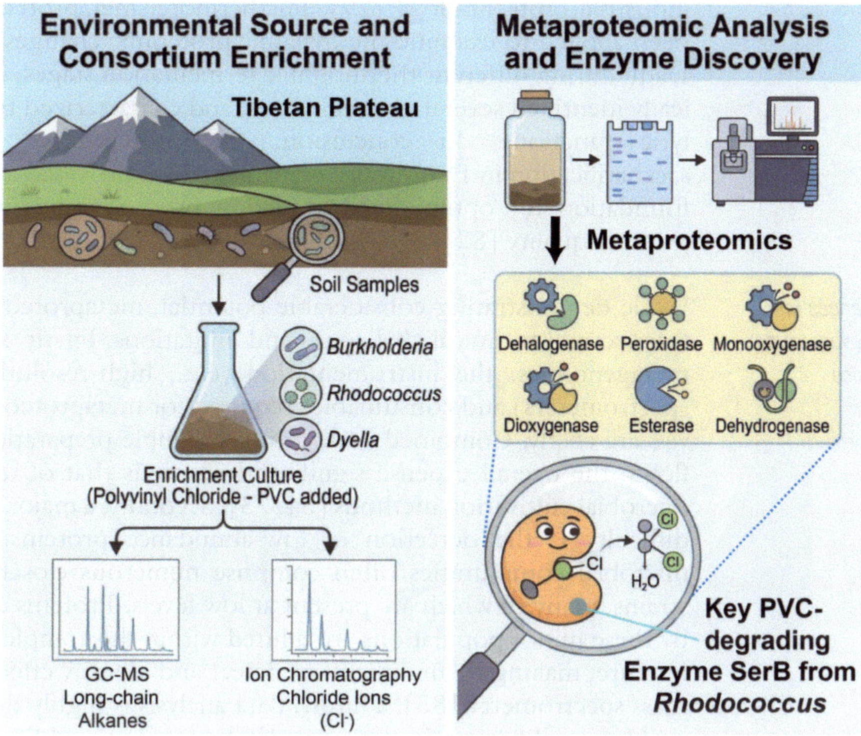

Fig. 5 Identification of PVC-degrading enzymes from alpine grassland soil via metaproteomics

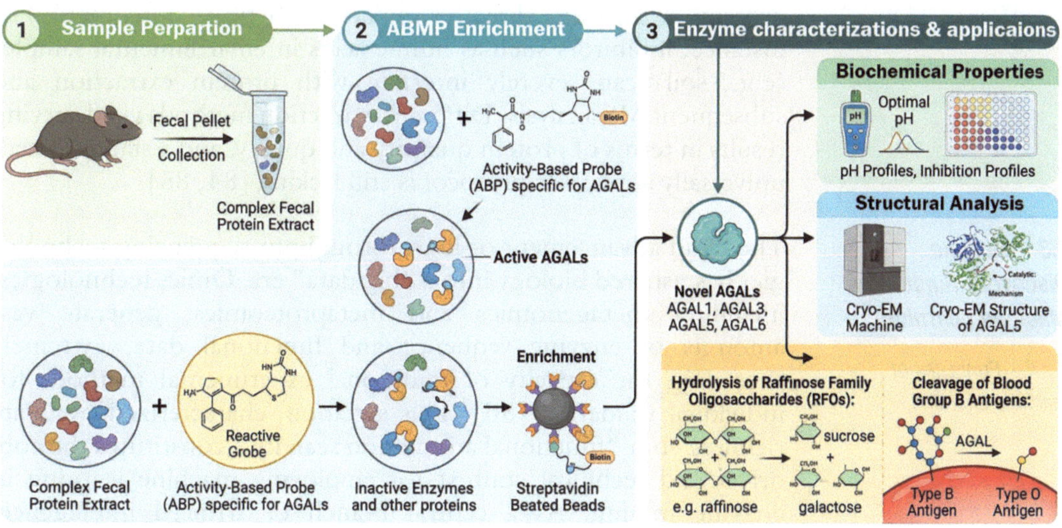

Fig. 6 Identification of α-galactosidase protein groups in the gut microbiome using activity-based metaproteomics (ABMP) screening

industrial biotechnology [79]. Furthermore, metaproteomics has been applied to examine the dynamic proteomic changes in Baijiu Daqu during different thermophilic fermentation stages, systematically identified several glucoamylases, and characterized their catalytic functions. In conclusion, the protein-level basis of saccharification in Daqu was elucidated, which provided a scientific foundation for optimizing production processes and improving product quality [82].

2.2.4 Challenges in Enzyme Discovery from Metaproteomics

While demonstrating considerable potential, metaproteomics also faces several technical challenges and limitations. Firstly, similar to metagenomics, the instrumentation (e.g., high-resolution mass spectrometers) and consumables required for metaproteomic analysis are costly. Combined with complex sample preparation workflows, the overall expense significantly exceeds that of traditional microbial cultivation methods [34, 73]. Secondly, a major technical difficulty is the detection of low-abundance proteins. Natural microbial communities often comprise numerous closely related strains, many of which are present at low levels. Proteins expressed by these minor populations are diluted within the complex protein mixture, making them difficult to detect and identify effectively via mass spectrometry [83]. Thirdly, data analysis is highly dependent on high-quality protein sequence databases. Since public databases cannot encompass the full diversity of environmental microbes, a substantial proportion of identified peptide spectra remain unmatched, leading to low protein identification rates [84]. Furthermore, the complexity of the sample itself and the efficiency of protein extraction directly impact analytical outcomes. For instance, inhibitors such as humic acids in environmental samples (e.g., soil) can severely interfere with protein extraction and subsequent MS analysis. Different extraction methods yield varying results in terms of protein quantity and quality, and a standardized, universally applicable protocol is still lacking [84, 85].

2.3 Enzyme Discovery Guided by Machine Learning

2.3.1 Principles of Machine Learning

The rapid advancement of high-throughput sequencing technologies has ushered biology into a "big data" era. Omics technologies, including metagenomics and metaproteomics, generate vast amounts of enzyme sequence and functional data, extremely exceeding the capacity of traditional experimental methods for individual validation [86]. This situation, characterized by "data richness" but "functional annotation scarcity," constitutes the core driver and technical context for employing machine learning in enzyme mining. As a central branch of artificial intelligence, machine learning can automatically uncover complex, nonlinear patterns and construct predictive models from these large-scale datasets [87].

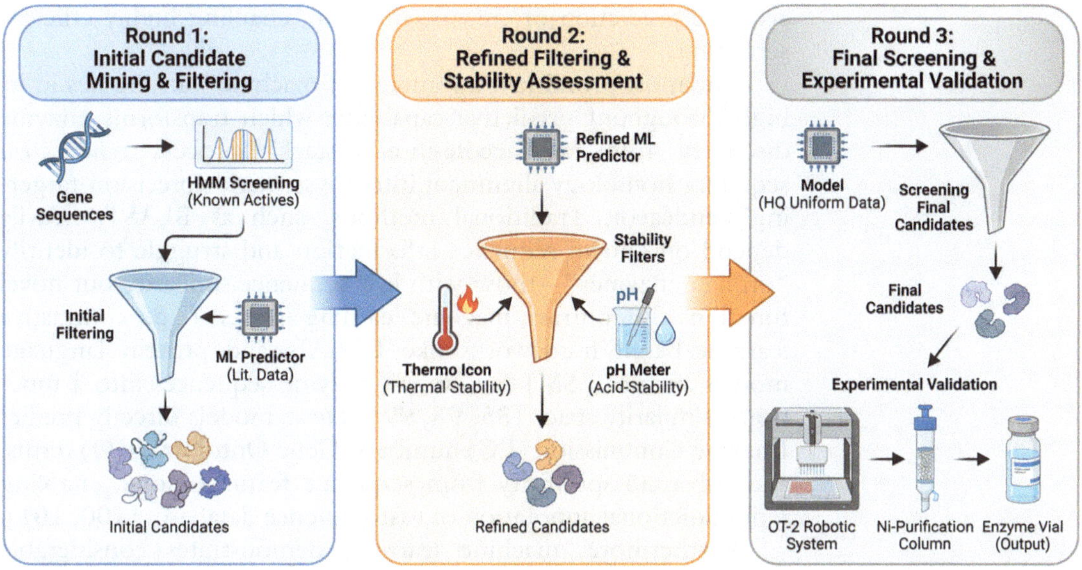

Fig. 7 Screening of novel hydrolases via integrating hidden Markov models (HMM) and supervised machine learning models

In the field of enzyme discovery and function prediction, machine learning models are now capable of high-throughput prediction of various key enzyme attributes (Fig. 7) [88]. These include functional classification, substrate specificity, and catalytic properties [89–92]. In enzyme design and engineering, the integration of machine learning has rendered traditional strategies such as directed evolution more intelligent and efficient [93]. By learning from limited experimental data to map the complex relationships between sequence and function, machine learning can predict which combinations of mutations are likely to yield beneficial traits, significantly reducing the experimental workload required for validation [87]. Furthermore, protein language models and generative AI (e.g., RFdiffusion) demonstrate the potential to generate novel enzyme sequences or structures, opening pathways for de novo enzyme design. Consequently, machine learning is increasingly becoming a powerful tool for rapidly navigating the sequence universe to pinpoint high-performance enzyme catalysts, substantially accelerating the process of enzyme discovery and design [94, 95].

2.3.2 Advantages of Machine Learning

Machine learning is revolutionizing enzyme discovery by leveraging its unique data-driven paradigm, offering substantial advantages over traditional approaches in terms of efficiency, scope, and precision [96]. Its principal strength is the capacity to infer complex functional attributes directly from sequence or structural data of uncharacterized proteins, such as those derived from metagenomic datasets [97]. This capability facilitates the high-throughput and accurate prediction of enzyme function, enabling highly efficient screening.

The most prominent advantage of machine learning lies in its high-throughput predictive capability, which transforms enzyme discovery from a "needle-in-a-haystack" process reliant on sequence homology alignment into a systematic "precision-targeting" endeavor. Traditional methods, such as BLAST, heavily depend on known sequence information and struggle to identify "orphan enzymes"—those with low sequence similarity but novel functions. In contrast, machine learning models (e.g., contrastive learning-based frameworks like CLEAN and protein language models such as ESM) can embed enzyme sequences into a functional similarity space [35, 98, 99]. These models directly predict Enzyme Commission (EC) numbers, Gene Ontology (GO) terms, and substrate specificity from sequence features alone, enabling rapid functional annotation of vast sequence databases [100, 101].

Furthermore, machine learning demonstrates considerable potential for the precise prediction of key enzymatic catalytic properties, providing crucial support for the rational selection of high-performance enzymes. Conventional experimental approaches for characterizing enzyme properties, including kinetic parameters (k_{cat}/K_M), optimum temperature, and pH, are typically low-throughput and resource-intensive [102, 103]. In contrast, deep learning models, such as CataPro, DLK$_{cat}$, and UniKP, integrate pretrained protein sequence representations with substrate molecular fingerprints to accurately predict these essential parameters. These predictions have exhibited strong agreement with experimental validations, substantially reducing the need for trial-and-error experimentation [104–106].

2.3.3 Enzyme Screening Via Machine Learning

As a powerful data mining tool, machine learning (ML) has become an indispensable approach for efficiently screening novel enzymes and bioactive compounds from vast sequence or chemical datasets. Its value is well demonstrated in two core application areas, including enzyme function prediction and active compound screening. The key research achievements are elaborated below. One study developed an ML model to predict the activity and substrate specificity of OleA enzymes in the thiolase superfamily. The established computational framework enabled the effective identification of enzyme variants with potential catalytic activity toward novel substrates. Experimental validation confirmed that the selected variants demonstrated significantly enhanced catalytic efficiency (k_{cat}/K_m) in specific applications. By developing a machine learning model, this study predicted the activity and substrate specificity of OleA enzymes in the thiolase superfamily. The established computational framework enabled the effective identification of enzyme variants with potential catalytic activity toward novel substrates.

Experimental validation confirmed that the selected variants demonstrated significantly enhanced efficiency in specific catalytic applications [107]. Another study integrated ML-based scoring functions with molecular dynamics simulations, successfully identifying 20 lead compounds from an initial pool of 9948 candidates. Among these, several exhibited nanomolar inhibitory activity against GH-20 β-N-acetylglucosaminidase, highlighting the considerable potential of computational biology in efficient discovery of targeted enzyme inhibitors. Notably, the most promising compound (1146525) serves as a viable scaffold for developing new antibacterial agents against *Vibrio* infections, providing a valuable starting point for novel therapeutics [108]. For natural product-based drug discovery, researchers integrated ML with in vitro chemical screening to efficiently identify compounds from indigenous Thai plants that inhibit α-amylase and α-glucosidase. The strategy involved two key steps: first, ML algorithms were used for preliminary screening and activity prediction of phytochemicals; second, in vitro assays validated the efficacy of candidate compounds. This workflow established an efficient approach for rapid discovery of potential anti-diabetic lead compounds [109]. Similarly, researchers adopted an integrated "machine learning-based virtual screening-molecular docking" strategy to identify potential anti-inflammatory cyclooxygenase (COX) inhibitors from the medicinal plant *Tinospora cordifolia*. Following the core logic of "activity prediction-binding validation," the strategy first used ML to predict the bioactivity of chemical constituents from *Tinospora cordifolia* leaf extracts, then verified their binding stability with COX-1 and COX-2 enzymes through molecular docking and molecular dynamics simulations. Results demonstrated that the compound 7,9-Di-tert-butyl-1-oxaspiro(4,5)deca-6,9-diene-2,8-dione exhibited strong binding affinity toward both COX-1 and COX-2, and 100 ns molecular dynamics simulations confirmed the formation of a stable enzyme-inhibitor complex. This integrated strategy provides an efficient computational-experimental framework for rapidly discovering novel COX inhibitors from natural products [110].

2.3.4 Challenges in Enzyme Discovery from Machine Learning

Despite its considerable potential in enzyme discovery, machine learning applications face significant challenges in data quality, model reliability, and experimental validation. The most critical bottleneck lies in the scarcity and inconsistency of high-quality data, because the functional annotations of most protein sequences remain unverified experimentally [88]. Data bias constitutes another major issue; for instance, data on certain enzyme families (e.g., hydrolases) are overrepresented, whereas many rare enzyme classes suffer from severe data paucity, resulting in prediction models often being biased toward well-represented enzyme families [111]. At the model level, significant limitations remain in generalization capability and interpretability. Many models perform well within the distribution of the training data, but their predictive accuracy declines sharply when applied to new enzyme families with distinct sequences or structures [111]. Furthermore, an efficient closed loop between machine learning predictions and experimental validation has yet to be established. Biocatalysts identified through virtual screening still require wet-lab validation, and the throughput and cost of high-throughput functional characterization techniques remain major bottlenecks [112]. Additionally, the accessibility and usability of computational tools pose another barrier. Beyond the aforementioned core challenges, another non-negligible barrier is the accessibility and usability of computational tools. Many advanced machine learning tools lack user-friendly interfaces, creating a high entry barrier for experimental scientists without extensive computational biology expertise, which in turn hinders the broader adoption of such technologies [111].

2.4 De Novo Enzyme Design

2.4.1 Principles of De Novo Enzyme Design

The core principle of de novo enzyme design involves computationally and AI-driven creation of novel, nonnatural enzyme molecules. This process starts with a thorough analysis of the target chemical reaction mechanism to identify key catalytic residues and their spatial configurations required for transition state stabilization and thereby constructs a theoretical minimal active site model (Fig. 8) [113]. Subsequently, researchers employ protein structure prediction and design algorithms, such as the matching algorithm within Rosetta, to scan extensive protein scaffold databases. This process aims to identify backbone structures that can precisely

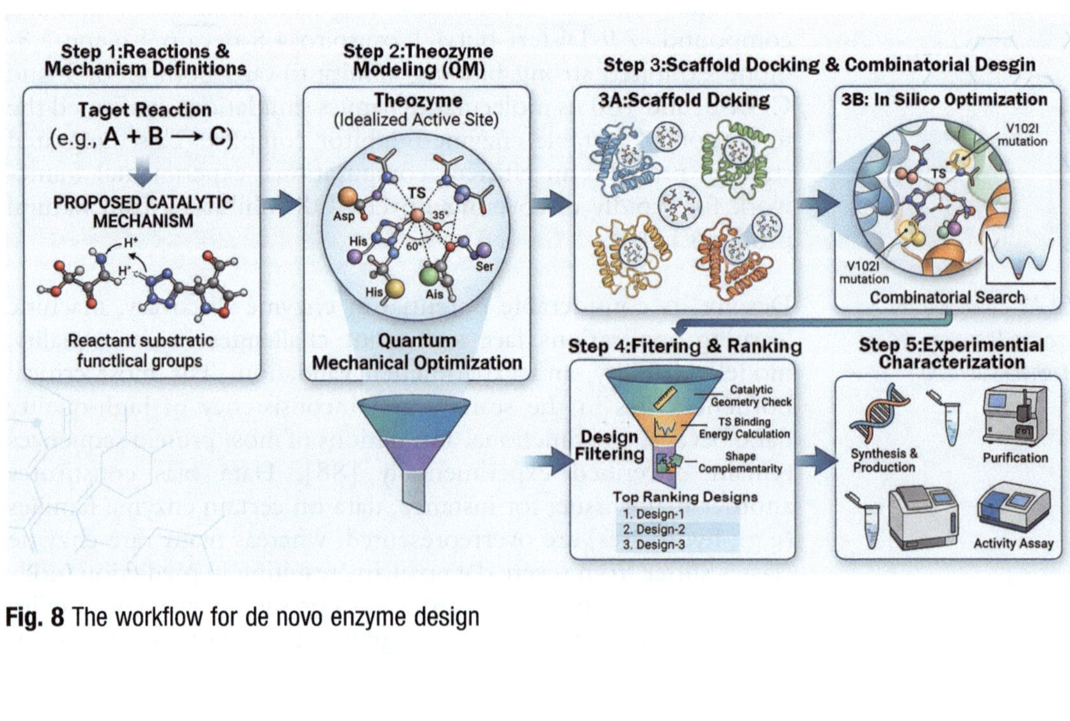

Fig. 8 The workflow for de novo enzyme design

accommodate the theoretical active site. The selected scaffolds are then computationally optimized by adjusting surrounding residues to maximize transition state stabilization and minimize the reaction activation energy [114]. Recently, artificial intelligence has significantly advanced this field. For instance, the diffusion model-based RFdiffusion tool enables the de novo generation of entirely novel enzyme protein structures, while the deep neural network PLACER allows for refined optimization of atomic arrangements within the active site and its geometric conformations throughout the catalytic cycle [95, 115]. Meanwhile, protein language models (such as the ESM series) and inverse folding algorithms (e.g., ProteinMPNN) are capable of "inverting" a designed three-dimensional structure to generate amino acid sequences with high foldability and stability [116, 117].

The creation of de novo enzymes is primarily achieved through three distinct technological pathways. The first approach relies exclusively on computational protein design methods. This strategy utilizes computational protein design (CPD) technology to automatically redesign amino acid sequences based on specific protein backbone coordinates, enabling precise structural control and the assignment of novel catalytic functions [118]. The second approach employs transition-state analogs (TSAs) as haptens to harness the diversity of the immune system for generating catalytic antibodies (abzymes) [38]. These catalytic antibodies effectively bind substrates and significantly accelerate reaction rates, demonstrating catalytic properties analogous to natural enzymes [119]. The third approach involves the systematic screening of large protein libraries using mRNA display technology. In this method, the reaction substrate is linked to the protein expressed by mRNA display [120]. This connection is achieved by reverse transcribing the mRNA into cDNA, which hybridizes with an oligonucleotide primer pre-immobilized on the substrate. During the reaction, catalytically active proteins convert the substrate into a product, enabling the isolation of the cDNA encoding the active enzyme based on product formation [121]. The retrieved cDNA can be directly sequenced or used for subsequent screening rounds after amplification [122]. Additionally, site-directed mutagenesis can be performed on the cDNA to further optimize the enzyme's catalytic properties [119].

2.4.2 Advantages of De Novo Enzyme Design

The rapid advancement of synthetic chemistry has led to the creation of numerous artificial compounds with novel structures and corresponding new reaction types [123]. However, naturally evolved enzymes lack the inherent capability to catalyze these non-natural transformations, as they were not shaped by evolutionary pressures to accommodate such synthetic substrates [124]. Consequently, the existing repository of natural enzymes falls short in meeting the demand for highly efficient and selective catalysis of these artificial reactions [125]. To achieve precise catalysis of chemical transformations not found in nature, researchers must move beyond reliance on natural enzyme templates and turn to de novo design and synthesis of entirely artificial enzyme catalysts. De novo enzyme design offers revolutionary advantages over traditional enzyme engineering methods. Its core innovation lies in completely liberating enzyme design from dependence on natural templates, instead utilizing computational simulation and artificial intelligence to create enzyme proteins not found in nature, thereby enabling true customization of catalytic function [38]. This design concept, which breaks free from the constraints of natural templates, endows it with the unique advantage of integrating multiple desirable traits into enzyme molecules, such as significantly enhanced thermostability, improved tolerance in organic solvents, and even the ability to utilize nonnatural cofactors. This approach thereby overcomes the inherent limitation of natural enzymes, which are typically functional only within a narrow range of environmental conditions [118]. Compared to rational design or directed evolution techniques, which rely on modifying natural enzyme structures, the most significant advantage of de novo enzyme design lies in its unparalleled freedom for functional creation [119]. This structural minimalism not only facilitates the processes of heterologous expression and industrial-scale production but also simplifies the investigation of structure-function relationships [126].

2.4.3 Case Studies in De Novo Enzyme Design

De novo enzyme design has emerged as a powerful methodology for creating novel enzymatic molecules. Leveraging computational biology and artificial intelligence technologies, this approach enables the precise construction of enzyme structures and functions, achieving multiple breakthroughs in expanding reaction types, enhancing stability, and improving catalytic efficiency. The following elaborates on these achievements through typical cases. In a pioneering study, 72 proteins were computationally designed, among which 32 exhibited retro-aldol reaction activity, representing a milestone in developing novel biocatalysts. Experimental characterization confirmed that the designed enzyme significantly accelerated the reaction rate compared to the uncatalyzed baseline. Although its catalytic efficiency remains lower than that of natural enzymes, this work successfully demonstrates the substantial potential of computational protein design for expanding the repertoire of enzyme-catalyzed reactions [127]. Another study focused on improving enzyme stability: researchers created a de novo hyperthermostable heme-containing peroxidase through computational design. Experimental validation confirmed the efficacy of this strategy, demonstrating successful intracellular assembly and high catalytic efficiency. The designed enzyme exhibited superior performance, including significantly enhanced thermal stability (enabling

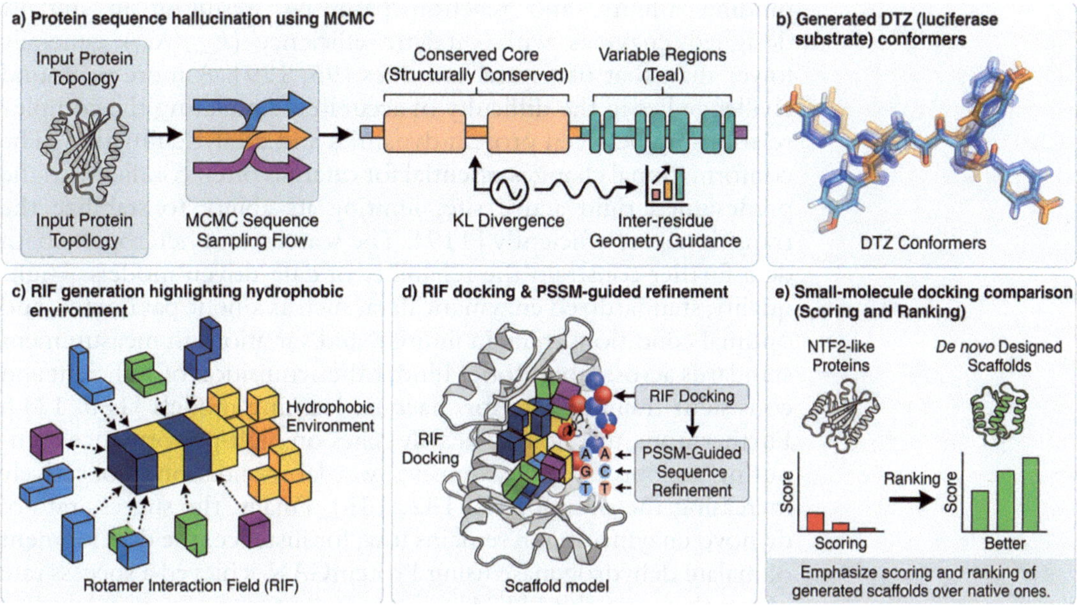

Fig. 9 Computational design of de novo luciferases

activity under high temperatures and in organic solvents), functional promiscuity, and high catalytic efficiency [40]. Beyond protein enzymes, another study successfully designed and synthesized a novel RNA enzyme using mRNA display-directed in vitro evolution. Experimental results demonstrated that this artificial enzyme exhibits remarkably high catalytic efficiency, with a reaction rate 10^6 times faster than the uncatalyzed reaction. Its product yield surpassed that of previously reported ligase ribozymes, while displaying substrate saturation kinetics analogous to natural enzymes [41]. Researchers also developed a de novo artificial luciferase, and experimental validation demonstrated its ability to effectively catalyze luminescence reactions with various luciferin substrates. The most efficient catalyst was a small (13.9 kDa) thermostable enzyme with a melting temperature exceeding 95 °C. It exhibited luminescence intensity comparable to that of natural luciferase while possessing significantly higher substrate specificity. This finding provides a novel strategy for developing bioluminescent probes (Fig. 9) [128].

2.4.4 Challenges in De Novo Enzyme Design

Despite its considerable potential, de novo enzyme design still faces multiple fundamental challenges. The accuracy of computational design remains the primary bottleneck. Although tools such as AlphaFold2 and RFdiffusion can predict or generate protein structures, the precise design of highly efficient catalytic sites remains exceptionally challenging. Current computational models have limited precision in predicting transition-state stabilization, substrate binding affinity, and reaction pathways, resulting in initially designed enzymes with catalytic efficiency (k_{cat}/K_m) generally lower than that of natural enzymes [95, 129]. A more profound challenge lies in the difficulty of accurately simulating the complex relationship between protein dynamics and catalytic function. The conformational changes essential for catalysis often conflict with the predesigned rigid active site, limiting its ability to stabilize the transition state efficiently [119]. The scarcity and heterogeneity of data further constrain the reliability of data-driven models. High-quality, standardized enzymatic data, such as kinetic parameters and optimal conditions, remain limited, and variations in measurement standards across laboratories hinder the acquisition of sufficient and consistent training data for machine learning models [130, 131]. Furthermore, technology heavily relies on high-performance computing resources and extensive wet-lab validation, collectively increasing the overall cost [132, 133]. Finally, the success rate of de novo enzyme design remains low; for instance, the development of malate dehydrogenase using ProteinGAN achieved a success rate of less than one-third [134].

3 Discovery and Application of Novel Enzymes

3.1 Conversion of Lignocellulose

Lignocellulose, the Earth's most abundant renewable feedstock, is primarily sourced from the tens of billions of tons of agricultural and forestry waste generated globally each year [135]. Despite its potential, the massive accumulation of this biomass not only represents a critical failure in resource utilization but also poses a significant threat to environmental pollution [136, 137]. Compared to traditional physicochemical methods, enzymatic hydrolysis using cellulases offers significant advantages, such as milder operating conditions, higher substrate specificity, and better environmental compatibility [138]. Cellulose decomposition is a critical process for converting biomass into fuels and chemicals [139]. Through metagenomic analysis of soil samples covered with bagasse, we identified a novel metalloenzyme, CelOCE, with oxidative cellulose-lysing activity from unclassified DNA of unknown function within a lignocellulose-degrading microbial community. CelOCE cleaves cellulose via an exo-acting mechanism with C1 region selectivity, exclusively generating cellobionic acid as the product. The CelOCE gene was integrated into *Trichoderma reesei* Br_Tr03 via CRISPR-Cas9 to develop a strain for lignocellulose biorefineries, enabling the production of enzyme blends under industrial-like conditions. Under industrial operational settings, glucose release from pretreated biomass increased by 20%. The promoting effect of CelOCE significantly surpassed that of other enzyme genes, demonstrating its strong potential in cellulose degradation. Furthermore, CelOCE-containing enzyme mixtures were successfully produced in 65 L and 300 L pilot-scale bioreactors, confirming their feasibility for industrial application. This study enhances our understanding of cellulose decomposition mechanisms and provides innovative solutions for the biomass conversion of agricultural and industrial residues [140]. Another study identified a novel lytic polysaccharide monooxygenase (LPMO), MtLPMO9V, from a thermophilic fungus, which promotes cellulose hydrolysis by disrupting crystalline regions of cellulose. After successful expression of MtLPMO9V, computational protein design and molecular dynamics simulations were combined to introduce disulfide bonds and point mutations, resulting in an engineered variant with enhanced catalytic activity and thermostability. The optimized mutant M3 exhibited 88% higher enzyme activity than the wild-type MtLPMO9V, with an optimal temperature of 90 °C and a catalytic efficiency (k_{cat}/K_m) nearly twice that of the unmodified enzyme. Using a commercial cellulase preparation (Celluclast 1.5 L) alone, the glucose yield from microcrystalline cellulose (MCC) reached 58% at 96 h. With the addition of MtLPMO9V at 0.4 mg/g dry substrate, the glucose yield increased to 70.6% at 96 h, representing a 22% improvement compared to cellulase alone. The novel thermostable MtLPMO9V demonstrated strong synergistic effects with various cellulases across different cellulose substrates, indicating significant potential for industrial development of enzymatic lignocellulose biorefining processes [141]. Moreover, another study employed a metagenomic approach to identify a purified cellulase, CelB, from a novel strain of *Bacillus amyloliquefaciens* OKB3—a soil-derived isolate exhibiting high cellulolytic activity. CelB demonstrated strong substrate affinity and low activation energy, indicative of high catalytic efficiency [142]. The enzyme achieved optimal activity at 60 °C and pH 6, while maintaining stability under pH 5.5–6.0 and temperatures of 0–4 °C. Its cellulolytic function operated independently of metal cofactors. Compared to other substrates, CelB showed higher specificity toward carboxymethyl cellulose (CMC). These findings suggest the potential of CelB for efficient bioconversion of agricultural waste into biofuels, which could yield significant economic and environmental benefits in industrial biotechnology applications [142].

3.2 Degradation and Recycle of Plastics

The pervasive environmental crisis of white pollution, caused by synthetic polymers such as polyethylene terephthalate (PET), has become a global concern [143]. These plastics exhibit extreme persistence in natural environments, resisting microbial breakdown and persisting for centuries [144]. Consequently, the development of efficient plastic-degrading enzymes has emerged as an essential strategy for achieving sustainable plastic waste recycling and upcycling (Fig. 10) [145]. However, existing PET hydrolases (PETases) are constrained by limited sequence diversity and suboptimal

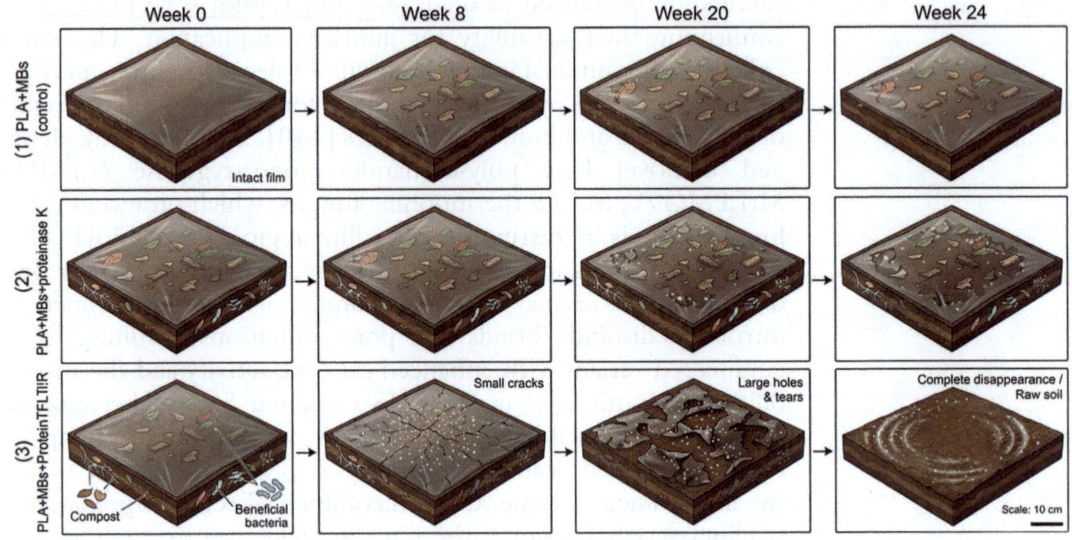

Fig. 10 Disintegration of PLA films with incorporated engineered protease PAM under home-compost conditions

performance. To overcome the limitations of existing PET-degrading enzymes, researchers have developed high-efficiency degrading enzymes through three core technical routes: machine learning-assisted novel enzyme mining, traditional microbial isolation and screening, and machine learning-aided enzyme engineering. A recent study integrated protein language models (PLMs) with representation tree ensembles and employed machine learning to identify PETases based on structural similarity inferred from sequence information. Using the crystal structure of *Is*PETase as a template, VenusMine identified 34 target proteins, 14 of which were candidate PET-degrading enzymes. Among these, a PET hydrolase from *Kibdelosporangium banguiense* (*Kb*PETase) exhibited a melting temperature (T_m) 32 °C higher than that of *Is*PETase. Moreover, *Kb*PETase demonstrated the highest PET degradation efficiency among all characterized PET-degrading enzymes under optimal catalytic conditions and also showed significant degradation activity against a commercial PET film (AF-PET, Goodfellow Cambridge Ltd., Cat. No. ES301445) [146]. Another study isolated a membrane-anchored PET esterase (OQN32_06240) from *Rhodococcus pyridinivorans* P23 through traditional microbial culture and genomic sequencing, using PET as the sole carbon source. Beyond PET depolymerization, OQN32_06240 hydrolyzed MHET to TPA under acidic conditions, elucidating the PET degradation pathway in strain P23 and the enzyme's catalytic mechanism. Based on these findings, a biofilm-based model for marine PET biodegradation was proposed: P23 colonizes PET surfaces via biofilm formation, enabling continuous degradation by retaining degraded products as a sustained carbon source while isolating acidic microenvironments from alkaline seawater to optimize OQN32_06240's MHET hydrolase activity. In this study, *R. pyridinivorans* P23 reduced the weight of PET by 4.28% over a 35-day cultivation period. In contrast, four other marine strains achieved only 1.3–1.8% weight loss (30 days). Notably, nearly complete degradation of 60 mg of PET film was observed within 42 days of incubation with P23. Although the degradation efficiency of *R. pyridinivorans* P23 does not yet meet industrial application standards, its marine origin—coupled with the severe plastic pollution in ocean environments—positions this bacterium, along with similar PET-degrading and assimilating microbial communities, play a vital role in the natural bioremediation of marine plastic pollution [147]. A separate study developed a novel PET hydrolase through machine learning-assisted enzyme engineering. Researchers first employed molecular docking and enzyme contact surface analysis to investigate the binding mode of the substrate 2-HE (MHET)$_3$ in the active site of the wild-type LCC enzyme. Eleven key amino acid residues were selected for site-directed saturation mutagenesis, generating 209 variants. Screening identified mutants with enhanced activity, such as F243I and F243W. To improve thermostability, disulfide bridges including D238C and S283C were introduced, conferring greater stability under high temperatures. The final optimized variant, ICCG, achieved 90% PET depolymerization within 10 h, with a productivity of 16.7 g TA/L/h, significantly surpassing known enzymes. This enzyme efficiently degrades postconsumer PET waste, producing high-purity terephthalic acid suitable for synthesizing new PET bottles. The mechanical properties of the recycled PET are comparable to those of petroleum-based PET, while the enzyme cost constitutes only 4% of the virgin PET price, advancing the development of a circular plastics economy [148]. Another study leveraged the GRASE (Graph Neural Network-based Recommendation of Active and Stable Enzymes) framework, which integrates self-supervised and supervised learning, to identify highly efficient and glycolytically compatible urethanases. A standout enzyme demonstrated exceptional performance in polyurethane degradation, maintaining stability in 6M diethylene glycol and achieving near-complete depolymerization (98% conversion in 8 h) of kilogram-scale commercial PU foam, highlighting its significant potential for industrial recycling [149].

4 Future Direction

Multi-omics and machine learning have accelerated the discovery of novel enzymes, such as PET hydrolases, haloalkane dehalogenases, and bile acid metabolic enzymes [150]. A critical bottleneck in computational enzyme discovery is the reliance on sequence homology, which often fails to predict function accurately as it overlooks the complex interplay between sequence, structure, and dynamics, leading to high false-positive rates and functionally misannotated candidates. Recent studies have confirmed that protein language models (PLMs), including ProtGPT2 [151], ProGEN [152], ESM-1b [153], ESM-2 [154], MSA transformer [155], and ProTrans [156], have been successfully applied to protein engineering, antibody design, and enzyme functional assessment. Recent successes include ESM-Ezy, which fine-tuned the ESM-1b model to mine for novel multicopper oxidases with superior properties by navigating the protein semantic space [157]. More advanced strategies integrate PLMs with structural retrieval. For example, VenusMine leveraged this approach to discover *Kb*PETase, a highly efficient and thermostable PET hydrolase that outperforms established benchmarks. The frontier of this field lies in models like Saprot, which directly encode 3D structural information as tokens alongside amino acid sequences, creating a truly structure-aware language model [146]. This integrated approach enables powerful zero-shot prediction capabilities. In a striking example, the Saprot-based tool ColabSaprot predicted mutations for a xylanase, with 13 of 20 tested variants showing enhanced activity without any task-specific training. The enzymes discovered and optimized through these advanced PLMs provide functionally reliable, high-performance genetic parts for engineering high-efficiency cell factories, promising broad applications in synthetic biology.

Conflict of Interest

The authors have no conflicts of interest to declare that are relevant to the content of this chapter.

References

1. Knowles JR (1991) Enzyme catalysis: not different, just better. Nature 350(6314): 121–124. https://doi.org/10.1038/350121a0
2. Agarwal PK (2006) Enzymes: an integrated view of structure, dynamics and function. Microb Cell Factories 5:2. https://doi.org/10.1186/1475-2859-5-2
3. Chatzinikolaou PN, Margaritelis NV, Paschalis V, Theodorou AA, Vrabas IS, Kyparos A et al (2024) Erythrocyte metabolism. Acta Physiol 240(3). https://doi.org/10.1111/apha.14081

4. Gao J, Ma S, Major DT, Nam K, Pu J, Truhlar DG (2006) Mechanisms and free energies of enzymatic reactions. Chem Rev 106(8): 3188–3209. https://doi.org/10.1021/cr050293k
5. Zingg JM (2015) Vitamin E: a role in signal transduction. Annu Rev Nutr 35:135–173. https://doi.org/10.1146/annurev-nutr-071714-034347
6. Crick F (1970) Central dogma of molecular biology. Nature 227(5258):561–563. https://doi.org/10.1038/227561a0
7. Bar-Even A, Noor E, Milo R (2012) A survey of carbon fixation pathways through a quantitative lens. J Exp Bot 63(6):2325–2342. https://doi.org/10.1093/jxb/err417
8. Anbu P, Gopinath SC, Chaulagain BP, Tang TH, Citartan M (2015) Microbial enzymes and their applications in industries and medicine 2014. Biomed Res Int 2015:816419. https://doi.org/10.1155/2015/816419
9. Kirk O, Borchert TV, Fuglsang CC (2002) Industrial enzyme applications. Curr Opin Biotechnol 13(4):345–351. https://doi.org/10.1016/s0958-1669(02)00328-2
10. Chapman J, Ismail AE, Dinu CZ (2018) Industrial applications of enzymes: recent advances, techniques, and outlooks. Catalysts 8(6). https://doi.org/10.3390/catal8060238
11. Singh R, Kumar M, Mittal A, Mehta PK (2016) Microbial enzymes: industrial progress in 21st century. 3 Biotech 6(2):174. https://doi.org/10.1007/s13205-016-0485-8
12. Jiang F, Doudna JA (2017) CRISPR-Cas9 structures and mechanisms. Annu Rev Biophys 46:505–529. https://doi.org/10.1146/annurev-biophys-062215-010822
13. Kuchner O, Arnold FH (1997) Directed evolution of enzyme catalysts. Trends Biotechnol 15(12):523–530. https://doi.org/10.1016/s0167-7799(97)01138-4
14. Turner NJ (2003) Directed evolution of enzymes for applied biocatalysis. Trends Biotechnol 21(11):474–478. https://doi.org/10.1016/j.tibtech.2003.09.001
15. Cheng F, Zhu L, Schwaneberg U (2015) Directed evolution 2.0: improving and deciphering enzyme properties. Chem Commun (Cambridge, England). 51(48):9760–9772. https://doi.org/10.1039/c5cc01594d
16. Eijsink VG, Gåseidnes S, Borchert TV, van den Burg B (2005) Directed evolution of enzyme stability. Biomol Eng 22(1–3): 21–30. https://doi.org/10.1016/j.bioeng.2004.12.003
17. Jumper J, Evans R, Pritzel A, Green T, Figurnov M, Ronneberger O et al (2021) Highly accurate protein structure prediction with AlphaFold. Nature 596(7873): 583–589. https://doi.org/10.1038/s41586-021-03819-2
18. Varadi M, Anyango S, Deshpande M, Nair S, Natassia C, Yordanova G et al (2022) AlphaFold protein structure database: massively expanding the structural coverage of protein-sequence space with high-accuracy models. Nucleic Acids Res 50(D1):D439–Dd44. https://doi.org/10.1093/nar/gkab1061
19. Hult K, Berglund P (2007) Enzyme promiscuity: mechanism and applications. Trends Biotechnol 25(5):231–238. https://doi.org/10.1016/j.tibtech.2007.03.002
20. Yang G, Miton CM, Tokuriki N (2020) A mechanistic view of enzyme evolution. Protein Sci 29(8):1724–1747. https://doi.org/10.1002/pro.3901
21. Babtie A, Tokuriki N, Hollfelder F (2010) What makes an enzyme promiscuous? Curr Opin Chem Biol 14(2):200–207. https://doi.org/10.1016/j.cbpa.2009.11.028
22. Pandya C, Farelli JD, Dunaway-Mariano D, Allen KN (2014) Enzyme promiscuity: engine of evolutionary innovation. J Biol Chem 289(44):30229–30236. https://doi.org/10.1074/jbc.R114.572990
23. Jensen RA (1976) Enzyme recruitment in evolution of new function. Ann Rev Microbiol 30:409–425. https://doi.org/10.1146/annurev.mi.30.100176.002205
24. Khersonsky O, Roodveldt C, Tawfik DS (2006) Enzyme promiscuity: evolutionary and mechanistic aspects. Curr Opin Chem Biol 10(5):498–508. https://doi.org/10.1016/j.cbpa.2006.08.011
25. Khersonsky O, Tawfik DS (2010) Enzyme promiscuity: a mechanistic and evolutionary perspective. Annu Rev Biochem 79:471–505. https://doi.org/10.1146/annurev-biochem-030409-143718
26. Bridgham JT, Ortlund EA, Thornton JW (2009) An epistatic ratchet constrains the direction of glucocorticoid receptor evolution. Nature 461(7263):515–519. https://doi.org/10.1038/nature08249
27. Aharoni A (2009) Mining for new enzymes. Microb Biotechnol 2(2):128–129. https://doi.org/10.1111/j.1751-7915.2009.00090_1.x
28. Thapa S, Li H, OHair J, Bhatti S, Chen FC, Nasr KA et al (2019) Biochemical characteristics of microbial enzymes and their significance from industrial perspectives. Mol

Biotechnol 61(8):579–601. https://doi.org/10.1007/s12033-019-00187-1

29. Ferrer M, Golyshina O, Beloqui A, Golyshin PN (2007) Mining enzymes from extreme environments. Curr Opin Microbiol 10(3): 207–214. https://doi.org/10.1016/j.mib.2007.05.004
30. Schloss PD, Handelsman J (2003) Biotechnological prospects from metagenomics. Curr Opin Biotechnol 14(3):303–310. https://doi.org/10.1016/s0958-1669(03)00067-3
31. Hugenholtz P, Tyson GW (2008) Metagenomics. Nature 455(7212):481–483. https://doi.org/10.1038/455481a
32. Garg R, Srivastava R, Brahma V, Verma L, Karthikeyan S, Sahni G (2016) Biochemical and structural characterization of a novel halotolerant cellulase from soil metagenome. Sci Rep 6:39634. https://doi.org/10.1038/srep39634
33. Yin YR, Li XW, Long CH, Li L, Hang YY, Rao MD et al (2023) Characterization of a GH10 extremely thermophilic xylanase from the metagenome of hot spring for prebiotic production. Sci Rep 13(1):16053. https://doi.org/10.1038/s41598-023-42920-6
34. Wilmes P, Heintz-Buschart A, Bond PL (2015) A decade of metaproteomics: where we stand and what the future holds. Proteomics 15(20):3409–3417. https://doi.org/10.1002/pmic.201500183
35. Yang J, Li FZ, Arnold FH (2024) Opportunities and challenges for machine learning-assisted enzyme engineering. ACS Central Sci 10(2):226–241. https://doi.org/10.1021/acscentsci.3c01275
36. Gado JE, Knotts M, Shaw AY, Marks D, Gauthier NP, Sander C et al (2025) Machine learning prediction of enzyme optimum pH. Nat Mach Intell 7(5):716–729. https://doi.org/10.1038/s42256-025-01026-6
37. Nana Teukam YG, Kwate Dassi L, Manica M, Probst D, Schwaller P, Laino T (2024) Language models can identify enzymatic binding sites in protein sequences. Comput Struct Biotechnol J 23:1929–1937. https://doi.org/10.1016/j.csbj.2024.04.012
38. Bolon DN, Voigt CA, Mayo SL (2002) De novo design of biocatalysts. Curr Opin Chem Biol 6(2):125–129. https://doi.org/10.1016/s1367-5931(02)00303-4
39. Cui XC, Zheng Y, Liu Y, Yuchi Z, Yuan YJ (2025) AI-driven de novo enzyme design: strategies, applications, and future prospects. Biotechnol Adv 82:108603. https://doi.org/10.1016/j.biotechadv.2025.108603
40. Watkins DW, JMX J, Grayson KJ, Wood N, Steventon JW, Le Vay KK et al (2017) Construction and in vivo assembly of a catalytically proficient and hyperthermostable de novo enzyme. Nat Commun 8(1):358. https://doi.org/10.1038/s41467-017-00541-4
41. Ikawa Y, Tsuda K, Matsumura S, Inoue T (2004) De novo synthesis and development of an RNA enzyme. Proc Natl Acad Sci USA 101(38):13750–13755. https://doi.org/10.1073/pnas.0405886101
42. Braun M, Tripp A, Chakatok M, Kaltenbrunner S, Fischer C, Stoll D et al (2024) Computational enzyme design by catalytic motif scaffolding. bioRxiv. https://doi.org/10.1101/2024.08.02.606416
43. Klein AS, Leiss-Maier F, Mühlhofer R, Boesen B, Mustafa G, Kugler H et al (2024) A de novo metalloenzyme for cerium photoredox catalysis. J Am Chem Soc 146(38): 25976–25985. https://doi.org/10.1021/jacs.4c04618
44. Ravindran R, Jaiswal AK (2016) Microbial enzyme production using Lignocellulosic food industry wastes as feedstock: a review. Bioengineering (Basel, Switzerland). 3(4). https://doi.org/10.3390/bioengineering3040030
45. Prakash O, Shouche Y, Jangid K, Kostka JE (2013) Microbial cultivation and the role of microbial resource centers in the omics era. Appl Microbiol Biotechnol 97(1):51–62. https://doi.org/10.1007/s00253-012-4533-y
46. Tanaka T, Kawasaki K, Daimon S, Kitagawa W, Yamamoto K, Tamaki H et al (2014) A hidden pitfall in the preparation of agar media undermines microorganism cultivability. Appl Environ Microbiol 80(24): 7659–7666. https://doi.org/10.1128/aem.02741-14
47. Mirete S, Morgante V, González-Pastor JE (2016) Functional metagenomics of extreme environments. Curr Opin Biotechnol 38: 143–149. https://doi.org/10.1016/j.copbio.2016.01.017
48. Lorenz P, Schleper C (2002) Metagenome—a challenging source of enzyme discovery. J Mol Catal B Enzym 19-20:13–19. https://doi.org/10.1016/S1381-1177(02)00147-9
49. Alma'abadi AD, Gojobori T, Mineta K (2015) Marine metagenome as a resource for novel enzymes. Genomics Proteomics Bioinformatics 13(5):290–295. https://doi.org/10.1016/j.gpb.2015.10.001
50. Patrascu O, Béguet-Crespel F, Marinelli L, Le Chatelier E, Abraham A-L, Leclerc M et al

(2017) A fibrolytic potential in the human ileum mucosal microbiota revealed by functional metagenomic. Sci Rep 7(1):40248. https://doi.org/10.1038/srep40248

51. Madhavan A, Sindhu R, Parameswaran B, Sukumaran RK, Pandey A (2017) Metagenome analysis: a powerful tool for enzyme bioprospecting. Appl Biochem Biotechnol 183(2):636–651. https://doi.org/10.1007/s12010-017-2568-3
52. Xing MN, Zhang XZ, Huang H (2012) Application of metagenomic techniques in mining enzymes from microbial communities for biofuel synthesis. Biotechnol Adv 30(4): 920–929. https://doi.org/10.1016/j.biotechadv.2012.01.021
53. Ufarté L, Laville É, Duquesne S, Potocki-Veronese G (2015) Metagenomics for the discovery of pollutant degrading enzymes. Biotechnol Adv 33(8):1845–1854. https://doi.org/10.1016/j.biotechadv.2015.10.009
54. Sleator RD, Shortall C, Hill C (2008) Metagenomics. Lett Appl Microbiol 47(5): 361–366. https://doi.org/10.1111/j.1472-765X.2008.02444.x
55. Suenaga H (2015) Targeted metagenomics unveils the molecular basis for adaptive evolution of enzymes to their environment. Front Microbiol 6:1018. https://doi.org/10.3389/fmicb.2015.01018
56. Uchiyama T, Miyazaki K (2009) Functional metagenomics for enzyme discovery: challenges to efficient screening. Curr Opin Biotechnol 20(6):616–622. https://doi.org/10.1016/j.copbio.2009.09.010
57. Ngara TR, Zhang H (2018) Recent advances in function-based metagenomic screening. Genomics Proteomics Bioinform 16(6): 405–415. https://doi.org/10.1016/j.gpb.2018.01.002
58. Prayogo FA, Budiharjo A, Kusumaningrum HP, Wijanarka W, Suprihadi A, Nurhayati N (2020) Metagenomic applications in exploration and development of novel enzymes from nature: a review. J Genet Eng Biotechnol 18(1):39. https://doi.org/10.1186/s43141-020-00043-9
59. Parachin NS, Gorwa-Grauslund MF (2011) Isolation of xylose isomerases by sequence- and function-based screening from a soil metagenomic library. Biotechnol Biofuels 4(1):9. https://doi.org/10.1186/1754-6834-4-9
60. Yun J, Ryu S (2005) Screening for novel enzymes from metagenome and SIGEX, as a way to improve it. Microb Cell Factories 4(1): 8. https://doi.org/10.1186/1475-2859-4-8
61. Hess M, Sczyrba A, Egan R, Kim T-W, Chokhawala H, Schroth G et al (2011) Metagenomic discovery of biomass-degrading genes and genomes from cow rumen. Science 331(6016):463–467. https://doi.org/10.1126/science.1200387
62. Jeon JH, Kim JT, Lee HS, Kim SJ, Kang SG, Choi SH et al (2011) Novel lipolytic enzymes identified from metagenomic library of deep-sea sediment. Evid-Based Complement Alternat Med: eCAM 2011:271419. https://doi.org/10.1155/2011/271419
63. Rahfeld P, Sim L, Moon H, Constantinescu I, Morgan-Lang C, Hallam SJ et al (2019) An enzymatic pathway in the human gut microbiome that converts A to universal O type blood. Nat Microbiol 4(9):1475–1485. https://doi.org/10.1038/s41564-019-0469-7
64. Hou X, Gu J, Chen J, Ullah H, Mamtimin T, Wang X et al (2025) Identification and functional characterization of a novel CaSrpA enzyme for selenite reduction and selenium nanoparticle formation. J Hazard Mater 494: 138486. https://doi.org/10.1016/j.jhazmat.2025.138486
65. Mamtimin T, Ouyang X, Wu W-M, Zhou T, Hou X, Khan A et al (2024) Novel Feruloyl esterase for the degradation of Polyethylene Terephthalate (PET) screened from the gut microbiome of plastic-degrading mealworms (Tenebrio Molitor Larvae). Environ Sci Technol 58(40):17717–17731. https://doi.org/10.1021/acs.est.4c01495
66. Fadiji AE, Babalola OO (2020) Metagenomics methods for the study of plant-associated microbial communities: a review. J Microbiol Methods 170:105860. https://doi.org/10.1016/j.mimet.2020.105860
67. Dulanto Chiang A, Dekker JP (2020) From the pipeline to the bedside: advances and challenges in clinical metagenomics. J Infect Dis 221(Supplement_3):S331–SS40. https://doi.org/10.1093/infdis/jiz151
68. Wang WL, Xu SY, Ren ZG, Tao L, Jiang JW, Zheng SS (2015) Application of metagenomics in the human gut microbiome. World J Gastroenterol 21(3):803–814. https://doi.org/10.3748/wjg.v21.i3.803
69. Siggins A, Gunnigle E, Abram F (2012) Exploring mixed microbial community functioning: recent advances in metaproteomics. FEMS Microbiol Ecol 80(2):265–280. https://doi.org/10.1111/j.1574-6941.2011.01284.x
70. Schiebenhoefer H, Van Den Bossche T, Fuchs S, Renard BY, Muth T, Martens L (2019) Challenges and promise at the

interface of metaproteomics and genomics: an overview of recent progress in metaproteogenomic data analysis. Expert Rev Proteomics 16(5):375–390. https://doi.org/10.1080/14789450.2019.1609944

71. Van Den Bossche T, Armengaud J, Benndorf D, Blakeley-Ruiz JA, Brauer M, Cheng K et al (2025) The microbiologist's guide to metaproteomics. Imeta 4(3): e70031. https://doi.org/10.1002/imt2.70031
72. Wilmes P, Bond PL (2006) Metaproteomics: studying functional gene expression in microbial ecosystems. Trends Microbiol 14(2): 92–97. https://doi.org/10.1016/j.tim.2005.12.006
73. Speda J, Jonsson BH, Carlsson U, Karlsson M (2017) Metaproteomics-guided selection of targeted enzymes for bioprospecting of mixed microbial communities. Biotechnol Biofuels 10:128. https://doi.org/10.1186/s13068-017-0815-z
74. Hanson BT, Hewson I, Madsen EL (2014) Metaproteomic survey of six aquatic habitats: discovering the identities of microbial populations active in biogeochemical cycling. Microb Ecol 67(3):520–539. https://doi.org/10.1007/s00248-013-0346-5
75. Wang Y, Zhou Y, Xiao X, Zheng J, Zhou H (2020) Metaproteomics: a strategy to study the taxonomy and functionality of the gut microbiota. J Proteome 219:103737. https://doi.org/10.1016/j.jprot.2020.103737
76. Zhang X, Li L, Butcher J, Stintzi A, Figeys D (2019) Advancing functional and translational microbiome research using meta-omics approaches. Microbiome. 7(1):154. https://doi.org/10.1186/s40168-019-0767-6
77. Wang H-B, Zhang Z-X, Li H, He H-B, Fang C-X, Zhang A-J et al (2011) Characterization of metaproteomics in crop rhizospheric soil. J Proteome Res 10(3):932–940. https://doi.org/10.1021/pr100981r
78. Bastida F, Moreno JL, Nicolás C, Hernández T, García C (2009) Soil metaproteomics: a review of an emerging environmental science. Significance, methodology and perspectives. Eur J Soil Sci 60(6):845–859. https://doi.org/10.1111/j.1365-2389.2009.01184.x
79. Jiang J, Czuchry D, Ru Y, Peng H, Shen J, Wang T et al (2024) Activity-based metaproteomics driven discovery and enzymological characterization of potential α-galactosidases in the mouse gut microbiome. Commun Chem 7(1):184. https://doi.org/10.1038/s42004-024-01273-5
80. Sukul P, Schäkermann S, Bandow JE, Kusnezowa A, Nowrousian M, Leichert LI (2017) Simple discovery of bacterial biocatalysts from environmental samples through functional metaproteomics. Microbiome 5(1):28. https://doi.org/10.1186/s40168-017-0247-9
81. Jiang Y, Fu B, Wu W-M, Zhou C, Khan A, Zhang G et al (2025) Soil microbes in the Tibetan Plateau degrade polyvinyl chloride and harbor novel dehalogenase SerB. Environ Int 202:109708. https://doi.org/10.1016/j.envint.2025.109708
82. Xia Y, Zhu M, Du Y, Wu Z, Gomi K, Zhang W (2022) Metaproteomics reveals protein composition of multiple saccharifying enzymes in nongxiangxing daqu and jiangxiangxing daqu under different thermophilic temperatures. Int J Food Sci Technol 57(8): 5102–5113. https://doi.org/10.1111/ijfs.15818
83. Blackburn JM, Martens L (2016) The challenge of metaproteomic analysis in human samples. Expert Rev Proteomics 13(2): 135–138. https://doi.org/10.1586/14789450.2016.1135058
84. Abiraami TV, Singh S, Nain L (2020) Soil metaproteomics as a tool for monitoring functional microbial communities: promises and challenges. Rev Environ Sci Biotechnol 19(1):73–102. https://doi.org/10.1007/s11157-019-09519-8
85. Pan H, Wattiez R, Gillan D (2024) Soil metaproteomics for microbial community profiling: methodologies and challenges. Curr Microbiol 81(8):257. https://doi.org/10.1007/s00284-024-03781-y
86. Wassan JT, Wang H, Browne F, Zheng H (2019) A comprehensive study on predicting functional role of metagenomes using machine learning methods. IEEE/ACM Trans Comput Biol Bioinform 16(3): 751–763. https://doi.org/10.1109/tcbb.2018.2858808
87. Mazurenko S, Prokop Z, Damborsky J (2020) Machine learning in enzyme engineering. ACS Catal 10(2):1210–1223. https://doi.org/10.1021/acscatal.9b04321
88. Markus B, C GC, Andreas K, Arkadij K, Stefan L, Gustav O et al (2023) Accelerating biocatalysis discovery with machine learning: a paradigm shift in enzyme engineering, discovery, and design. ACS Catal 13(21): 14454–14469. https://doi.org/10.1021/acscatal.3c03417

89. Tan JX, Lv H, Wang F, Dao FY, Chen W, Ding H (2019) A survey for predicting enzyme family classes using machine learning methods. Curr Drug Targets 20(5):540–550. https://doi.org/10.2174/1389450119666181002143355
90. Li G, Rabe KS, Nielsen J, Engqvist MKM (2019) Machine learning applied to predicting microorganism growth temperatures and enzyme catalytic optima. ACS Synth Biol 8(6):1411–1420. https://doi.org/10.1021/acssynbio.9b00099
91. Bonk BM, Weis JW, Tidor B (2019) Machine learning identifies chemical characteristics that promote enzyme catalysis. J Am Chem Soc 141(9):4108–4118. https://doi.org/10.1021/jacs.8b13879
92. Goldman S, Das R, Yang KK, Coley CW (2022) Machine learning modeling of family wide enzyme-substrate specificity screens. PLoS Comput Biol 18(2):e1009853. https://doi.org/10.1371/journal.pcbi.1009853
93. Singh N, Malik S, Gupta A, Srivastava KR (2021) Revolutionizing enzyme engineering through artificial intelligence and machine learning. Emerg Top Life Sci 5(1):113–125. https://doi.org/10.1042/ETLS20200257
94. Ferruz N, Höcker B (2022) Controllable protein design with language models. Nat Mach Intell 4(6):521–532. https://doi.org/10.1038/s42256-022-00499-z
95. Watson JL, Juergens D, Bennett NR, Trippe BL, Yim J, Eisenach HE et al (2023) De novo design of protein structure and function with RFdiffusion. Nature 620(7976):1089–1100. https://doi.org/10.1038/s41586-023-06415-8
96. Vanella R, Kovacevic G, Doffini V, Fernández de Santaella J, Nash MA (2022) High-throughput screening, next generation sequencing and machine learning: advanced methods in enzyme engineering. Chem Commun (Cambridge, England) 58(15):2455–2467. https://doi.org/10.1039/d1cc04635g
97. Robinson SL, Piel J, Sunagawa S (2021) A roadmap for metagenomic enzyme discovery. Nat Prod Rep 38(11):1994–2023. https://doi.org/10.1039/d1np00006c
98. Kroll A, Ranjan S, Engqvist MKM, Lercher MJ (2023) A general model to predict small molecule substrates of enzymes based on machine and deep learning. Nat Commun 14(1):2787. https://doi.org/10.1038/s41467-023-38347-2
99. Hirota K, Salim F, Yamada T (2025) DeepES: deep learning-based enzyme screening to identify orphan enzyme genes. Bioinformatics 41(3):btaf053. https://doi.org/10.1093/bioinformatics/btaf053
100. Zhou J, Huang M (2024) Navigating the landscape of enzyme design: from molecular simulations to machine learning. Chem Soc Rev 53(16):8202–8239. https://doi.org/10.1039/d4cs00196f
101. Ryu JY, Kim HU, Lee SY (2019) Deep learning enables high-quality and high-throughput prediction of enzyme commission numbers. Proc Natl Acad Sci 116(28):13996–14001. https://doi.org/10.1073/pnas.1821905116
102. Pan RJ, Wu YC, Wang QP, Hong Y (2009) Preparation and catalytic properties of platinum dioxide nanoparticles: a comparison between conventional heating and microwave-assisted method. Chem Eng J 153(1–3):206–210. https://doi.org/10.1016/j.cej.2009.06.009
103. Koshland DE (2002) The application and usefulness of the ratio kcat/KM. Bioorg Chem 30(3):211–213. https://doi.org/10.1006/bioo.2002.1246
104. Liu C, Wu JX, Chen YB, Liu YH, Zheng YJ, Liu L et al (2025) Advances in zero-shot prediction-guided enzyme engineering using machine learning. ChemCatChem 17(5). https://doi.org/10.1002/cctc.202401542
105. Kroll A, Lercher MJ (2024) DLKcat cannot predict meaningful k (cat) values for mutants and unfamiliar enzymes. Biol Methods Protoc 9(1):bpae061. https://doi.org/10.1093/biomethods/bpae061
106. Wang Z, Xie D, Wu D, Luo X, Wang S, Li Y et al (2025) Robust enzyme discovery and engineering with deep learning using CataPro. Nat Commun 16(1):2736. https://doi.org/10.1038/s41467-025-58038-4
107. Robinson SL, Smith MD, Richman JE, Aukema KG, Wackett LP (2020) Machine learning-based prediction of activity and substrate specificity for OleA enzymes in the thiolase superfamily. Synth Biol 5(1):ysaa004. https://doi.org/10.1093/synbio/ysaa004
108. Phengsakun G, Boonyarit B, Rungrotmongkol T, Suginta W (2023) Structure-based virtual screening for potent inhibitors of GH-20 β-N-acetylglucosaminidase: classical and machine learning scoring functions, and molecular dynamics simulations. Comput Biol Chem 104:107856. https://doi.org/10.1016/j.compbiolchem.2023.107856

109. Srisongkram T, Waithong S, Thitimetharoch T, Weerapreeyakul N (2022) Machine learning and in vitro chemical screening of potential α-amylase and α-glucosidase inhibitors from thai indigenous plants. Nutrients 14(2). https://doi.org/10.3390/nu14020267
110. Mathe A, Mulpuru V, Katari SK, Karlapudi AP, Venkateswarulu TC (2024) Virtual screening and invitro evaluation of cyclooxygenase inhibitors from Tinospora cordifolia using the machine learning tool. J Biomol Struct Dyn 42(23):13275–13289. https://doi.org/10.1080/07391102.2023.2275175
111. Vornholt T, Stockinger P, Mutný M, Jeschek M, Nestl B, Oberdorfer G et al (2025) Of revolutions and roadblocks: the emerging role of machine learning in biocatalysis. ACS Central Sci 11(10):1828–1838. https://doi.org/10.1021/acscentsci.5c00949
112. Landwehr GM, Bogart JW, Magalhaes C, Hammarlund EG, Karim AS, Jewett MC (2025) Accelerated enzyme engineering by machine-learning guided cell-free expression. Nat Commun 16(1):865. https://doi.org/10.1038/s41467-024-55399-0
113. Kries H, Blomberg R, Hilvert D (2013) De novo enzymes by computational design. Curr Opin Chem Biol 17(2):221–228. https://doi.org/10.1016/j.cbpa.2013.02.012
114. Richter F, Leaver-Fay A, Khare SD, Bjelic S, Baker D (2011) De novo enzyme design using Rosetta3. PLoS One 6(5):e19230. https://doi.org/10.1371/journal.pone.0019230
115. Xuan Q, Dehang W, Zhenkun S, Xiaoping L, Hongwu M (2025) Generative artificial intelligence for function-driven de novo enzyme design. Synthetic Biol Eng 3(3):10015. https://doi.org/10.70322/sbe.2025.10015
116. Hua C, Lu J, Liu Y, Zhang O, Tang J, Ying R et al (2024) Reaction-conditioned de novo enzyme design with GENzyme. Arxiv. https://doi.org/10.48550/arXiv.2411.16694
117. Li Z, Lou M, Sun C, Li Z (2024) Engineering a robust UDP-glucose pyrophosphorylase for enhanced biocatalytic synthesis via ProteinMPNN and ancestral sequence reconstruction. J Agric Food Chem 72(27):15284–15292. https://doi.org/10.1021/acs.jafc.4c03126
118. Zanghellini A (2014) de novo computational enzyme design. Curr Opin Biotechnol 29:132–138. https://doi.org/10.1016/j.copbio.2014.03.002
119. Golynskiy MV, Seelig B (2010) De novo enzymes: from computational design to mRNA display. Trends Biotechnol 28(7):340–345. https://doi.org/10.1016/j.tibtech.2010.04.003
120. Seelig B (2011) mRNA display for the selection and evolution of enzymes from in vitro-translated protein libraries. Nat Protoc 6(4):540–552. https://doi.org/10.1038/nprot.2011.312
121. Valencia CA, Cotten SW, Dong B, Liu R (2008) mRNA-display-based selections for proteins with desired functions: a protease-substrate case study. Biotechnol Prog 24(3):561–569. https://doi.org/10.1021/bp070473a
122. Kamalinia G, Grindel BJ, Takahashi TT, Millward SW, Roberts RW (2021) Directing evolution of novel ligands by mRNA display. Chem Soc Rev 50(16):9055–9103. https://doi.org/10.1039/d1cs00160d
123. Keasling JD (2008) Synthetic biology for synthetic chemistry. ACS Chem Biol 3(1):64–76. https://doi.org/10.1021/cb7002434
124. Gao L, Ding Q, Lei X (2024) Hunting for the intermolecular diels–alderase. Acc Chem Res 57(15):2166–2183. https://doi.org/10.1021/acs.accounts.4c00315
125. Hossack EJ, Hardy FJ, Green AP (2023) Building enzymes through design and evolution. ACS Catal 13(19):12436–12444. https://doi.org/10.1021/acscatal.3c02746
126. Hou K, Huang W, Qi M, Tugwell TH, Alturaifi TM, Chen Y et al (2025) De novo design of porphyrin-containing proteins as efficient and stereoselective catalysts. Science 388(6747):665–670. https://doi.org/10.1126/science.adt7268
127. Jiang L, Althoff EA, Clemente FR, Doyle L, Röthlisberger D, Zanghellini A et al (2008) De novo computational design of Retro-Aldol enzymes. Science 319(5868):1387–1391. https://doi.org/10.1126/science.1152692
128. Yeh AH-W, Norn C, Kipnis Y, Tischer D, Pellock SJ, Evans D et al (2023) De novo design of luciferases using deep learning. Nature 614(7949):774–780. https://doi.org/10.1038/s41586-023-05696-3
129. Baker D (2010) An exciting but challenging road ahead for computational enzyme design. Protein Sci 19(10):1817–1819. https://doi.org/10.1002/pro.481
130. Savitski MM, Nielsen ML, Kjeldsen F, Zubarev RA (2005) Proteomics-grade de novo sequencing approach. J Proteome Res

4(6):2348–2354. https://doi.org/10.1021/pr050288x

131. Ma B, Johnson R (2012) De Novo sequencing and homology searching. Mol Cell Proteomics 11(2). https://doi.org/10.1074/mcp.O111.014902
132. Marshall LR, Zozulia O, Lengyel-Zhand Z, Korendovych IV (2019) Minimalist de Novo design of protein catalysts. ACS Catal 9(10): 9265–9275. https://doi.org/10.1021/acscatal.9b02509
133. Puranik R, Quan G, Werner J, Zhou R, Xu Z (2015) A pipeline for completing bacterial genomes using in silicoand wet lab approaches. BMC Genomics 16(3):S7. https://doi.org/10.1186/1471-2164-16-S3-S7
134. Wang X, Xu K, Tan Y, Liu S, Zhou J (2023) Possibilities of using De novo design for generating diverse functional food enzymes. Int J Mol Sci 24(4). https://doi.org/10.3390/ijms24043827
135. Hendriks ATWM, Zeeman G (2009) Pretreatments to enhance the digestibility of lignocellulosic biomass. Bioresour Technol 100(1):10–18. https://doi.org/10.1016/j.biortech.2008.05.027
136. Thapa S, Mishra J, Arora N, Mishra P, Li H, O'Hair J et al (2020) Microbial cellulolytic enzymes: diversity and biotechnology with reference to lignocellulosic biomass degradation. Rev Environ Sci Biotechnol 19(3): 621–648. https://doi.org/10.1007/s11157-020-09536-y
137. Demirbas A (2008) Products from lignocellulosic materials via degradation processes. Energy Sources Part A-Recovery Util Environ Effects 30(1):27–37. https://doi.org/10.1080/00908310600626705
138. Baldrian P, Šnajdr J (2011) Lignocellulose-degrading enzymes in soils. In: Shukla G, Varma A (eds) Soil enzymology. Springer Berlin Heidelberg, Berlin, Heidelberg, pp 167–186
139. Béguin P, Aubert J-P (1994) The biological degradation of cellulose. FEMS Microbiol Rev 13(1):25–58. https://doi.org/10.1111/j.1574-6976.1994.tb00033.x
140. Santos CA, Morais MAB, Mandelli F, Lima EA, Miyamoto RY, Higasi PMR et al (2025) A metagenomic 'dark matter' enzyme catalyses oxidative cellulose conversion. Nature 639(8056):1076–1083. https://doi.org/10.1038/s41586-024-08553-z
141. Li W, Yuan H, Hu Y, Dou S, Zhang E, Wu Q et al (2025) Construction of new thermostable MtLPMO9V in synergism with cellulases for efficient lignocellulosic hydrolysis. Bioresour Technol 430:132594. https://doi.org/10.1016/j.biortech.2025.132594
142. Bhatt B, Bhatt K, Lal S, R S, Bhatt V (2024) Production of a novel cellulase by Bacillus amyloliquefaciens OKB3 isolated from soil: purification and characterization. Int J Biol Macromol 282:137454. https://doi.org/10.1016/j.ijbiomac.2024.137454
143. Benavides Fernández CD, Guzmán Castillo MP, Quijano Pérez SA, Carvajal Rodríguez LV (2022) Microbial degradation of polyethylene terephthalate: a systematic review. SN Appl Sci 4(10):263. https://doi.org/10.1007/s42452-022-05143-4
144. Barnard E, Arias JJR, Thielemans W (2021) Chemolytic depolymerisation of PET: a review. Green Chem 23(11):3765–3789. https://doi.org/10.1039/d1gc00887k
145. Viljakainen VR, Hug LA (2021) New approaches for the characterization of plastic-associated microbial communities and the discovery of plastic-degrading microorganisms and enzymes. Comput Struct Biotechnol J 19:6191–6200. https://doi.org/10.1016/j.csbj.2021.11.023
146. Wu B, Zhong B, Zheng L, Huang R, Jiang S, Li M et al (2025) Harnessing protein language model for structure-based discovery of highly efficient and robust PET hydrolases. Nat Commun 16(1):6211. https://doi.org/10.1038/s41467-025-61599-z
147. Guo W, Duan J, Shi Z, Yu X, Shao Z (2023) Biodegradation of PET by the membrane-anchored PET esterase from the marine bacterium Rhodococcus pyridinivorans P23. Commun Biol 6(1):1090. https://doi.org/10.1038/s42003-023-05470-1
148. Tournier V, Topham CM, Gilles A, David B, Folgoas C, Moya-Leclair E et al (2020) An engineered PET depolymerase to break down and recycle plastic bottles. Nature 580(7802): 216–219. https://doi.org/10.1038/s41586-020-2149-4
149. Chen Y, Sun J, Shi K, Zhu T, Li R, Li R et al (2025) Glycolysis-compatible urethanases for polyurethane recycling. Science 390(6772): 503–509. https://doi.org/10.1126/science.adw4487
150. Vasina M, Vanacek P, Hon J, Kovar D, Faldynova H, Kunka A et al (2022) Advanced database mining of efficient haloalkane dehalogenases by sequence and structure bioinformatics and microfluidics. Chem Catal 2(10): 2704–2725. https://doi.org/10.1016/j.checat.2022.09.011
151. Ferruz N, Schmidt S, Höcker B (2022) ProtGPT2 is a deep unsupervised language model for protein design. Nat Commun

13(1):4348. https://doi.org/10.1038/s41467-022-32007-7

152. Wang M, Yang N, Forcade-Perkins NJ, Weng N (2024) ProGen: projection-based adversarial attack generation against network intrusion detection. IEEE Trans Inf Forensics Secur 19:5476–5491. https://doi.org/10.1109/TIFS.2024.3402155
153. Li Y, Wei Y, Xu S, Tan Q, Zong L, Wang J et al (2023) AcrNET: predicting anti-CRISPR with deep learning. Bioinformatics 39(5). https://doi.org/10.1093/bioinformatics/btad259
154. Zhang F, Li J, Wen Z, Fang C (2024) FusPB-ESM2: fusion model of ProtBERT and ESM-2 for cell-penetrating peptide prediction. Comput Biol Chem 111:108098. https://doi.org/10.1016/j.compbiolchem.2024.108098
155. Hong Y, Lee J, Ko J (2022) A-Prot: protein structure modeling using MSA transformer. BMC Bioinform 23(1):93. https://doi.org/10.1186/s12859-022-04628-8
156. Lu J, Chen H, Qiu J (2024) PhosHSGN: deep neural networks combining sequence and protein spatial information to improve protein phosphorylation site prediction. Ieee Access 12:100611–100627. https://doi.org/10.1109/access.2024.3427792
157. Qian H, Wang Y, Zhou X, Gu T, Wang H, Lyu H et al (2025) ESM-Ezy: a deep learning strategy for the mining of novel multicopper oxidases with superior properties. Nat Commun 16(1):3274. https://doi.org/10.1038/s41467-025-58521-y

Chapter 8

Omics Approaches to Unraveling the Complexity of the Gut-Lung Axis

Bichen Peng and Xiao Chang

Abstract

The complex, bidirectional communication between the gut and the lungs, known as the "gut-lung axis," profoundly influences host immune homeostasis and the pathogenesis of respiratory diseases. In recent years, multi-omics approaches, including metagenomics, metabolomics, and metatranscriptomics, have emerged as the core driving force for unraveling the complexity of this interorgan cross talk network. This review aims to systematically summarize the current omics-based evidence in the field of the gut-lung axis. We highlight key communication mechanisms discovered through multi-omics integration, particularly how gut microbiota-derived metabolites, exemplified by short-chain fatty acids (SCFAs), mediate distal immune regulation. Concurrently, we consolidate omics evidence from the contexts of respiratory infectious diseases, chronic lung disorders, and aging, systematically delineating the impact of gut dysbiosis on pulmonary pathophysiology via the gut-lung axis and emphasizing the feasibility of disease management in patients with lung diseases by modulating the gut microbiota. Although omics technologies have significantly advanced our understanding of this field, the challenge of effectively integrating vast, heterogeneous data and transitioning from "correlation" to "causation" remains a primary hurdle. By reviewing and discussing the current omics evidence in the gut-lung axis, this paper aims to provide new perspectives for future mechanistic explorations and clinical translation strategies.

Key words Gut-lung axis, Microbiome, SCFAs, Omics, Lung disease

1 Introduction

The human microbiome, composed of diverse microbial communities including bacteria, viruses, and fungi, plays a pivotal role in maintaining host homeostasis [1]. Among these communities, the gut microbiome represents the largest and most extensively studied microbial ecosystem in the human body [2, 3]. Its collective genome contains a gene repertoire that vastly exceeds that of the human genome, estimated to be more than 200-fold larger [4]. With advancing research into the gut microbiome, it has become widely accepted that the commensal gut microbiota can modulate immune responses not only within the gastrointestinal

Tao Huang (ed.), *Microbial 'Omics: Methods and Protocols*, Methods in Molecular Biology, vol. 3033,
https://doi.org/10.1007/978-1-0716-5264-0_8,

system but also in distal organs, establishing it as a dominant factor influencing human health [3, 5, 6].

Indeed, a growing body of evidence highlights the existence of the "gut-lung axis" (GLA), a critical molecular signaling cross talk between the gut microbiota and the lungs [7–9]. Although the gut and lungs are anatomically distinct organs, they share fundamental similarities: both are exposed to the external environment, possess a protective mucus layer colonized by microorganisms [10], and even share a common embryological origin [11–13].

The clinical co-occurrence of pulmonary and intestinal diseases also provides preliminary evidence for gut-lung interactions [14]. One study, utilizing 16S rRNA gene sequencing of gastrointestinal and respiratory microbiota in a murine sepsis model and of bronchoalveolar lavage fluid from patients with acute respiratory distress syndrome (ARDS), demonstrated gut-to-lung bacterial translocation in both experimental sepsis and human ARDS. This finding suggests a potential mechanism by which the gut microbiome contributes to critical illness [15]. Furthermore, a wide spectrum of respiratory diseases, including asthma [16], chronic obstructive pulmonary disease (COPD) [17], pulmonary hypertension, COVID-19, and even lung cancer [6, 18], have been shown to be influenced by the gut microbiota, affecting lung immunity via the gut-lung axis.

Driven by the advancement of next-generation sequencing (NGS) technologies and multi-omics studies, the primary communication pathways of the gut-lung axis are gradually being elucidated, and our understanding of gut microbiota-host interactions has significantly improved. Nevertheless, the downstream analysis and interpretation of complex, integrated omics data remain challenging. This review aims to provide an overview of the current omics-based findings on the gut-lung axis, explore the role of the microbiome in respiratory and intestinal health, and further discuss how to better integrate omics data pertaining to the gut-lung axis.

2 The Lungs and Their Microbiome

The term "microbiome," first proposed by Joshua Lederberg in 2001, has been widely adopted to define "the entire habitat, including the microorganisms (bacteria, archaea, lower and higher eukaryotes, and viruses), their genomes (i.e., genes), and the surrounding environmental conditions" [19, 20]. Subsequently, the Human Microbiome Project (HMP) was launched in 2007 [21]. However, due to the long-held belief that the healthy lung is a sterile environment [22], the lungs were not included among the 18 body sites sampled by the HMP. In 2010, culture-independent microbial identification techniques based on 16S rRNA gene sequencing were first applied to the airways of healthy subjects, revealing the existence of a microbial community in the lower respiratory tract. This community was found to have a composition that differed significantly between states of health and disease [23]. After subsequent studies further validated that sampling via bronchoscopy accurately reflects the microbiome of healthy lungs [24, 25], the number of publications in this field surged.

With advancing research, the composition of the lung microbiome has been more comprehensively characterized. According to 16S rRNA studies, the bacteriome of healthy lungs is characterized at the family level by *Firmicutes*, *Bacteroidetes*, *Proteobacteria*, *Fusobacteria*, and *Actinobacteria* and at the operational taxonomic unit (OTU) level by the genera *Prevotella*, *Veillonella*, and *Streptococcus* [24–27]. The bacterial load in lung tissue is relatively low compared to the gut bacteriome, with an estimated 20–1252 bacterial cells per 1000 human cells [28]. The virome of healthy lungs contains a variety of bacteriophages and members of the *Circoviridae* family [29, 30], with abundance and complexity increasing in states of immunosuppression [31]. The mycobiome of healthy lungs primarily includes the genera *Eremothecium*, *Systenostrema*, and *Malassezia* and the family *Davidiellaceae*, with its composition differing significantly from that in diseased lungs [32, 33].

Overall, a close relationship exists between the diversity and relative abundance of the lung microbial community and the state of lung health [34, 35]. On one hand, findings from two cross-sectional studies revealed a negative correlation between the diversity of microbial exposure and the risk of childhood asthma, suggesting that certain microbiomes are protective factors against asthma [36]. On the other hand, dysbiosis of the lung microbiome may increase the risk of developing respiratory diseases [37–39]. Of course, most studies emphasize that the microbiome of a healthy lung is typically low-density, relatively transient, and constantly renewed. Dysbiosis occurs when conditions arise that favor the forced replication, persistence, or even colonization of certain microorganisms [16, 40, 41]. For instance, acute exacerbations of chronic obstructive pulmonary disease are often due to the acquisition of new bacterial strains [42]; the diversity of the airway microbiome in cystic fibrosis decreases significantly as the disease worsens [43, 44]; and an increased abundance of *Proteobacteria* is associated with acute exacerbations of asthma [45].

An interesting observation is that the variation in the lung microbiome within an individual (compared to other body sites) is significantly less than the variation between individuals [46]. Although the host microbiome is dynamic, it undoubtedly possesses unique and individualized characteristics (Fig. 1) [47]. While the origins of the lung microbiome are not yet fully understood, evidence suggests that the development of the

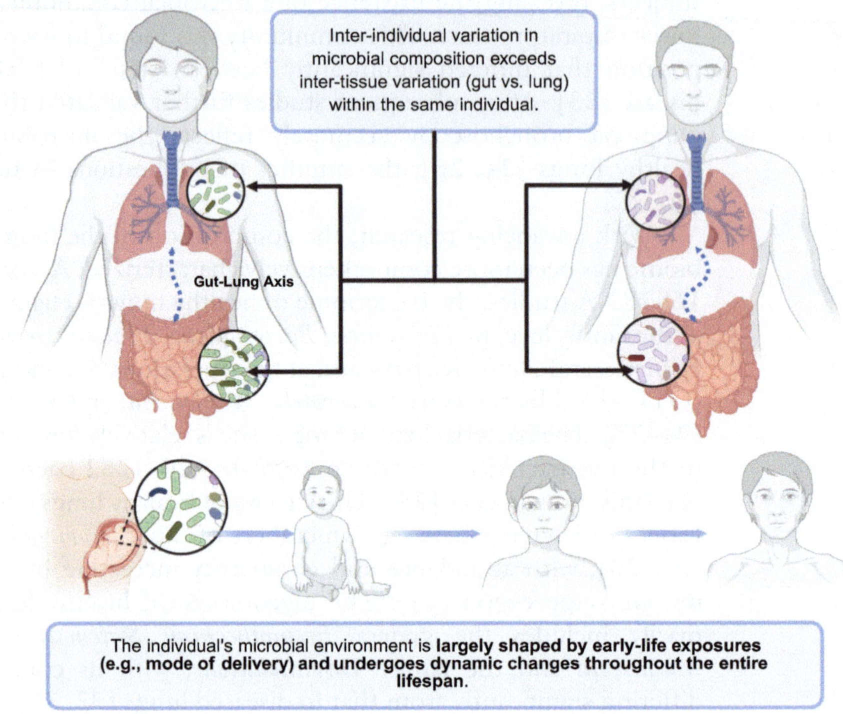

Fig. 1 The human microbiome: Individualized characteristics, gut–lung connectivity, and lifelong evolution. Panel **a** (Individual Microbial Signatures) demonstrates that the differences between the lung and gut microbiomes within the same individual are significantly smaller than the differences between the microbiomes of different individuals. Despite the dynamic nature of the host microbiome, it possesses distinct, individualized signatures that differentiate one host from another. The bidirectional dashed line represents the gut–lung axis, indicating the systemic connection and microbial cross talk between these two distinct mucosal sites. Panel **b** (Developmental and Lifecycle Dynamics) demonstrates that the establishment of the respiratory microbiome is largely determined by early-life exposures. As illustrated by the timeline from the fetal stage to old age, an individual's microbial environment undergoes continuous dynamic changes throughout the lifespan yet remains profoundly influenced by early-life events

respiratory microbiome is largely dependent on early-life exposures, including the mode of delivery [47–50].

Concurrently, murine models suggest that the formation of the early-life airway microbiome is a critical parameter in the process of immune maturation. A lack of pulmonary microbial exposure in early life (within the first 10 days after birth) in mice may lead to insufficient immune training of the host, thereby increasing the risk of allergic airway inflammation [51]. Moreover, certain features of the early-life gut microbiota/microbiome, such as reduced diversity or an abnormal composition, have also been shown to be associated with susceptibility to allergic airway inflammation. Multiple large-scale studies have indicated that low alpha-diversity of the gut microbiota in the first year of life is linked to subsequent respiratory diseases, particularly asthma [52–54]. Low diversity of the gut microbiome within the first month of a neonate's life is associated with an increased susceptibility to school-age asthma [55], and colonization with *Clostridium difficile* within the first month of life is also linked to school-age asthma [56]. Data from the Canadian Healthy Infant Longitudinal Development (CHILD) Study demonstrated that infants at risk for asthma exhibit a transient dysbiosis of the gut microbiota within the first 100 days of life, characterized by a significant reduction in the relative abundance of the genera *Lachnospira*, *Veillonella*, *Faecalibacterium*, and *Rothia* [57].

However, the relationship between microbiome and lung-specific immune development is not confined to early life but is a dynamic process that spans the entire life course [58]. This dynamic nature presents challenges for studying the relationship between the lung microbiome and pulmonary diseases.

3 The Gut-Lung Axis

The host microbiome is indeed not static; its composition continuously evolves throughout the host's life course, reflecting changes in external factors such as diet, medication use, and environmental exposures [59]. Advances in high-throughput sequencing technologies have laid the foundation for systematically studying the interactions between distal organs across the lifespan. Against this backdrop, the bidirectional regulatory network between the gut microbiota and the respiratory system, termed the "gut-lung axis," has increasingly become a focus of research [9]. The theoretical basis for this axis stems from the "gut-lymph" hypothesis proposed by Samuelson et al. [60].

It has become a scientific consensus that the gut and lungs engage in distal communication via the gut-lung axis to collectively maintain host immune homeostasis and modulate disease susceptibility. The functional state of this axis is highly dependent on the health status of the gut microbiota [61, 62].

3.1 Short-Chain Fatty Acids: Fundamental Mediators of the Gut-Lung Axis

Although the complete regulatory network of the gut-lung axis has not been fully elucidated, existing evidence clearly points to the central role of gut microbiota-derived metabolites. As key effector molecules, SCFAs (including acetate, propionate, and butyrate), which are primarily produced by the anaerobic fermentation of dietary fiber by the gut microbiota, can systemically influence the homeostasis, susceptibility, and outcomes of various lung diseases [8, 62, 63]. Preliminary evidence shows that patients with gastrointestinal diseases have lower SCFA levels and are more prone to developing pulmonary lesions [8]. Furthermore, compared to healthy individuals, patients with respiratory infections exhibit reduced activity in metabolic pathways related to SCFA fermentation and a significant decrease in the abundance of SCFA-producing bacteria [64]. Corroborating evidence also comes from disease-specific association studies; a metagenome-wide association study (MWAS) revealed alterations in the SCFA metabolic potential of the gut microbiota in asthma patients [65]. Evidence from hospital settings demonstrates that antibiotic therapy in hospitalized patients is associated with decreased diversity of the gut microbiome and depletion of SCFA producers, and this antibiotic-induced dysbiosis impairs pulmonary defense against multidrug-resistant *Klebsiella pneumoniae* [66].

Based on the evidence above, targeting the gut microbiota to enhance host resistance to respiratory diseases is emerging as a highly attractive strategy [67]. The core of this strategy lies in gut microbiota-derived metabolites, particularly SCFAs, whose protective effects have been confirmed at multiple research levels. Integrated analysis of metagenomics and metabolomics has revealed that SCFAs can reduce the risk of both primary and secondary respiratory infections. This is attributed to their ability to modulate allergic airway inflammation and improve overall epithelial defense against pathogens [68, 69].

To further investigate causal relationships and underlying mechanisms, animal models have provided more direct evidence. For instance, acetate supplementation can effectively reduce the bacterial load in the lungs and decrease the systemic dissemination of lung-derived bacteria, thereby improving the survival rate of mice with secondary respiratory infections [70]. Similarly, modulating the endogenous microbiota through dietary interventions (e.g., a high-fiber diet) or pharmacological agents (e.g., baicalin) can exert comparable effects by increasing circulating SCFA levels [63, 71, 72]. Specifically, these SCFAs can, on one hand, directly act on the bone marrow to promote hematopoiesis and, on the other, suppress the function of dendritic cells (DCs), thereby attenuating their capacity to induce TH2 cell-mediated allergic airway inflammation [63]. Acetate that circulates to the lungs can initiate local immune defense by activating its specific receptor, FFAR2 [71]. Furthermore, the coadministration of acetate, propionate, and butyrate has been shown to counteract the impairment of gut barrier function and secondary infections following influenza infection [73]. A recent study also provided evidence of how commensal gut protozoa can influence respiratory diseases by affecting the lung immune environment. Intestinal colonization with *Tritrichomonas musculis* (*T.mu*) induces an immune network composed of gut-derived iILC2s, lung-resident T cells, and B cells. This immune network, by driving a significant expansion of pulmonary eosinophils, not only exacerbates the severity of allergic airway inflammation but also inhibits the systemic dissemination of *Mycobacterium tuberculosis.*

Although the immunomodulatory effects of SCFAs have been extensively demonstrated in rodent models, their role in humans remains to be fully elucidated, and relevant clinical cohort studies are particularly scarce. Nevertheless, the limited number of existing human studies has begun to reveal the association between SCFAs and respiratory health in various disease contexts. In a multi-omics study of a high-risk, immunosuppressed population (hematopoietic stem cell transplant recipients), patients with a higher abundance of butyrate-producing bacteria were found to have a fivefold lower risk of developing viral respiratory infections [74]. Concurrently, a study of Taiwanese children showed that reduced fecal butyrate levels were associated with elevated dust mite-specific immunoglobulin E (IgE) levels and an increased risk of early-childhood asthma [75]. For patients with obstructive airway diseases, lower host SCFA levels are significantly associated with higher mucus plug scores [76]. Moreover, in patients with cachexia, including those with lung cancer, all fecal SCFA levels showed a decreasing trend, with the reduction in acetate concentration being particularly significant [77, 78]. The scarcity of human data necessitates further research to determine whether increasing systemic SCFA levels through dietary or nutritional interventions can prevent or alleviate immune-inflammatory responses in humans and to explore strategies and pathways for supporting gut and respiratory health in specific populations by targeting systemic SCFA levels.

3.2 The Gut-Lung Axis and Respiratory Infectious Diseases

Antibiotic treatment can disrupt gut microbiota homeostasis, thereby increasing the susceptibility of mice to respiratory pathogens, such as *Streptococcus pneumoniae* and *Klebsiella pneumoniae*. Importantly, this susceptible phenotype can be reversed by fecal microbiota transplantation (FMT) [79]. To further elucidate the specific molecular mechanisms by which the microbiota regulates pulmonary immunity, studies have linked antibiotic-induced susceptibility to specific cytokine changes. It was found that the enhanced susceptibility of antibiotic-treated mice to *S. pneumoniae* and *K. pneumoniae* infection was significantly associated with reduced production of IL-17A and granulocyte-macrophage colony-stimulating factor (GM-CSF) in the lungs [80]. This susceptibility can be reversed by transplanting either local (respiratory) or distal (intestinal) microbiota. Critically, this protective effect is dependent on GM-CSF, as its neutralization with an antibody abrogates the protective effect. IL-17A, in turn, was identified as an upstream inducer of GM-CSF. This provides strong evidence that both gut and respiratory microbial communities can act on a key pulmonary cytokine axis to maintain immune homeostasis.

In addition to bacterial respiratory infections, viral respiratory infections are also influenced by the gut-lung axis. A study combining metagenomics and targeted metabolomics demonstrated that indole-3-propionic acid (IPA), a key metabolite produced by the gut microbiota, plays a crucial protective role in combating influenza A virus (IAV) infection. A decline in IPA levels was directly correlated with disease exacerbation, whereas IPA supplementation effectively treated influenza [69]. Furthermore, respiratory viral infections can lead to significant changes in the host's fecal metabolome and lipid metabolism. Following respiratory syncytial virus (RSV) infection, the abundance of sphingolipids, polyunsaturated fatty acids (PUFAs), and the SCFA valerate was all increased in the host's fecal metabolome [81]. Metagenomic studies of COVID-19 patients have also revealed the feedback regulation of respiratory diseases on the gut microbiota. Compared to healthy controls, the fecal microbiome of hospitalized COVID-19 patients underwent persistent changes, characterized by the enrichment of opportunistic pathogens and the depletion of beneficial commensals. These changes in the fecal microbiota were correlated with fecal SARS-CoV-2 load and the severity of COVID-19 [82]. Additional clinical studies have further shown that oral administration of probiotics (mainly *Bifidobacterium* and *Lactobacillus* species) can induce host antiviral activity, positively affect the composition and diversity of the host gut microbiota, and is associated with improved clinical outcomes in COVID-19 patients [83]. This, to some extent, demonstrates the feasibility of strategies aimed at mitigating the severity of respiratory infectious diseases by modulating the gut microbiota.

3.3 The Gut-Lung Axis and Chronic Lung Diseases

A substantial body of multi-omics evidence indicates that various chronic lung diseases, including bronchiectasis [84], cystic fibrosis [43], COPD [27, 28], and asthma [23], are closely associated with a dysbiotic airway microbiome. Unlike the gastrointestinal tract, the alveolar space under normal conditions provides an environment that is generally unfavorable for most bacteria, resulting in minimal bacterial growth [85]. However, in patients with chronic lung diseases, structural damage to the airways and disruption of local immune homeostasis create conditions conducive to the colonization of opportunistic pathogens. In this context, not only does the airway microbiome change, but alterations in the composition of the gut microbiome have also been noted.

A multi-omics study has revealed the clinical relevance of inter-organ microbial interactions. Through unsupervised clustering analysis based on "gut-lung interaction" patterns, this study classified patients with stable bronchiectasis into two subgroups with significantly different clinical phenotypes: a "high-interaction group," characterized by the co-occurrence of pulmonary *Pseudomonas aeruginosa* with gut *Bacteroides* and *Saccharomyces*, which was associated with more frequent exacerbations and greater disease severity, and a "low-interaction group," enriched with pulmonary commensals and gut *Candida*, whose disease state was relatively stable. To investigate the causality of this association, the lung *Pseudomonas*-gut *Bacteroides* relationship observed in the "high-interaction group" was further validated in a murine model of pulmonary *P. aeruginosa* infection. This interaction was abrogated following pretreatment with the antibiotic imipenem, confirming the relevance and therapeutic potential of targeting the gut microbiome to influence the gut-lung axis [84].

Cystic fibrosis (CF) is a hereditary respiratory disease characterized by bronchiectasis and obstructive lesions. The disease manifests with increased viscosity of secretions, altered pulmonary microbial diversity, and infections resulting from the progressive colonization of pathogenic microorganisms in the airways [86]. Similarly, alterations in the gut microbiota structure have also been reported in CF patients [7, 87].

Numerous epidemiological studies have shown that COPD is associated with an increased incidence of changes in gut health, and vice versa [9, 88–90]. Gut microbiota dysbiosis has been observed in COPD patients compared to healthy individuals [7]. Concurrently, studies have found that the composition of the gut microbiota significantly influences the progression of COPD induced by cigarette smoke exposure and that fecal microbiota transplantation (with *Parabacteroides goldsteinii*) can reverse the pathological process of COPD [91].

Existing sequencing data clearly indicate a significant correlation between reduced gut microbial diversity in early infancy and the asthma phenotype in childhood [55, 92]. Notably, these differences in diversity disappear by 12 months of age, long before the onset of disease symptoms [55]. Further research has found that in children at high risk for asthma, the relative abundance of bacterial genera, such as *Lachnospira*, *Veillonella*, *Faecalibacterium*, and *Rothia*, is significantly reduced. This reduction in bacterial taxa is accompanied by lower levels of acetate in the host's feces and dysregulation of enterohepatic metabolites. Inoculating germ-free mice with these four bacterial taxa was shown to ameliorate airway inflammation in their adult offspring, suggesting a causal role for these bacterial taxa in preventing the development of asthma [57].

4 Aging and the Microbiome

The aging process is accompanied by a natural decline in physiological functions and cumulative exposure to environmental toxins. These factors collectively lead to the deterioration of lung tissue and function, thereby significantly increasing susceptibility to a variety of pulmonary diseases [93]. This heightened risk encompasses a range of chronic non-infectious diseases, such as COPD, idiopathic pulmonary fibrosis (IPF), and lung cancer, as well as infectious diseases like community-acquired pneumonia. Among these, age-related changes in the immune system are considered a primary driver of asthma exacerbation in the elderly [94].

Although the impact of aging on lung health is receiving increasing attention, direct omics-based evidence on the lower respiratory tract microbiome in the elderly population is still lacking due to challenges in sample acquisition. However, murine models provide important indirect evidence, revealing that age influences the recovery process following pulmonary infection with *Streptococcus pneumoniae*. Compared to young mice, aged mice exhibit lower efficiency in clearing the pathogen, along with a significant increase in the abundance of Firmicutes operational taxonomic units (OTUs) [95]. More importantly, the composition of the upper respiratory tract microbiome in aged mice fails to return to its initial state for up to four weeks post-infection. This phenomenon provides strong evidence that aging leads to a significant loss of "resilience" in the microbial community, meaning its capacity for self-repair and restoration of homeostasis after an external perturbation (such as infection) is severely impaired [96].

Existing evidence indicates that the composition and function of the gut microbiome are profoundly regulated by dietary patterns. This regulation is not only manifested in macroscopic dietary types (e.g., diets rich in animal protein, plant protein, or dietary fiber) but also depends on the microscopic food matrix, that is, the composition of food components and their complex interactions [97]. Therefore, it is reasonable to infer that the key biological mechanism by which certain dietary interventions can influence age-related molecular and cellular processes [98] is likely achieved through the remodeling of the gut microbiota.

Furthermore, aging also increases the prevalence of gastroesophageal reflux [99], suggesting that the transfer of microbiota from the stomach to the lungs may occur more easily in older individuals. This, in turn, could provide an enhanced and unchecked communication pathway between the gut and the lungs.

5 Conclusion and Future Perspectives

The association between compositional changes in the gut and airway microbiota and a variety of chronic lung diseases and respiratory infections has become an area of increasing scholarly focus. Although the "cross talk" between the gut and lungs is widely considered to be bidirectional, the vast majority of current research has focused on the "gut-to-lung" axis of regulation. The "lung-to-gut" axis, however, remains relatively under-explored, representing a significant direction for future research.

Currently, mechanistic investigations of the "gut-to-lung" axis are primarily concentrated on two potential pathways. The first is metabolite-mediated systemic communication, wherein metabolites produced by the gut microbiota (particularly under the influence of factors like diet), such as SCFAs, reach the lungs via the bloodstream to exert their broad-spectrum anti-inflammatory and immunomodulatory functions. The second is the direct translocation of microbes or their components, where bacteria or their products from the gut microbiota may be directly "seeded" into the airways, acting on local immune cells to shape their response patterns. However, the relative contributions of these two pathways in different disease contexts, as well as their interactions, remain to be thoroughly investigated.

In recent years, a foundational understanding of the gut-lung axis in the context of respiratory diseases has been established, yet more in-depth mechanistic studies are required to delineate its specific pathways and mediating molecules. Some studies have already explored the feasibility of using multi-class models based on the fecal microbiome for disease diagnosis [100]. The success of such research paradigms, however, critically depends on high-quality datasets and superior, reproducible machine learning methods with high clinical relevance.

To bridge the gap from "correlation" to "causation," future research must integrate multi-dimensional omics data with advanced computational biology methods. The combined application of methods such as metagenomics, metabolomics, metatranscriptomics, and metaproteomics, particularly within a longitudinal cohort design, can provide a dynamic perspective for exploring functional changes in the lung microbiome during disease progression, microbe-microbe interactions, microbe-host cell interaction networks, and the key regulatory circuits of inflammation initiation and resolution.

However, to identify the core biological processes driving phenotypic differences from such vast and heterogeneous multi-omics data, the use of machine learning methods is essential. These approaches can integrate data from different levels into a unified probabilistic model, learning a series of latent factors to disentangle the main sources of variation in the data. By analyzing the weights of these latent factors on different omics features (such as specific microbial species, metabolites, or gene expression), we can more accurately infer the key regulatory axes connecting the gut microbiota, circulating metabolites, and the host's pulmonary response. This will enable the elucidation of underlying disease mechanisms at a systemic level, truly transforming data correlation into biological insight.

Conflicts of Interest

The authors have no conflicts of interest to declare that are relevant to the content of this chapter.

References

1. Zhang Y, Thomas JP, Korcsmaros T, Gul L (2024) Integrating multi-omics to unravel host-microbiome interactions in inflammatory bowel disease. Cell Rep Med 5(9): 101738. https://doi.org/10.1016/j.xcrm.2024.101738
2. Zhu Y, Michelle Luo T, Jobin C, Young HA (2011) Gut microbiota and probiotics in colon tumorigenesis. Cancer Lett 309(2): 119–127. https://doi.org/10.1016/j.canlet.2011.06.004
3. Chotirmall SH, Gellatly SL, Budden KF, Mac Aogain M, Shukla SD, Wood DL, Hugenholtz P, Pethe K, Hansbro PM (2017) Microbiomes in respiratory health and disease: an Asia-Pacific perspective. Respirology 22(2):240–250. https://doi.org/10.1111/resp.12971
4. Metwaly A, Reitmeier S, Haller D (2022) Microbiome risk profiles as biomarkers for inflammatory and metabolic disorders. Nat Rev Gastroenterol Hepatol 19(6):383–397. https://doi.org/10.1038/s41575-022-00581-2
5. Backhed F, Ley RE, Sonnenburg JL, Peterson DA, Gordon JI (2005) Host-bacterial mutualism in the human intestine. Science 307(5717):1915–1920. https://doi.org/10.1126/science.1104816
6. Ma PJ, Wang MM, Wang Y (2022) Gut microbiota: a new insight into lung diseases. Biomed Pharmacother 155:113810. https://doi.org/10.1016/j.biopha.2022.113810
7. Marsland BJ, Trompette A, Gollwitzer ES (2015) The gut-lung axis in respiratory disease. Ann Am Thorac Soc 12(Suppl 2):S150–S156. https://doi.org/10.1513/AnnalsATS.201503-133AW
8. Dang AT, Marsland BJ (2019) Microbes, metabolites, and the gut-lung axis. Mucosal Immunol 12(4):843–850. https://doi.org/10.1038/s41385-019-0160-6
9. Budden KF, Gellatly SL, Wood DL, Cooper MA, Morrison M, Hugenholtz P, Hansbro PM (2017) Emerging pathogenic links between microbiota and the gut-lung axis. Nat Rev Microbiol 15(1):55–63. https://doi.org/10.1038/nrmicro.2016.142
10. Eladham MW, Selvakumar B, Saheb Sharif-Askari N, Saheb Sharif-Askari F, Ibrahim SM, Halwani R (2024) Unraveling the gut-lung axis: exploring complex mechanisms in disease interplay. Heliyon 10(1):e24032. https://doi.org/10.1016/j.heliyon.2024.e24032
11. Yiangou L, Ross ADB, Goh KJ, Vallier L (2018) Human pluripotent stem cell-derived endoderm for modeling development and clinical applications. Cell Stem Cell 22(4): 485–499. https://doi.org/10.1016/j.stem.2018.03.016
12. Shu W, Lu MM, Zhang Y, Tucker PW, Zhou D, Morrisey EE (2007) Foxp2 and Foxp1 cooperatively regulate lung and esophagus development. Development 134(10): 1991–2000. https://doi.org/10.1242/dev.02846
13. Ramalho-Santos M, Melton DA, McMahon AP (2000) Hedgehog signals regulate multiple aspects of gastrointestinal development. Development 127(12):2763–2772. https://doi.org/10.1242/dev.127.12.2763
14. Kraft SC, Earle RH, Roesler M, Esterly JR (1976) Unexplained bronchopulmonary disease with inflammatory bowel disease. Arch Intern Med 136(4):454–459
15. Dickson RP, Singer BH, Newstead MW, Falkowski NR, Erb-Downward JR, Standiford TJ, Huffnagle GB (2016) Enrichment of the lung microbiome with gut bacteria in sepsis and the acute respiratory distress syndrome. Nat Microbiol 1(10):16113. https://doi.org/10.1038/nmicrobiol.2016.113
16. Hufnagl K, Pali-Scholl I, Roth-Walter F, Jensen-Jarolim E (2020) Dysbiosis of the gut and lung microbiome has a role in asthma. Semin Immunopathol 42(1):75–93. https://doi.org/10.1007/s00281-019-00775-y
17. Wang L, Cai Y, Garssen J, Henricks PAJ, Folkerts G, Braber S (2023) The bidirectional gut-lung axis in chronic obstructive pulmonary disease. Am J Respir Crit Care Med 207(9):1145–1160. https://doi.org/10.1164/rccm.202206-1066TR
18. Wypych TP, Wickramasinghe LC, Marsland BJ (2019) The influence of the microbiome on respiratory health. Nat Immunol 20(10): 1279–1290. https://doi.org/10.1038/s41590-019-0451-9
19. Dickson RP, Erb-Downward JR, Martinez FJ, Huffnagle GB (2016) The microbiome and the respiratory tract. Annu Rev Physiol 78: 481–504. https://doi.org/10.1146/annurev-physiol-021115-105238
20. Marchesi JR, Ravel J (2015) The vocabulary of microbiome research: a proposal. Microbiome 3:31. https://doi.org/10.1186/s40168-015-0094-5
21. Turnbaugh PJ, Ley RE, Hamady M, Fraser-Liggett CM, Knight R, Gordon JI (2007)

The human microbiome project. Nature 449(7164):804–810. https://doi.org/10.1038/nature06244

22. Natalini JG, Singh S, Segal LN (2023) The dynamic lung microbiome in health and disease. Nat Rev Microbiol 21(4):222–235. https://doi.org/10.1038/s41579-022-00821-x
23. Hilty M, Burke C, Pedro H, Cardenas P, Bush A, Bossley C, Davies J, Ervine A, Poulter L, Pachter L, Moffatt MF, Cookson WO (2010) Disordered microbial communities in asthmatic airways. PLoS One 5(1):e8578. https://doi.org/10.1371/journal.pone.0008578
24. Charlson ES, Bittinger K, Haas AR, Fitzgerald AS, Frank I, Yadav A, Bushman FD, Collman RG (2011) Topographical continuity of bacterial populations in the healthy human respiratory tract. Am J Respir Crit Care Med 184(8):957–963. https://doi.org/10.1164/rccm.201104-0655OC
25. Dickson RP, Erb-Downward JR, Freeman CM, McCloskey L, Falkowski NR, Huffnagle GB, Curtis JL (2017) Bacterial topography of the healthy human lower respiratory tract. mBio 8(1). https://doi.org/10.1128/mBio.02287-16
26. Segal LN, Alekseyenko AV, Clemente JC, Kulkarni R, Wu B, Gao Z, Chen H, Berger KI, Goldring RM, Rom WN, Blaser MJ, Weiden MD (2013) Enrichment of lung microbiome with supraglottic taxa is associated with increased pulmonary inflammation. Microbiome 1(1):19. https://doi.org/10.1186/2049-2618-1-19
27. Erb-Downward JR, Thompson DL, Han MK, Freeman CM, McCloskey L, Schmidt LA, Young VB, Toews GB, Curtis JL, Sundaram B, Martinez FJ, Huffnagle GB (2011) Analysis of the lung microbiome in the "healthy" smoker and in COPD. PLoS One 6(2):e16384. https://doi.org/10.1371/journal.pone.0016384
28. Sze MA, Dimitriu PA, Hayashi S, Elliott WM, McDonough JE, Gosselink JV, Cooper J, Sin DD, Mohn WW, Hogg JC (2012) The lung tissue microbiome in chronic obstructive pulmonary disease. Am J Respir Crit Care Med 185(10):1073–1080. https://doi.org/10.1164/rccm.201111-2075OC
29. Abbas AA, Diamond JM, Chehoud C, Chang B, Kotzin JJ, Young JC, Imai I, Haas AR, Cantu E, Lederer DJ, Meyer KC, Milewski RK, Olthoff KM, Shaked A, Christie JD, Bushman FD, Collman RG (2017) The perioperative lung transplant virome: torque Teno viruses are elevated in donor lungs and show divergent dynamics in primary graft dysfunction. Am J Transplant 17(5):1313–1324. https://doi.org/10.1111/ajt.14076
30. Willner D, Furlan M, Haynes M, Schmieder R, Angly FE, Silva J, Tammadoni S, Nosrat B, Conrad D, Rohwer F (2009) Metagenomic analysis of respiratory tract DNA viral communities in cystic fibrosis and non-cystic fibrosis individuals. PLoS One 4(10):e7370. https://doi.org/10.1371/journal.pone.0007370
31. Young JC, Chehoud C, Bittinger K, Bailey A, Diamond JM, Cantu E, Haas AR, Abbas A, Frye L, Christie JD, Bushman FD, Collman RG (2015) Viral metagenomics reveal blooms of anelloviruses in the respiratory tract of lung transplant recipients. Am J Transplant 15(1):200–209. https://doi.org/10.1111/ajt.13031
32. Charlson ES, Diamond JM, Bittinger K, Fitzgerald AS, Yadav A, Haas AR, Bushman FD, Collman RG (2012) Lung-enriched organisms and aberrant bacterial and fungal respiratory microbiota after lung transplant. Am J Respir Crit Care Med 186(6):536–545. https://doi.org/10.1164/rccm.201204-0693OC
33. van Woerden HC, Gregory C, Brown R, Marchesi JR, Hoogendoorn B, Matthews IP (2013) Differences in fungi present in induced sputum samples from asthma patients and non-atopic controls: a community based case control study. BMC Infect Dis 13:69. https://doi.org/10.1186/1471-2334-13-69
34. Budden KF, Shukla SD, Rehman SF, Bowerman KL, Keely S, Hugenholtz P, Armstrong-James DPH, Adcock IM, Chotirmall SH, Chung KF, Hansbro PM (2019) Functional effects of the microbiota in chronic respiratory disease. Lancet Respir Med 7(10):907–920. https://doi.org/10.1016/S2213-2600(18)30510-1
35. He Y, Wen Q, Yao F, Xu D, Huang Y, Wang J (2017) Gut-lung axis: the microbial contributions and clinical implications. Crit Rev Microbiol 43(1):81–95. https://doi.org/10.1080/1040841X.2016.1176988
36. Ege MJ, Mayer M, Normand AC, Genuneit J, Cookson WO, Braun-Fahrlander C, Heederik D, Piarroux R, von Mutius E, Group GTS (2011) Exposure to environmental microorganisms and childhood asthma. N Engl J Med 364(8):701–709. https://doi.org/10.1056/NEJMoa1007302
37. Beck JM, Young VB, Huffnagle GB (2012) The microbiome of the lung. Transl Res

160(4):258–266. https://doi.org/10.1016/j.trsl.2012.02.005

38. de Steenhuijsen Piters WA, Sanders EA, Bogaert D (2015) The role of the local microbial ecosystem in respiratory health and disease. Philos Trans R Soc Lond Ser B Biol Sci 370(1675). https://doi.org/10.1098/rstb.2014.0294
39. Segal LN, Rom WN, Weiden MD (2014) Lung microbiome for clinicians. New discoveries about bugs in healthy and diseased lungs. Ann Am Thorac Soc 11(1):108–116. https://doi.org/10.1513/AnnalsATS.201310-339FR
40. Barcik W, Boutin RCT, Sokolowska M, Finlay BB (2020) The role of lung and gut microbiota in the pathology of asthma. Immunity 52(2):241–255. https://doi.org/10.1016/j.immuni.2020.01.007
41. Levy M, Kolodziejczyk AA, Thaiss CA, Elinav E (2017) Dysbiosis and the immune system. Nat Rev Immunol 17(4):219–232. https://doi.org/10.1038/nri.2017.7
42. Sethi S, Evans N, Grant BJB, Murphy TF (2002) New strains of bacteria and exacerbations of chronic obstructive pulmonary disease. N Engl J Med 347(7):465–471. https://doi.org/10.1056/NEJMoa012561
43. Zhao J, Schloss PD, Kalikin LM, Carmody LA, Foster BK, Petrosino JF, Cavalcoli JD, VanDevanter DR, Murray S, Li JZ, Young VB, LiPuma JJ (2012) Decade-long bacterial community dynamics in cystic fibrosis airways. Proc Natl Acad Sci 109(15):5809–5814. https://doi.org/10.1073/pnas.1120577109
44. Fodor AA, Klem ER, Gilpin DF, Elborn JS, Boucher RC, Tunney MM, Wolfgang MC (2012) The adult cystic fibrosis airway microbiota is stable over time and infection type, and highly resilient to antibiotic treatment of exacerbations. PLoS One 7(9):e45001. https://doi.org/10.1371/journal.pone.0045001
45. Marri PR, Stern DA, Wright AL, Billheimer D, Martinez FD (2013) Asthma-associated differences in microbial composition of induced sputum. J Allergy Clin Immunol 131(2):346–352 e341–343. https://doi.org/10.1016/j.jaci.2012.11.013
46. Dickson RP, Erb-Downward JR, Freeman CM, McCloskey L, Beck JM, Huffnagle GB, Curtis JL (2015) Spatial variation in the healthy human lung microbiome and the adapted Island model of lung biogeography. Ann Am Thorac Soc 12(6):821–830. https://doi.org/10.1513/AnnalsATS.201501-029OC
47. Integrative HMPRNC (2019) The integrative human microbiome project. Nature 569(7758):641–648. https://doi.org/10.1038/s41586-019-1238-8
48. Biesbroek G, Tsivtsivadze E, Sanders EA, Montijn R, Veenhoven RH, Keijser BJ, Bogaert D (2014) Early respiratory microbiota composition determines bacterial succession patterns and respiratory health in children. Am J Respir Crit Care Med 190(11):1283–1292. https://doi.org/10.1164/rccm.201407-1240OC
49. Teo SM, Mok D, Pham K, Kusel M, Serralha M, Troy N, Holt BJ, Hales BJ, Walker ML, Hollams E, Bochkov YA, Grindle K, Johnston SL, Gern JE, Sly PD, Holt PG, Holt KE, Inouye M (2015) The infant nasopharyngeal microbiome impacts severity of lower respiratory infection and risk of asthma development. Cell Host Microbe 17(5):704–715. https://doi.org/10.1016/j.chom.2015.03.008
50. Pattaroni C, Watzenboeck ML, Schneidegger S, Kieser S, Wong NC, Bernasconi E, Pernot J, Mercier L, Knapp S, Nicod LP, Marsland CP, Roth-Kleiner M, Marsland BJ (2018) Early-life formation of the microbial and immunological environment of the human airways. Cell Host Microbe 24(6):857–865.e854. https://doi.org/10.1016/j.chom.2018.10.019
51. Gollwitzer ES, Saglani S, Trompette A, Yadava K, Sherburn R, McCoy KD, Nicod LP, Lloyd CM, Marsland BJ (2014) Lung microbiota promotes tolerance to allergens in neonates via PD-L1. Nat Med 20(6):642–647. https://doi.org/10.1038/nm.3568
52. Boutin RCT, Sbihi H, Dsouza M, Malhotra R, Petersen C, Dai D, Sears MR, Moraes TJ, Becker AB, Azad MB, Mandhane PJ, Subbarao P, Finlay BB, Turvey SE (2020) Mining the infant gut microbiota for therapeutic targets against atopic disease. Allergy 75(8):2065–2068. https://doi.org/10.1111/all.14244
53. Patrick DM, Sbihi H, Dai DLY, Al Mamun A, Rasali D, Rose C, Marra F, Boutin RCT, Petersen C, Stiemsma LT, Winsor GL, Brinkman FSL, Kozyrskyj AL, Azad MB, Becker AB, Mandhane PJ, Moraes TJ, Sears MR, Subbarao P, Finlay BB, Turvey SE (2020) Decreasing antibiotic use, the gut microbiota, and asthma incidence in children: evidence from population-based and prospective cohort studies. Lancet Respir Med 8(11):1094–1105. https://doi.org/10.1016/S2213-2600(20)30052-7

54. Depner M, Taft DH, Kirjavainen PV, Kalanetra KM, Karvonen AM, Peschel S, Schmausser-Hechfellner E, Roduit C, Frei R, Lauener R, Divaret-Chauveau A, Dalphin JC, Riedler J, Roponen M, Kabesch M, Renz H, Pekkanen J, Farquharson FM, Louis P, Mills DA, von Mutius E, group Ps, Ege MJ (2020) Maturation of the gut microbiome during the first year of life contributes to the protective farm effect on childhood asthma. Nat Med 26(11):1766–1775. https://doi.org/10.1038/s41591-020-1095-x
55. Abrahamsson TR, Jakobsson HE, Andersson AF, Bjorksten B, Engstrand L, Jenmalm MC (2014) Low gut microbiota diversity in early infancy precedes asthma at school age. Clin Exp Allergy 44(6):842–850. https://doi.org/10.1111/cea.12253
56. van Nimwegen FA, Penders J, Stobberingh EE, Postma DS, Koppelman GH, Kerkhof M, Reijmerink NE, Dompeling E, van den Brandt PA, Ferreira I, Mommers M, Thijs C (2011) Mode and place of delivery, gastrointestinal microbiota, and their influence on asthma and atopy. J Allergy Clin Immunol 128(5):948–955 e941–943. https://doi.org/10.1016/j.jaci.2011.07.027
57. Arrieta MC, Stiemsma LT, Dimitriu PA, Thorson L, Russell S, Yurist-Doutsch S, Kuzeljevic B, Gold MJ, Britton HM, Lefebvre DL, Subbarao P, Mandhane P, Becker A, McNagny KM, Sears MR, Kollmann T, Investigators CS, Mohn WW, Turvey SE, Finlay BB (2015) Early infancy microbial and metabolic alterations affect risk of childhood asthma. Sci Transl Med 7(307):307ra152. https://doi.org/10.1126/scitranslmed.aab2271
58. Huang YJ, Boushey HA (2015) The microbiome in asthma. J Allergy Clin Immunol 135(1):25–30. https://doi.org/10.1016/j.jaci.2014.11.011
59. Quercia S, Candela M, Giuliani C, Turroni S, Luiselli D, Rampelli S, Brigidi P, Franceschi C, Bacalini MG, Garagnani P, Pirazzini C (2014) From lifetime to evolution: timescales of human gut microbiota adaptation. Front Microbiol 5:587. https://doi.org/10.3389/fmicb.2014.00587
60. Samuelson DR, Welsh DA, Shellito JE (2015) Regulation of lung immunity and host defense by the intestinal microbiota. Front Microbiol 6:1085. https://doi.org/10.3389/fmicb.2015.01085
61. Nicholson JK, Wilson ID (2003) Opinion: understanding 'global' systems biology: metabonomics and the continuum of metabolism. Nat Rev Drug Discov 2(8):668–676. https://doi.org/10.1038/nrd1157
62. Ashique S, De Rubis G, Sirohi E, Mishra N, Rihan M, Garg A, Reyes RJ, Manandhar B, Bhatt S, Jha NK, Singh TG, Gupta G, Singh SK, Chellappan DK, Paudel KR, Hansbro PM, Oliver BG, Dua K (2022) Short Chain Fatty Acids: fundamental mediators of the gut-lung axis and their involvement in pulmonary diseases. Chem Biol Interact 368:110231. https://doi.org/10.1016/j.cbi.2022.110231
63. Trompette A, Gollwitzer ES, Yadava K, Sichelstiel AK, Sprenger N, Ngom-Bru C, Blanchard C, Junt T, Nicod LP, Harris NL, Marsland BJ (2014) Gut microbiota metabolism of dietary fiber influences allergic airway disease and hematopoiesis. Nat Med 20(2):159–166. https://doi.org/10.1038/nm.3444
64. Woodall CA, McGeoch LJ, Hay AD, Hammond A (2022) Respiratory tract infections and gut microbiome modifications: a systematic review. PLoS One 17(1):e0262057. https://doi.org/10.1371/journal.pone.0262057
65. Wang Q, Li F, Liang B, Liang Y, Chen S, Mo X, Ju Y, Zhao H, Jia H, Spector TD, Xie H, Guo R (2018) A metagenome-wide association study of gut microbiota in asthma in UK adults. BMC Microbiol 18(1):114. https://doi.org/10.1186/s12866-018-1257-x
66. Dorner PJ, Anandakumar H, Rowekamp I, Fiocca Vernengo F, Millet Pascual-Leone B, Krzanowski M, Sellmaier J, Bruning U, Fritsche-Guenther R, Pfannkuch L, Kurth F, Milek M, Igbokwe V, Lober U, Gutbier B, Holstein M, Heinz GA, Mashreghi MF, Schulte LN, Klatt AB, Caesar S, Wienhold SM, Offermanns S, Mack M, Witzenrath M, Jordan S, Beule D, Kirwan JA, Forslund SK, Wilck N, Bartolomaeus H, Heimesaat MM, Opitz B (2024) Clinically used broad-spectrum antibiotics compromise inflammatory monocyte-dependent antibacterial defense in the lung. Nat Commun 15(1):2788. https://doi.org/10.1038/s41467-024-47149-z
67. Rowland SN, Green CG, Halliwill JR, Singanayagam A, Heaney LM (2025) Gut feelings on short-chain fatty acids to regulate respiratory health. Trends Endocrinol Metab. https://doi.org/10.1016/j.tem.2024.12.007
68. Brown JA, Sanidad KZ, Lucotti S, Lieber CM, Cox RM, Ananthanarayanan A, Basu S, Chen J, Shan M, Amir M, Schmidt F,

Weisblum Y, Cioffi M, Li T, Rowdo FM, Martin ML, Guo CJ, Lyssiotis C, Layden BT, Dannenberg AJ, Bieniasz PD, Lee B, Inohara N, Matei I, Plemper RK, Zeng MY (2022) Gut microbiota-derived metabolites confer protection against SARS-CoV-2 infection. Gut Microbes 14(1):2105609. https://doi.org/10.1080/19490976.2022.2105609
69. Heumel S, de Rezende RV, Urien C, Specque F, Brito Rodrigues P, Robil C, Delval L, Sencio V, Descat A, Deruyter L, Ferreira S, Gomes Machado M, Barthelemy A, Angulo FS, Haas JT, Goosens JF, Wolowczuk I, Grangette C, Rouille Y, Grimaud G, Lenski M, Hennart B, Ramirez Vinolo MA, Trottein F (2024) Shotgun metagenomics and systemic targeted metabolomics highlight indole-3-propionic acid as a protective gut microbial metabolite against influenza infection. Gut Microbes 16(1): 2325067. https://doi.org/10.1080/19490976.2024.2325067
70. Sencio V, Barthelemy A, Tavares LP, Machado MG, Soulard D, Cuinat C, Queiroz-Junior CM, Noordine ML, Salome-Desnoulez S, Deryuter L, Foligne B, Wahl C, Frisch B, Vieira AT, Paget C, Milligan G, Ulven T, Wolowczuk I, Faveeuw C, Le Goffic R, Thomas M, Ferreira S, Teixeira MM, Trottein F (2020) Gut dysbiosis during influenza contributes to pulmonary pneumococcal superinfection through altered short-chain fatty acid production. Cell Rep 30(9):2934–2947 e2936. https://doi.org/10.1016/j.celrep.2020.02.013
71. Peng LY, Shi HT, Tan YR, Shen SY, Yi PF, Shen HQ, Fu BD (2021) Baicalin inhibits APEC-induced lung injury by regulating gut microbiota and SCFA production. Food Funct 12(24):12621–12633. https://doi.org/10.1039/d1fo02407h
72. Jang YO, Lee SH, Choi JJ, Kim DH, Choi JM, Kang MJ, Oh YM, Park YJ, Shin Y, Lee SW (2020) Fecal microbial transplantation and a high fiber diet attenuates emphysema development by suppressing inflammation and apoptosis. Exp Mol Med 52(7): 1128–1139. https://doi.org/10.1038/s12276-020-0469-y
73. Sencio V, Gallerand A, Gomes Machado M, Deruyter L, Heumel S, Soulard D, Barthelemy J, Cuinat C, Vieira AT, Barthelemy A, Tavares LP, Guinamard R, Ivanov S, Grangette C, Teixeira MM, Foligne B, Wolowczuk I, Le Goffic R, Thomas M, Trottein F (2021) Influenza virus infection impairs the gut's barrier properties and Favors secondary enteric bacterial infection through reduced production of short-chain fatty acids. Infect Immun 89(9): e0073420. https://doi.org/10.1128/IAI.00734-20
74. Haak BW, Littmann ER, Chaubard J-L, Pickard AJ, Fontana E, Adhi F, Gyaltshen Y, Ling L, Morjaria SM, Peled JU, van den Brink MR, Geyer AI, Cross JR, Pamer EG, Taur Y (2018) Impact of gut colonization with butyrate producing microbiota on respiratory viral infection following allo-HCT. Blood. https://doi.org/10.1182/blood-2018-01-828996
75. Chiu CY, Cheng ML, Chiang MH, Kuo YL, Tsai MH, Chiu CC, Lin G, Genuneit J (2019) Gut microbial-derived butyrate is inversely associated with IgE responses to allergens in childhood asthma. Pediatr Allergy Immunol 30(7):689–697. https://doi.org/10.1111/pai.13096
76. Tanabe N, Matsumoto H, Morimoto C, Hirai T (2025) Sputum short-chain fatty acids, microbiome, inflammation, and mucus plugging in obstructive airway disease. J Allergy Clin Immunol 155(5):1675–1680. https://doi.org/10.1016/j.jaci.2025.01.031
77. Ubachs J, Ziemons J, Soons Z, Aarnoutse R, van Dijk DPJ, Penders J, van Helvoort A, Smidt ML, Kruitwagen R, Baade-Corpelijn L, Olde Damink SWM, Rensen SS (2021) Gut microbiota and short-chain fatty acid alterations in cachectic cancer patients. J Cachexia Sarcopenia Muscle 12(6):2007–2021. https://doi.org/10.1002/jcsm.12804
78. Jiang H, Zeng W, Zhang X, Li Y, Wang Y, Peng A, Cao D (2023) Gut microbiota and its metabolites in non-small cell lung cancer and brain metastasis: from alteration to potential microbial markers and drug targets. Front Cell Infect Microbiol 13:1211855. https://doi.org/10.3389/fcimb.2023.1211855
79. Schuijt TJ, Lankelma JM, Scicluna BP, de Sousa e Melo F, Roelofs JJ, de Boer JD, Hoogendijk AJ, de Beer R, de Vos A, Belzer C, de Vos WM, van der Poll T, Wiersinga WJ (2016) The gut microbiota plays a protective role in the host defence against pneumococcal pneumonia. Gut 65(4):575–583. https://doi.org/10.1136/gutjnl-2015-309728
80. Brown RL, Sequeira RP, Clarke TB (2017) The microbiota protects against respiratory infection via GM-CSF signaling. Nat Commun 8(1):1512. https://doi.org/10.1038/s41467-017-01803-x
81. Groves HT, Higham SL, Moffatt MF, Cox MJ, Tregoning JS (2020) Respiratory viral infection alters the gut microbiota by

inducing inappetence. mBio 11(1). https://doi.org/10.1128/mBio.03236-19

82. Zuo T, Zhang F, Lui GCY, Yeoh YK, Li AYL, Zhan H, Wan Y, Chung ACK, Cheung CP, Chen N, Lai CKC, Chen Z, Tso EYK, Fung KSC, Chan V, Ling L, Joynt G, Hui DSC, Chan FKL, Chan PKS, Ng SC (2020) Alterations in gut microbiota of patients with COVID-19 during time of hospitalization. Gastroenterology 159(3):944–955 e948. https://doi.org/10.1053/j.gastro.2020.05.048
83. Zhang F, Lau RI, Liu Q, Su Q, Chan FKL, Ng SC (2023) Gut microbiota in COVID-19: key microbial changes, potential mechanisms and clinical applications. Nat Rev Gastroenterol Hepatol 20(5):323–337. https://doi.org/10.1038/s41575-022-00698-4
84. Narayana JK, Aliberti S, Mac Aogain M, Jaggi TK, Ali N, Ivan FX, Cheng HS, Yip YS, Vos MIG, Low ZS, Lee JXT, Amati F, Gramegna A, Wong SH, Sung JJY, Tan NS, Tsaneva-Atanasova K, Blasi F, Chotirmall SH (2023) Microbial dysregulation of the gut-lung axis in bronchiectasis. Am J Respir Crit Care Med 207(7):908–920. https://doi.org/10.1164/rccm.202205-0893OC
85. Dickson RP (2016) The microbiome and critical illness. Lancet Respir Med 4(1):59–72. https://doi.org/10.1016/S2213-2600(15)00427-0
86. Wang J, Li F, Tian Z (2017) Role of microbiota on lung homeostasis and diseases. Sci China Life Sci 60(12):1407–1415. https://doi.org/10.1007/s11427-017-9151-1
87. Madan JC, Koestler DC, Stanton BA, Davidson L, Moulton LA, Housman ML, Moore JH, Guill MF, Morrison HG, Sogin ML, Hampton TH, Karagas MR, Palumbo PE, Foster JA, Hibberd PL, O'Toole GA (2012) Serial analysis of the gut and respiratory microbiome in cystic fibrosis in infancy: interaction between intestinal and respiratory tracts and impact of nutritional exposures. mBio 3(4). https://doi.org/10.1128/mBio.00251-12
88. Rutten EP, Spruit MA, Franssen FM, Buurman WA, Wouters EF, Lenaerts K (2014) GI symptoms in patients with COPD. Chest 145(6):1437–1438. https://doi.org/10.1378/chest.14-0285
89. Brassard P, Vutcovici M, Ernst P, Patenaude V, Sewitch M, Suissa S, Bitton A (2015) Increased incidence of inflammatory bowel disease in Quebec residents with airway diseases. Eur Respir J 45(4):962–968. https://doi.org/10.1183/09031936.00079414
90. Keely S, Hansbro PM (2014) Lung-gut cross talk: a potential mechanism for intestinal dysfunction in patients with COPD. Chest 145(2):199–200. https://doi.org/10.1378/chest.13-2077
91. Lai HC, Lin TL, Chen TW, Kuo YL, Chang CJ, Wu TR, Shu CC, Tsai YH, Swift S, Lu CC (2022) Gut microbiota modulates COPD pathogenesis: role of anti-inflammatory Parabacteroides goldsteinii lipopolysaccharide. Gut 71(2):309–321. https://doi.org/10.1136/gutjnl-2020-322599
92. Bisgaard H, Li N, Bonnelykke K, Chawes BL, Skov T, Paludan-Muller G, Stokholm J, Smith B, Krogfelt KA (2011) Reduced diversity of the intestinal microbiota during infancy is associated with increased risk of allergic disease at school age. J Allergy Clin Immunol 128(3):646–652 e641–645. https://doi.org/10.1016/j.jaci.2011.04.060
93. Saint-Criq V, Lugo-Villarino G, Thomas M (2021) Dysbiosis, malnutrition and enhanced gut-lung axis contribute to age-related respiratory diseases. Ageing Res Rev 66:101235. https://doi.org/10.1016/j.arr.2020.101235
94. Cho SJ, Stout-Delgado HW (2020) Aging and lung disease. Annu Rev Physiol 82:433–459. https://doi.org/10.1146/annurev-physiol-021119-034610
95. Krone CL, Biesbroek G, Trzcinski K, Sanders EA, Bogaert D (2014) Respiratory microbiota dynamics following Streptococcus pneumoniae acquisition in young and elderly mice. Infect Immun 82(4):1725–1731. https://doi.org/10.1128/IAI.01290-13
96. Thevaranjan N, Whelan FJ, Puchta A, Ashu E, Rossi L, Surette MG, Bowdish DM (2016) Streptococcus pneumoniae colonization disrupts the microbial community within the upper respiratory tract of aging mice. Infect Immun 84(4):906–916. https://doi.org/10.1128/IAI.01275-15
97. Aguilera JM (2019) The food matrix: implications in processing, nutrition and health. Crit Rev Food Sci Nutr 59(22):3612–3629. https://doi.org/10.1080/10408398.2018.1502743
98. Derrien M, Veiga P (2017) Rethinking diet to aid human-microbe symbiosis. Trends Microbiol 25(2):100–112. https://doi.org/10.1016/j.tim.2016.09.011
99. Poh CH, Navarro-Rodriguez T, Fass R (2010) Review: treatment of gastroesophageal reflux disease in the elderly. Am J Med 123(6):496–501. https://doi.org/10.1016/j.amjmed.2009.07.036

100. Su Q, Liu Q, Lau RI, Zhang J, Xu Z, Yeoh YK, Leung TWH, Tang W, Zhang L, Liang JQY, Yau YK, Zheng J, Liu C, Zhang M, Cheung CP, Ching JYL, Tun HM, Yu J, Chan FKL, Ng SC (2022) Faecal microbiome-based machine learning for multi-class disease diagnosis. Nat Commun 13(1):6818. https://doi.org/10.1038/s41467-022-34405-3

Chapter 9

Gut Microbial Metabolism in Colorectal Cancer

Yaowei Qin, Zhen Feng, Zhonghua Cheng, Danfeng Wen, Jiufeng Li, and Zuoyun Wang

Abstract

The human gut is colonized by a diverse array of microorganisms, including bacteria, fungi, viruses, and archaea. They produce various metabolites such as bile acids, tryptophan derivatives, fatty acids, and vitamins, which play crucial roles in maintaining gut homeostasis and are also associated with a variety of intestinal diseases, including colorectal cancer (CRC). CRC is one of the most common malignant tumors and the second leading cause of cancer-related death worldwide, with a high tendency to metastasize to distant organs such as the liver and lungs. Extensive research has demonstrated that gut microbiota and their metabolites influence colorectal carcinogenesis through multiple mechanisms. This review summarizes the mechanisms underlying the association between gut microbial metabolism and CRC and outlines the clinical translational prospects of gut microbiota and their metabolites, including their roles as biomarkers and therapeutic agents. Additionally, the gaps in current research and potential directions for future studies are identified and discussed.

Key words Colorectal cancer, Gut microbial metabolism, Fecal immunochemical test, Cancer-associated fibroblasts, Intestinal microbiota

1 Introduction

Colorectal cancer (CRC) is one of the most common malignancies and the second leading cause of cancer-related deaths worldwide [1], largely due to its tendency to metastasize to distant organs such as the liver and lungs, with a significant potential for peritoneal carcinomatosis and other metastatic sites including bones and the brain[2]. According to recent estimates, approximately 154, 270 people in the United States will be diagnosed with colorectal cancer, and about 52, 900 will die from the disease in 2025 [3]. Given this substantial disease burden, early screening and diagnosis of CRC are critical for improving treatment outcomes. Currently,

Yaowei Qin, Zhonghua Cheng and Danfeng Wen contributed equally with all other contributors.

Tao Huang (ed.), *Microbial 'Omics: Methods and Protocols*, Methods in Molecular Biology, vol. 3033, https://doi.org/10.1007/978-1-0716-5264-0_9, the most widely used CRC screening method globally is the fecal immunochemical test (FIT). Colonoscopy remains the most common screening modality in the United States and is also frequently employed in several other regions including parts of Germany and Poland. Although a one-time FIT is less sensitive than colonoscopy, colonoscopy invasive procedure leads to relatively lower screening compliance [4].

A recent study demonstrated that *Campylobacter jejuni*-derived cytolethal distending toxin (CDT) promotes CRC metastasis via the JAK2-STAT3-MMP9 pathway in mouse models of liver or lung metastasis. These findings were further validated in human colon tissues infected with *C. jejuni* and in patient-derived colonoid tumors exposed to CDT [5]. Numerous studies have demonstrated that microorganisms within and outside tumors exert significant impacts on both the immune system and immunotherapy [6]. These evidences underscore a key dimension of CRC pathogenesis: the intimate involvement of gut microbiota as key components of the tumor microenvironment. Over the past decade, research has shown that ecological disruptions among gut bacteria, fungi, viruses, and archaea are consistently associated with colorectal tumorigenesis and may contribute to its pathogenesis.

Mechanistic studies have indicated that gut microbiota influence colorectal cancer development by releasing various metabolites, proteins, and macromolecules that interact with host colonic epithelial and immune cells. For example, introducing *Fusobacterium nucleatum* or *Bacteroides fragilis* into *Apc*$^{\mathrm{Min}/+}$ mice accelerates intestinal tumorigenesis. *Fusobacterium nucleatum* has also been found to interact with cancer-associated fibroblasts (CAFs), inducing a pro-inflammatory CAF (iCAF) phenotype. Experiments show that cancer cells exposed to these iCAFs display enhanced invasive capacity [7]. Additionally, it's reported that *pks*$^{+}$ *E. coli*, *Peptostreptococcus anaerobius*, *Clostridium* spp., and *Bacteroides fragilis* promote colon tumorigenesis through colibactin production [8, 9], PI3K-AKT pathway activation [10], secondary bile acid production [11, 12], and *Bacteroides fragilis* toxin production, respectively [13]. Conversely, under conditions of mild colonic inflammation, the gut microbiome has also been shown to exert protective effects by suppressing the expression of the long noncoding RNA (lncRNA)Snhg9, thereby inhibiting tumor development in mice [7].

Therefore, we summarize the current understanding of how gut microbiota and their key metabolic processes influence CRC, including bile acid metabolism, tryptophan metabolism, fatty acid metabolism, urea cycle, and vitamin synthesis. We also summarize research on gut microbiota and their metabolites as diagnostic biomarkers, as well as the potential of developing clinical therapeutic strategies leveraging gut microbiota. Additionally, we summarize the current research focuses, analyze the core mechanisms and biomarkers, and identify the challenges, research gaps, and future development directions in this field. To present the relevant content more clearly, the figures in this paper were created using BioGDP.com [14].

2 An Overview of Gut Microbiota

The colonization of the human gut with microbes begins immediately at birth [15]. The human gut microbiota is an ecosystem populated by bacteria, fungi, viruses, archaea, and parasites, which collectively typically outnumber human cells [16]. These microorganisms play fundamental roles in fermenting indigestible carbohydrates, promoting nutrient absorption, synthesizing vitamins, and modulating the host's immune system [17], the reproductive system [18], and even the nervous system [19]. Through these metabolisms, gut microbiota can exert a significant impact on intestinal homeostasis and intestinal diseases. Here we provide a concise introduction to gut microbes.

2.1 Gut Bacteria: Composition, Metabolites, and Impacts on Gut Health

The gut bacterial colonies are predominantly composed of strict anaerobes, with minor populations of facultative anaerobes and aerobes [20, 21]. Although over 50 bacterial phyla have been identified, the human gut microbiota is dominated by only two of them: Bacteroidetes, like *Bacteroides* and *Prevotella*, and Firmicutes, like *Lactobacillus*, *Bifidobacterium*, and *Clostridium*. Other phyla, such as Proteobacteria, Verrucomicrobia, Actinobacteria, Fusobacteria, and Cyanobacteria, are relatively minor [15]. Though the numbers of these bacteria differ, they all have significant impacts on gut health. Certain species directly modulate physiological and pathological processes. For example, *Hhep* colonization promotes the expansion of *Hhep*-specific $CD4^{+}$ T follicular helper cells, thereby exerting antitumor effects [22]. Conversely, *Fusobacterium nucleatum*, a bacterium of the *Fusobacterium* genus, accelerates colorectal tumorigenesis by driving the over-proliferation and transformation of $LY6A^{+}$ regenerative stem cells (RSCs) [23].

Gut bacteria also affect host physiology through the production of metabolites. N-acetylmuramic acid (NAM), a core component of bacterial peptidoglycan, is depleted in CRC patients and appears to protect against tumor development. Studies indicate that NAM directly binds to protein kinase B (AKT1), curbing cancer cell proliferation and inducing cell death during intestinal tumorigenesis. It also prevents intestinal tumorigenesis in various CRC mouse models and patient-derived organoids [24]. 4-hydroxyphenylacetic acid, a gut microbial metabolite primarily from bacteria, stimulates CRC cells to secrete CXCL3, which mediates MDSC recruitment via the CXCL3/CXCR2 axis, leading to T-cell suppression and CRC progression [25].

Additionally, certain specific metabolites are produced by some particular bacteria. Reuterin, a metabolite of *Lactobacillus reuteri* (a gut probiotic), is beneficial to gut health [26]. Colibactin, a metabolite produced by *E. coli*, has been shown to promote colorectal carcinogenesis via mechanisms such as inducing DNA breaks and senescence [8, 27–29]. Indolamines, a substance produced by *M. morganii*, have been demonstrated to promote the development of colorectal cancer (CRC)[30]. *Bacteroides fragilis* facilitates CRC progression through the production of enterotoxigenic factors [31]. In summary, intestinal bacteria and their metabolites play a crucial role in gut health and are closely linked to colorectal cancer.

2.2 Gut Fungi: Composition and Impacts on Intestinal Health Via Metabolites and Immunity

The most frequently reported gut fungal groups include *Candida*, *Saccharomyces*, and *Malassezia* [32]. Alterations in fungal microbiome composition and function have been associated with various diseases [33–36]. In inflammatory bowel disease (IBD), fungal dysbiosis is characterized by an elevated Basidiomycota/Ascomycota ratio, reduced abundance of *Saccharomyces cerevisiae*, and an increased proportion of *Candida albicans* compared to healthy controls [33]. The metabolic activities of fungi also exert a significant impact on intestinal health. *Candida* species can produce carcinogenic metabolites, such as acetaldehyde [37, 38] and N-nitrobenzylmethylamine [39]. Furthermore, fungi engage in close interactions with bacteria. In Crohn's disease patients, *Candida tropicalis*, *Escherichia coli*, and *Serratia marcescens* have been found to co-associate and potentially form synergistic biofilms, which may contribute to inflammatory responses [32].

Gut fungi also have important connections with the immune system. A study has shown that mucosa-associated fungi (MAF) enhance intestinal epithelial function and protect mice against intestinal injury and bacterial infection. Notably, intestinal colonization with a defined consortium of mucosa-associated fungi was also shown to promote social behavior in mice. The protective effects of these fungi on the gut barrier function were found to depend on IL-22 production by $CD4^+$ T helper cells, whereas their behavioral effects were mediated through IL-17R-dependent signaling in neurons [40]. In summary, gut fungal metabolism and its interactions with bacteria jointly influence host gut health and are associated with various diseases.

2.3 Other Gut Microorganisms: Archaea, Viruses, and Parasites

Archaea, viruses, and parasites present in the human gut also play crucial roles in intestinal health. Phyla of Archaea, including Methanobacteriota, Thermoplasmatota, and Halobacteriota, are the three dominant taxa. Methanobacteriales and Methanomassiliicoccales are two major archaeal orders. At the species level, *Methanobrevibacter_A smithii* is the most abundant archaeal species, followed by *Methanobrevibacter_A sp900766745* and *Methanobrevibacter_A smithii_A* [41]. A recent study has shown that extracellular vesicles from human gut-derived methanogenic archaea are enriched in adhesin-like proteins. These proteins can be taken up by macrophages and induce responses in immune cells and epithelial cells [42], thus influencing gut health.

Recently, the role of viruses in the gut has sparked growing interest. Viruses exist in the gut at a tenfold higher abundance than prokaryotes. Gut viruses interact with bacteria, like bacteriophages, and host cells, like eukaryotic-targeting viruses, thereby influencing intestinal homeostasis [43]. Studies have shown that Crohn's disease (CD) and ulcerative colitis (UC) are associated with a significant increase of Caudovirales bacteriophages [44]. And an increasing number of studies have focused on using phage therapy for gut diseases [45–47], which shows clinical translation potential [48]. However, the relationship between gut viruses and colorectal cancer (CRC) requires further research. Finally, there may be parasites in the gut, such as roundworms, tapeworms, schistosomes [49], and cryptosporidia [50]. Parasites can disrupt gut homeostasis and can lead to significant morbidity [50, 51].

3 Gut Microbial Metabolism and Its Mechanistic Links to CRC

As previously described, the metabolites of gut microbiota can influence CRC through various mechanisms, which is a major focus in this field. This section systematically summarizes the mechanistic links between gut microbial metabolism and CRC, focusing on five core metabolic axes. In each section, we introduce the gut microbes involved in these metabolic processes (Table 1), describe the mechanisms by which gut microbial metabolites affect CRC, and mention other factors that influence these metabolic processes, such as diet and medications.

3.1 Gut Microbiota-Bile Acid Axis: Metabolic Pathways, Regulatory Factors, and CRC Implications

Bile acids are key regulators of the pro-inflammatory/anti-inflammatory balance. Bile acid metabolism plays crucial roles in inflammatory diseases and neoplastic diseases [67]. Notably, their metabolic products—secondary bile acids, which are mainly generated by gut microbiota-mediated transformation—exert a particularly important effect on CRC, including regulating CRC cell proliferation, apoptosis, and the tumor microenvironment. We first introduce the basic metabolic process of bile acids and the formation process of secondary bile acids, focusing on DCA and LCA (Fig. 1).

Intestinal microorganisms play crucial roles in the production of secondary bile acids. Strains of *Clostridium* cluster XIVa, like *Clostridium hylemonae*, can express bile acid deconjugating enzymes, such as BSH, to deconjugate CA. Strains of *Clostridium* cluster XIVa further convert free CA to DCA via the 7-α-dehydroxylation, serving as important auxiliary genera for intestinal DCA synthesis. Members of the genus *Lachnoclostridium*, like *Lachnoclostridium scindens*, efficiently convert free CA into DCA

Table 1
Overview of intestinal microbiota and related metabolites

Family	Genus	Species	Related metabolites
Bifidobacteriaceae	*Bifidobacterium*		NH_3, CO_2[52], VB [53]
Carnobacteriaceae	*Carnobacterium*	*Carnobacterium maltaromaticum*	VD [54]
Clostridiaceae	*Clostridium*	*Clostridium hylemonae* *Clostridium sporogenes* *Clostridium butyricum*	DCA [12] IPA [55] Butyrate [56]
Eggerthellaceae	*Eggerthella*	*Eggerthella lenta*	TDCA [57]
Eubacteriaceae	*Eubacterium*	*Eubacterium limosum*	LCA [12]
Lachnospiraceae			butyrate, VD[54]
Lachnospiraceae	*Faecalibacterium* *Lachnoclostridium* *Roseburia* *Ruminococcus*	*Faecalibacterium prausnitzii* *Lachnoclostridium scindens* *Roseburia intestinalis* *Ruminococcus gnavus*	butyrate[58], VD [54] DCA, LCA [12] butyrate [58] free CDCA [12], isoDCA [59]
Lactobacillaceae	*Lactobacillus*		ILA [60], VB [53]
Lactobacillaceae	*Lactobacillus*	*Lactobacillus reuteri* *Lactobacillus johnsonii* *Lactobacillus intestinalis*	IAA [61], I3A [62], Reuterin [26], NH_3, CO_2 [63] IPA [55] atRA, 13-cisRA [64]
Peptostreptococcaceae	*Peptostreptococcus*	*Peptostreptococcus anaerobius*	IDA [65]
Prevotellaceae	*Prevotella*	*Prevotella copri*	AA [66]

through 7α-dehydroxylation. The genus *Ruminococcus*, like *Ruminococcus gnavus*, can first deconjugate conjugated CDCA to produce free CDCA. Additionally, *Lachnoclostridium scindens* and *Eubacterium limosum* are capable of acting on free CDCA to convert it into LCA [12].

Research has shown that DCA inhibits Ca^{2+}-NFAT2 signaling in a PMCA-dependent manner, thus suppressing activation, proliferation, effector function, and cytotoxicity of $CD8^+$ T cells. DCA also affects $CD4^+$T cell similarly [70]. Besides, DCA can promote ferroptosis by upregulating hypoxia-inducible factor-2α (HIF2α) and divalent metal transporter-1 (DMT1), leading to the accumulation of ferrous iron. Conversely, CA, CDCA, and DCA can activate the farnesoid X receptor (FXR), upregulating GPX4 to resist ferroptosis [64]. This is similar to the way astaxanthin inhibits ferroptosis, both by upregulating GPX4[71]. Both DCA and LCA induce hyperproliferation, DNA damage, and oxidative stress.

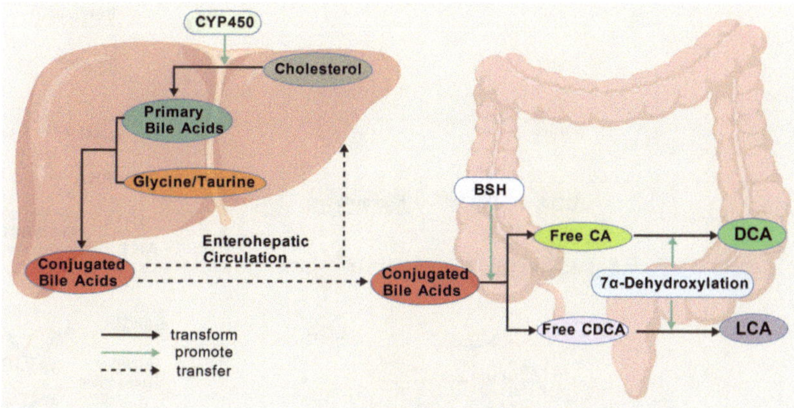

Fig. 1 The synthesis of bile acids and the formation process of secondary bile acids (DCA and LCA). Bile acids are synthesized in hepatocytes through cytochrome P450-mediated cholesterol oxidation [68]. In humans, primary bile acids are conjugated with glycine or taurine, forming bile salts that are stored in the gallbladder. Upon food intake, bile acids are released into the duodenum to facilitate the digestion and absorption of dietary lipids [69]. The majority of bile acids are reabsorbed in the terminal ileum and return to the liver via the enterohepatic circulation. The remainder enters the colon to participate in subsequent transformation. The conversion of primary to secondary bile acids involves two key microbial steps. Cholic acid (CA) firstly is deconjugated to its free form by bile salt hydrolase (BSH) and then undergoes 7α-dehydroxylation to form DCA. Similarly, chenodeoxycholic acid (CDCA) is converted into LCA through the same sequence

DCA and LCA can increase cyclin D1 expression, which is involved in cell cycle progression, via β-catenin signaling. Besides, DCA and LCA degrade the tumor suppressor p53 and enhance resistance to apoptosis [72] (Figs. 2a and 3a).

Additionally, it has been reported that the abundance of *E. lenta* in the intestine is positively correlated with taurodeoxycholic acid (TDCA). Excess TDCA can damage intestinal barrier integrity and also activate oncogenic MAPK/ERK signaling in colonic epithelial cells, thereby promoting the development of CRC. And cigarette smoke can increase TDCA and *E. lenta* [57]. Besides, the colonization of *Streptococcus gallolyticus subsp. gallolyticus* (SGG) is closely associated with CRC [74, 75]. Researches have shown that bile acids can enhance the activity of gallocin (a bacteriocin produced by SGG), thereby promoting the colonization of SGG [76]. The secondary bile acid 3β-hydroxydeoxycholic acid (isoDCA), produced by *Ruminococcus gnavus*, acts on dendritic cells (DCs) to reduce their immunostimulatory properties, thereby enhancing the induction of Foxp3 and subsequently regulating regulatory T (Treg) cells [59, 77].

Many factors can influence gut microbiota composition and bile acid metabolism, especially high-fat diet (HFD), which increases hepatic bile acid synthesis, resulting in greater colonic delivery and microbial conversion to secondary bile acids. These acids further reshape gut microbiota composition, creating a feed-forward cycle [72]. HFD enriches pathogenic *Alistipes* while

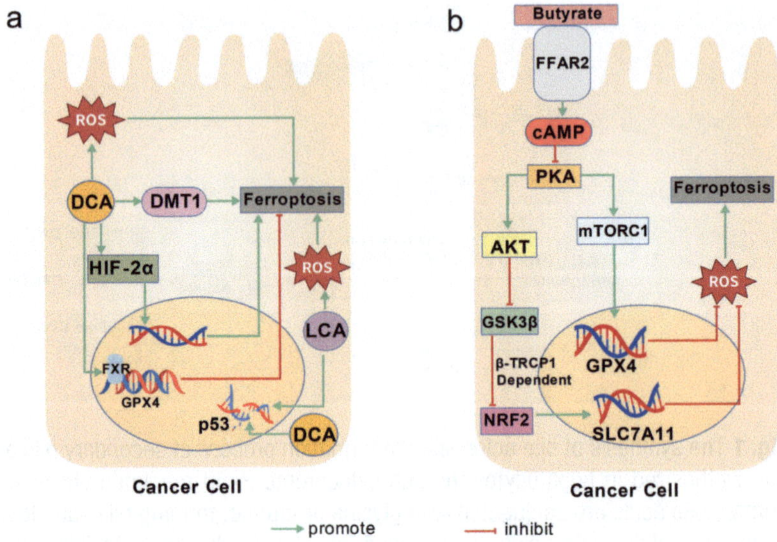

Fig. 2 The effects of DCA, LCA, and butyrate on cancer cells. Ferroptosis serves as a vital pathway to influence CRC. DCA, LCA, and butyrate can all increase ROS to promote ferroptosis. Additionally, DCA further facilitates ferroptosis by upregulating DMT1 and HIF-2α. Interestingly, DCA upregulates GPX4 to inhibit ferroptosis, whereas butyrate downregulates GPX4 and SLC7A11 to promote ferroptosis. Specifically, by binding to the FFAR2, butyrate can enhance the cyclic adenosine monophosphate (cAMP) in cells, thereby inhibiting the activity of protein kinase A (PKA). Subsequently, through AKT-GSK3β-β-TRCP1-NRF2 pathway and mammalian target of rapamycin complex 1 (mTORC1), butyrate downregulates the levels of SLC7A11 and GPX4, respectively. This leads to the accumulation of ROS, promoting ferroptosis [73]. Beyond the ferroptosis pathway, DCA and LCA also degrade the p53 gene to drive CRC progression

depleting *Parabacteroides distasonis*, causing damage to the intestinal epithelial barrier [78]. HFD also increases nonclassical amino acid conjugation of cholic acid (AA-CA), which affect stem cells growth. *Erysipelatoclostridium valens* and *Ruminococcus gnavus* are capable of synthesizing these AA-CA molecules [79]. *Lactobacillus johnsonii* increases CDCA in the intestine to suppress CRC induced by a high-fat diet [80]. Additionally, fibroblast growth factor 15 (FGF15) can activate the Hippo pathway to inhibit bile acid metabolism [81]. The Hippo pathway also influences the production of ROS [82].

3.2 Gut Microbiota-Mediated Tryptophan Metabolism: From Metabolite Synthesis to Regulatory Roles in CRC

Tryptophan (Trp) is the amino acid with the largest molecular weight among the 20 standard amino acids. It is also the rarest one and exhibits evolutionary conservation [83], and Trp serves as a precursor for a variety of metabolites [84]. Gut microbiota directly transform tryptophan into indole and its derivatives. It has been reported that *Lactobacillus* can metabolize tryptophan into indole-3-lactic acid (ILA), which enhances the expression of

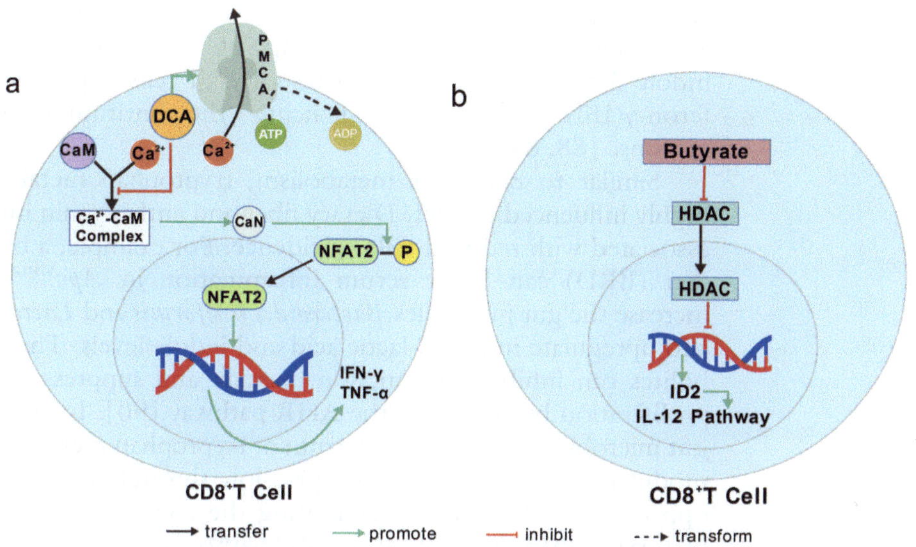

Fig. 3 The effects of DCA and butyrate on CD8$^+$ T cell. DCA and butyrate exert opposite effects on the activation of CD8$^+$ T cells. (**a**) By activating PMCA, DCA promotes the transport of Ca^{2+} out of the cell. This leads to a decrease in intracellular Ca^{2+} concentration, which in turn inhibits the formation of the Ca^{2+}-calmodulin (CaM) complex. Consequently, the activation of calcineurin (CaN) and the dephosphorylation of nuclear factor of activated T cells 2 (NFAT2) are suppressed, preventing NFAT2 from promoting the expression of relevant genes to activate CD8$^+$ T cells and finally inhibiting immune function. (**b**) As a HDAC inhibitor, butyrate regulates and enhances the expression of inhibitor of DNA binding 2 (ID2). This upregulation further activates the interleukin-12 (IL-12) pathway, thereby promoting the activation of CD8$^+$ T cells

key bacterial enzymes involved in tryptophan metabolism, thereby promoting the synthesis of other indole derivatives, including indole-3-propionic acid (IPA), indole-3-acetic acid (IAA), and indoleacrylic acid [60, 85]. *Lactobacillus reuteri* can also participate in tryptophan metabolism to synthesize indole-3-acetic acid (IAA) [61] and indole-3-aldehyde (I3A) [62]. It has been reported that *L. johnsonii* and *C. sporogenes* synergistically produce IPA [55]. IAA, IPA, and indoleacrylic acid are all ligands of aryl hydrocarbon receptor (AhR) [86, 87].

These tryptophan-derived metabolites exert diverse immunomodulatory and tumor-regulatory effects. IPA activates Tpex cells and sustains CD8$^+$ T-cell responses by modifying H3K27 acetylation in the super-enhancer region of the Tcf7 gene, thereby enhancing responsiveness to immune checkpoint blockade. Notably, IPA also augments the ICB response in pan-cancer and CRC-derived organoids [55]. In contrast, *P. anaerobius*-derived tryptophan metabolite trans-3-indolepropionic acid (IDA) can promote CRC development. Mechanistically, IDA, as an endogenous ligand of the aryl hydrocarbon receptor (AHR), transcriptionally upregulates ALDH1A3. ALDH1A3 uses retinaldehyde to produce NADH, which facilitates FSP1-mediated CoQ10 reduction, thereby inhibiting ferroptosis and promoting CRC progression

[65]. Tryptophan metabolites (such as L-KYN, 3-HK, and 3-HA) also promote ferroptosis by inhibiting SLC7A11 [64]. Additionally, indole-3-aldehyde (I3A) stimulates $CD8^+$ T cells to produce interferon-γ(IFN-γ), thereby enhancing the antitumor immune response [88, 89].

Similar to bile acids metabolism, tryptophan metabolism is highly influenced by diet. Dietary fiber and anthocyanin intake are associated with reduced CRC incidence. For example, a black rice diet (BRD) can lower serum inflammation in *Apc*$^{Min/+}$ mice, increase the gut probiotics *Bacteroides uniformis* and *Lactobacillus*, and upregulate indole-3-lactic acid and indole levels. These metabolites can inhibit cell junction damage and suppress CRC cell proliferation by activating the AHR pathway [90]. In conclusion, gut microbiota generate a spectrum of tryptophan metabolites that modulate colorectal cancer progression through multiple, often opposing, mechanisms—highlighting the complex role of microbial tryptophan metabolism in CRC pathogenesis.

3.3 Roles of Gut Microbiota-Mediated Fatty Acid Metabolism and Related Metabolites in CRC

Fatty acids are significant for energy metabolism [91]. Gut microbes actively participate in fatty acid metabolism, generating various metabolites with significant implications for CRC. For example, *Prevotella copri* can metabolize precursors of arachidonic acid (AA), such as dietary linoleic acid (LA), into arachidonic acid (AA) [66]. AA is an essential unsaturated fatty acid in CRC. Delta-5 desaturase (FADS1) promotes AA synthesis, enriching gram-negative microbes, like *E. coli.* These microbes activate the TLR4/MYD88 pathway, facilitating AA metabolism to PGE2 and promoting CRC progression [92].

Butyrate represents another crucial microbial metabolite, produced by species including *Clostridium butyricum* [56], *Faecalibacterium prausnitzii*, and *Roseburia intestinalis* [58]. Approximately half of butyrate-producing isolates employ the butyryl coenzyme A-acetyl coenzyme A transferase pathway, consuming acetate to produce butyrate [58]. Butyrate possesses tumor-suppressive properties, which may be attributed to its function as a histone deacetylase (HDAC) inhibitor [93]. Butyrate reduces the growth of normal intestinal stem cells and a series of cancer-derived cell lines [26]. Besides, butyrate can enhance antitumor treatment efficacy by regulating $CD8^+$ T-cell immunity in an ID2-dependent manner [94]. The activation of T cells is also influenced by MST1/2 in the Hippo signaling pathway [95]. Studies have shown that butyric acid can promote ferroptosis by inhibiting the GPX4-SLC7A11 axis and increasing ROS, thereby suppressing the progression of CRC [73]. However, recent evidence suggests that *Porphyromonas* species provoke cellular senescence via butyrate secretion, which is an oncogenic stress response [93] (Figs. 2b and 3b).

Other lipid-related metabolites also influence CRC. Reuterin, a compound produced by *Lactobacillus reuteri* during the anaerobic metabolism of glycerol, exhibits cytotoxicity and growth-inhibiting effects on colon cancer cells [26]. In addition, ketogenic diet (KD) promotes the production of stearic acid by gut microbiota, increasing the level of free stearate in the intestinal lumen, which in turn inhibits the growth of CRC cells [96]. In conclusion, via participating in fatty acid metabolism, gut microbiota produce a variety of metabolites that exert a dual effect on CRC. Among these metabolites, butyrate plays a particularly important role.

3.4 Insights into the Urea Cycle: Its Interaction with Gut Microbiota, Roles in CRC, and Antitumor Applications

The urea cycle stands as a vital metabolic pathway in humans, primarily functioning to convert toxic ammonia—a byproduct of protein catabolism—into urea, which is then excreted via the kidneys. This process is indispensable for preventing ammonia accumulation, as excessive ammonia can cause severe damage to the central nervous system and other organs, thereby safeguarding overall physiological homeostasis [97–99]. In the gut, certain microorganisms can secrete urease, which decomposes urea and inhibits the urea cycle. The most common ones are *Bifidobacterium* [52] and *Lactobacillus reuteri* [63].

Recently, emerging evidence has uncovered a notable link between this cycle and colorectal tumorigenesis: During the development of CRC, the host's urea cycle metabolism is significantly activated. This activation is often accompanied by distinct alterations in the gut microbiota composition—specifically, the depletion of beneficial bacteria with ureolytic activity (such as *Bifidobacterium*, which can break down urea) and a simultaneous overgrowth of pathogenic microbes lacking this ureolytic function, like *Fusobacterium nucleatum*. Notably, the urea produced through this activated cycle exerts pro-tumor effects. It can enter macrophages, a key component of the tumor microenvironment, and interfere with their normal antitumor function. Specifically, urea inhibits the binding efficiency of phosphorylated STAT1 (p-STAT1) to the promoter region of the SAT1 gene, a critical step in regulating immune responses. This interference further drives macrophages toward a pro-tumorigenic phenotype, marked by the accumulation of polyamines molecules that promote cell proliferation and tumor progression [52].

Regarding the urea cycle, pharmaceutical research has identified a potential therapeutic angle: metformin, a commonly used drug for type II diabetes, has been shown to inhibit the urea cycle. This inhibition leads to elevated ammonia concentrations in the tumor microenvironment [100], which in turn suppresses the growth of CRC cell lines such as HCT116 and SW480 [101]. Currently, reports exploring the direct connection between the urea cycle, microbial metabolism, and CRC remain relatively scarce, leaving room for further investigation. However, the urea cycle's relevance to cancer is not limited to CRC; it is associated with the progression of various malignancies and thus holds promising potential as a target for antitumor therapy. Take hepatocellular carcinoma (HCC) as an example: The inhibition or impairment of the urea cycle is a well-documented characteristic of HCC. In preclinical studies, strategies combining arginine deprivation with GCN2 inhibition and senescence-targeted therapy have proven effective in suppressing tumor growth in HCC models with urea cycle deficiency [102].

3.5 The Role of Gut Microbiota in Vitamin Synthesis, Metabolism, and Its Link to CRC

In the human digestive microecosystem, gut microbiota maintains intestinal homeostasis and plays a key role in vitamin synthesis, absorption, and metabolism. Its dysregulation links to diseases like CRC. Beneficial genera such as *Bifidobacteria* and *Lactobacilli* produce B vitamins (VB)[53], which affect CRC cell behavior via energy metabolism and signal transduction. Vitamin B_2 (VB_2) exerts a dual effect on CRC cell ferroptosis: It induces intracellular iron and reactive oxygen species (ROS) accumulation to promote ferroptosis [103] but also reduces ROS and inhibits SLC3A2 (a protein for iron homeostasis) to weaken ferroptosis [104]. Its final role depends on gut microbiota composition, VB_2 concentration, and CRC microenvironment.

Gut microbiota also regulate vitamin A (VA) metabolism. *Lactobacillus intestinalis* converts dietary VA into active atRA and 13-cisRA [64]; antibiotic-induced anaerobe elimination lowers their intestinal levels [105], suggesting this bacterium may regulate CRC cell ferroptosis sensitivity via VA metabolism. Additionally, *Carnobacterium maltaromaticum* (estrogen-dependently colonizing the gut) collaborates with other microbes, such as *F. prausnitzii* and *Lachnospiraceae bacterium*, to synthesize vitamin D [54]. Vitamin D binds to host VDR, activating pathways that regulate cell cycle, inhibit inflammation, and enhance apoptosis—ultimately inhibiting CRC development. These findings reveal a gut microbiota-vitamin D-CRC network, supporting microbiota modulation as a potential strategy for CRC prevention.

4 Clinical Translation of Gut Microbiota in CRC: Biomarkers and Therapeutic Strategies

The gut microbiota, host gene expression, and pathological images exhibit a strong correlation in CRC [106]. Different stages of CRC exhibit distinct gut microbial characteristics, and microbial profiles also vary among different population—this provides a direction for the clinical personalized diagnosis of CRC. Additionally, as mentioned earlier, certain gut microbial metabolites can inhibit CRC, which offers insights into microbial-based therapies for CRC. In this section, we summarize the CRC-related microbial biomarkers identified in current studies and introduce several therapeutic approaches leveraging gut microbial metabolism.

4.1 Signature Roles of Gut Microbiota and Metabolites

CRC typically develops from dysplastic precursor lesions through the accumulation of genetic alterations [107]. Adenoma is the most common precancerous lesion of colorectal cancer. Given the close link between CRC and gut microbiota and their metabolites, the latter show promise as CRC biomarkers for early detection. Numerous studies have identified distinct microbial signatures associated with adenomas and CRC. In adenomas patients, enriched microbiota include *Blautia obeum*, *Butyricicoccus faecihominis*, *Erysipelotrichaceae UCG-003* sp., *Dorea longicatena*, *Flavonifractor plautii*, *Bilophila wadsworthia*, *Bacteroides stercoris*, *Clostridium asparagiforme*, *Clostridium bolteae CAG 59*, and *Roseburia intestinalis* [108–110], while *Fusobacterium* is reduced [110]. In CRC patients, the Methanobacteriota phylum is enriched, and the Methanococcales is depleted [41]. At the genus level, *Peptostreptococcus*, *Parvimonas*, *Fusobacterium*, and *Campylobacter* are increased, while *Roseburia*, *Faecalibacterium*, *Eubacterium*, *Megamonas*, *Dorea*, *Ruminococcus*, and *Coprococcus* are decreased [108–110]. Tubular adenomas (TA) exhibit location-specific TA features: *Flavonifractor plautii* is more abundant in left-sided TAs, *Clostridium* sp. *CAG 58* in right-sided TAs, and *Bilophila wadsworthia* in both proximal and distal Tas [109]. Furthermore, the Wnt/β-catenin and PI3K/mTOR pathways are simultaneously overactivated in the lesions of familial adenomatous polyposis (FAP). In the fecal microbiota of FAP patients, the molecular markers of the Wnt/β-catenin pathway show a positive correlation with *Clostridium strictoides* type 1 and a negative correlation with *Bacteroides* spp. [111].

A series of studies have focused on exploring microbial markers associated with CRC. *Fusobacterium nucleatum* is significantly enriched in CRC and is considered a key marker microbe [26, 108, 112–114]. Moreover, the copy number of the *Fusobacterium* fadA gene is higher in CRC patients than in healthy individuals, suggesting it may serve as a marker for CRC risk [115]. In addition to *Fusobacterium nucleatum*, other microbes positively correlated with CRC include *Parvimonas_unclassified*, *Peptostreptococcus stomatis* [108, 116], *Selenomonas moorei*, *Prevotella asaccharolytica*, *Prevotella micra*, and *Prevotella stomatis* [117]. On the other hand, microbes negatively correlated with CRC are *Roseburia intestinalis* and *Faecalibacterium prausnitzii* [108], and *L. reuteri* is significantly reduced across all CRC stages [26]. At the phylum level, a general trend in CRC patients is a decrease in Bacteroidetes and an increase in Firmicutes [118]. A study also highlighted that at a meta-analysis FDR of 0.005, *Fusobacterium*, *Porphyromonas*, *Parvimonas*, *Peptostreptococcus*, *Gemella*, *Prevotella*, and *Solobacterium* have the potential to become markers [119].

Furthermore, regarding CRC-related microbial markers, Kong Chen et al. found that in LO-CRC, *Fusobacterium nucleatum* enrichment and short-chain fatty acid consumption are observed, including reduced microbial GABA biosynthesis and a shift in acetate/ethyl aldehyde metabolism toward acetyl-CoA production. In contrast, the multi-omics characteristics of EO-CRC are often associated with *Flavonifractor plautii* enrichment and increased tryptophan, bile acid, and choline metabolism [112]. Similarly, compared to old-onset CRC, young-onset CRC (yCRC) patients exhibit increased gut microbiota diversity. Metagenomic sequencing shows that genus *Streptococcus* remains a key microbe in old-onset CRC, while genus *Flavonifractor* dominates in yCRC [120]. Li Jinming et al. performed metagenomic sequencing and metabolomics analysis on 522 fecal samples and found differences in species and metabolites among CRC patients with different BMI levels. They identified features specific to obesity CRC, including enrichment of *Eubacterium sulci* and *Streptococcus oralis*, and depletion of *Clostridium butyricum* [118]. Additionally, a study has also shown that in patients with CRC, there is an enrichment of oleic acid accompanied by a depletion of allocholic acid [121]. In conclusion, while fully consistent results remain difficult to achieve in current research, microorganisms, as a noninvasive tool, hold great clinical significance for the diagnosis of CRC. The indicator role of microbiota is not limited to CRC. A study has shown that there are significant differences in the composition of oral microbiota between people with dental caries and those without. And oral microorganisms such as *Staphylococcus* and *Paludibacteraceae* F0058 may partially mediate the association between a high-sugar diet and dental caries [122]. *Neisseria oralis* and *Campylobacter gracilis* may increase the risk of CRC progression, while *Treponema medium* exerts a protective effect [123]. In conclusion, microorganisms hold great potential in the diagnosis of digestive tract-related diseases.

4.2 Treatment Methodologies

Owing to the close relationship between gut microbial metabolism and colorectal cancer, as well as the fact that these metabolites are linked to diet [124–126], many therapeutic ideas have emerged. For example, reducing dietary fat to control bile acid levels can prevent CRC. Additionally, NDGA and nervonic acid are depleted under HFD conditions, suggesting NDGA and nervonic acid may be potential supplements for obesity and CRC prevention [78]. Moreover, adopting an appropriate amount of black rice can prevent and treat CRC by improving gut microbiota and metabolites [90]. And studies have also shown that the proper supplementation of butyrate can enhance the efficacy of oxaliplatin [94]. The Mediterranean diet (MD), rich in fruits, vegetables, whole grains, nuts, and olive oil, can increase the proportion of intestinal Bacteroidetes and fecal short-chain fatty acid levels, improving the composition and diversity of the gut microbiota and reducing systemic inflammation [18].

Apart from dietary approaches, we can also use microbial metabolites to enhance immunotherapy sensitivity and effectiveness [55]. Also, targeting the IDA—AHR-ALDH1A3 axis to regulate ferroptosis for CRC treatment—is a possibility [65]. Besides, based on the close link between gut microbiota and the urea cycle, treating mouse models with urea cycle inhibitors or *Bifidobacterium*-based supplements can reduce urea-mediated tumorigenesis [52], providing a unique strategy for CRC treatment. Transplanting fecal microbiota from healthy individuals into patients to alleviate disease symptoms has been proven quite effective [127]. Therefore, many studies have explored leveraging gut probiotics and their metabolites to aid in CRC treatment. Notably, studies on *Lactobacillus reuteri* have shown promise in this regard. For instance, Xu Haiting et al. developed a nano-armed bacterium named LR-S-CD/CpG@LNP, which features optical responsiveness, immune-stimulating activity, and the ability to modulate gut microbiome metabolites. This engineered bacterium can regulate tryptophan metabolism to enhance immune responses [128]. Another study demonstrates that a bacteriaboat constructed using *Lactobacillus reuteri* as a carrier effectively enhances the therapeutic efficacy of 5-FU [129]. Moreover, studies have revealed that reuterin can activate macrophages to inhibit the progression of CRC [130].

5 Discussion

The above content presents the relevant findings of research on microbial metabolism in CRC from aspects such as molecular mechanisms and clinical applications. In this section, we analyze the current research focuses and promising clinical translation directions, summarize the challenges and unresolved issues faced by current studies on gut microbial metabolism in CRC, and identify the future development directions of this field through this approach.

5.1 Current Focus, Key Mechanisms, and Clinical Prospects in Gut Microbiota in CRC Research

In the field of gut microbiota research, current studies mainly focus on gut bacteria, fungi, and viruses, while investigations into archaea and their associations with other microorganisms remain limited. In research on the regulatory mechanisms of microbial metabolism in colorectal cancer (CRC), we have identified that the ferroptosis pathway and the regulation of host immune function are key targets through which microbiota influence CRC progression. Furthermore, the effects of microbial metabolites on CRC typically exhibit a dual role—the same metabolite can usually both promote and inhibit CRC development, depending on the specific pathway through which it acts on CRC.

In addition, due to the unique characteristics of the gut, gut microbiota and their metabolites are amenable to dietary intervention. Although the long-term effects of such interventions have not been fully elucidated, dietary modulation represents a highly feasible therapeutic approach with low risk. Among various metabolic processes, the functions and mechanisms of secondary bile acids have been most thoroughly studied, and butyrate has been repeatedly confirmed as a beneficial metabolite. Compared with bile acid metabolism, significant research gaps remain regarding the microbiota, metabolites, and mechanisms associated with lipid and vitamin metabolism. In contrast, studies on the urea cycle have primarily focused on liver diseases, with relatively few investigations into its relationship with CRC. In terms of clinical translational applications, microbial biomarkers hold promising prospects with the extensive application of artificial intelligence (AI) technology. Population-specific studies, including those stratified by BMI and other individual factors, suggest potential pathways toward personalized diagnostic and therapeutic strategies.

5.2 Challenges and Unresolved Questions in Gut Microbiota in CRC Research

Current research on gut microbiota in CRC faces several challenges. Many studies are limited by small sample sizes, which may compromise the statistical power and generalizability of findings [70, 57, 55]. And most research is based on animal models, and caution is needed when extrapolating these findings to humans [72, 57, 78, 52]. The specific molecular mechanisms and causal relationships between the gut microbiota, their metabolites, and CRC require further in-depth investigation [70, 72, 55, 52]. And the long-term impacts and clinical effectiveness and safety of interventions targeting the gut microbiota remain unclear [118]. Additionally, some studies have identified changes in bacterial composition and metabolism that may be a consequence of cancer rather than a cause, and this distinction needs to be clarified.

In biomarker identification, inconsistent results across studies present another challenge, despite employing similar methodologies and analytical tools, such as shotgun metagenomic sequencing, meta-analysis, and random forest models. This phenomenon may be attributed to the small sample size and is also closely associated with the sample sources. Besides, certain methodologies, such as random forest models, may be prone to overfitting. There is also a lack of large-scale population validation for some findings. Overcoming these challenges will require more extensive and well-designed studies.

5.3 Future Directions in Gut Microbiota Metabolism in CRC Research

Future research may focus on further investigating the mechanisms of action of classical metabolites—including identifying molecular-level targets and exploring the crosstalk between different mechanisms—and exploring the effects of other specific metabolites on CRC. Additionally, research on biomarkers needs to be more closely integrated with clinical practice to identify more accurate, highly specific, and consistent biomarkers. Research on microbial biomarkers associated with tumor metastasis also represents a highly promising area. Furthermore, alterations in metabolites of patients with CRC may be associated with gender, which warrants further investigation [131, 132]. The development of novel model algorithms to improve result accuracy is also warranted. In terms of treatment, it is necessary to investigate the clinical safety and long-term effects of microbial therapies while minimizing costs as much as possible. Moreover, with the development of AI technology, it has been widely applied in fields such as vaccine research and development [133], drug research and development [134], and so on. Therefore, the integration of emerging AI technologies holds promise for enabling precision treatment of CRC [135]. In conclusion, while gut microbiota and their metabolites are closely associated with CRC, this field remains substantial room for further research.

Acknowledgments

This research is supported by the National Key R&D Program of China (No. 2022YFA1106400 and No. 2020YFA0803201 to Z. W.), the National Natural Science Foundation of China (No. 32270886 to Z.W.), Young Talent Program of Shanghai Eastern Talent Plan (QNJY2024131 to Z.W), Shanghai Magnolia talent plan Pujiang project (Grant number: 24PJA020 to J.L.), the Natural Science Foundation of Shanghai (Grant number: 25ZR1401069 to J.L.), Shanghai 3-year Public Health Action Plan (Grant number: GWVI-11.1-39 to J.L.), Clinical Research Project of the Shanghai Municipal Health Commission (202140472 to Z.F), and 2023 peak discipline of key medical disciplines in Xuhui District(SHXHZDXK202309).

Conflicts of Interest The authors have no conflicts of interest to declare that are relevant to the content of this chapter.

References

1. Keum N, Giovannucci E (2019) Global burden of colorectal cancer: emerging trends, risk factors and prevention strategies. Nat Rev Gastroenterol Hepatol 16(12):713–732. https://doi.org/10.1038/s41575-019-0189-8
2. Cañellas-Socias A, Sancho E, Batlle E (2024) Mechanisms of metastatic colorectal cancer. Nat Rev Gastroenterol Hepatol 21(9): 609–625. https://doi.org/10.1038/s41575-024-00934-z
3. Surveillance E, and End Results Program (2025) Colorectal cancer stat facts. National Cancer Institute. https://seer.cancer.gov/statfacts/html/colorect.html. Accessed 10 09 2025
4. Shaukat A, Levin TR (2022) Current and future colorectal cancer screening strategies. Nat Rev Gastroenterol Hepatol 19(8): 521–531. https://doi.org/10.1038/s41575-022-00612-y

5. He Z, Yu J, Gong J, Wu J, Zong X, Luo Z, He X, Cheng WM, Liu Y, Liu C, Zhang Q, Dai L, Ding T, Gao B, Gharaibeh RZ, Huang J, Jobin C, Lan P (2024) Campylobacter jejuni-derived cytolethal distending toxin promotes colorectal cancer metastasis. Cell Host Microbe 32(12):2080–2091. e2086. https://doi.org/10.1016/j.chom. 2024.11.006
6. Du Y, Liu L, Ma W, Yan W, Mao W, Du Y, Cui K, Yu P, Li Z, Sansonetti PJ, Gao Y-Z, Zhao H (2023) The role of extratumoral and intratumoral microorganisms in cancer immunotherapy. Innov Life 1(2):100016. https://doi.org/10.59717/j.xinn-life.2023. 100016
7. Karta J, Meyers M, Rodriguez F, Koncina E, Gilson C, Klein E, Gabola M, Benzarti M, Pérez Escriva P, Molina Tijeras JA, Correia Tavares Bernardino C, Ponath F, Carpentier A, Pujabet MA, Schmoetten M, Tsenkova M, Saoud P, Gaigneaux A, Ternes D, Alonso L, Zügel N, Willemssen E, Koppes P, Léonard D, Casanova LP, Haan S, Mittelbronn M, Meiser J, Pozdeev VI, Vogel J, Nuciforo PG, Wilmes P, Letellier E (2025) Fusobacterium nucleatum interacts with cancer-associated fibroblasts to promote colorectal cancer. EMBO J 44(19): 5375–5393. https://doi.org/10.1038/ s44318-025-00542-w
8. Nougayrède JP, Homburg S, Taieb F, Boury M, Brzuszkiewicz E, Gottschalk G, Buchrieser C, Hacker J, Dobrindt U, Oswald E (2006) Escherichia coli induces DNA double-strand breaks in eukaryotic cells. Science 313(5788):848–851. https://doi.org/ 10.1126/science.1127059
9. Pleguezuelos-Manzano C, Puschhof J, Rosendahl Huber A, van Hoeck A, Wood HM, Nomburg J, Gurjao C, Manders F, Dalmasso G, Stege PB, Paganelli FL, Geurts MH, Beumer J, Mizutani T, Miao Y, van der Linden R, van der Elst S, Garcia KC, Top J, Willems RJL, Giannakis M, Bonnet R, Quirke P, Meyerson M, Cuppen E, van Boxtel R, Clevers H (2020) Mutational signature in colorectal cancer caused by genotoxic pks(+) E. coli. Nature 580(7802):269–273. https://doi.org/10.1038/s41586-020- 2080-8
10. Long X, Wong CC, Tong L, Chu ESH, Ho Szeto C, Go MYY, Coker OO, Chan AWH, Chan FKL, Sung JJY, Yu J (2019) Peptostreptococcus anaerobius promotes colorectal carcinogenesis and modulates tumour immunity. Nat Microbiol 4(12):2319–2330. https:// doi.org/10.1038/s41564-019-0541-3
11. Fu T, Coulter S, Yoshihara E, Oh TG, Fang S, Cayabyab F, Zhu Q, Zhang T, Leblanc M, Liu S, He M, Waizenegger W, Gasser E, Schnabl B, Atkins AR, Yu RT, Knight R, Liddle C, Downes M, Evans RM (2019) FXR regulates intestinal cancer stem cell proliferation. Cell 176(5):1098–1112.e1018. https://doi.org/10.1016/j.cell.2019. 01.036
12. Kiriyama Y, Nochi H (2023) The role of gut microbiota-derived lithocholic acid, deoxycholic acid and their derivatives on the function and differentiation of immune cells. Microorganisms 11(11). https://doi.org/ 10.3390/microorganisms11112730
13. Wu S, Morin PJ, Maouyo D, Sears CL (2003) Bacteroides fragilis enterotoxin induces c-Myc expression and cellular proliferation. Gastroenterology 124(2):392–400. https:// doi.org/10.1053/gast.2003.50047
14. Jiang S, Li H, Zhang L, Mu W, Zhang Y, Chen T, Wu J, Tang H, Zheng S, Liu Y, Wu Y, Luo X, Xie Y, Ren J (2025) Generic Diagramming Platform (GDP): a comprehensive database of high-quality biomedical graphics. Nucleic Acids Res 53(D1):D1670– D1676. https://doi.org/10.1093/nar/ gkae973
15. Sekirov I, Russell SL, Antunes LC, Finlay BB (2010) Gut microbiota in health and disease. Physiol Rev 90(3):859–904. https://doi. org/10.1152/physrev.00045.2009
16. Wong CC, Yu J (2023) Gut microbiota in colorectal cancer development and therapy. Nat Rev Clin Oncol 20(7):429–452. https://doi.org/10.1038/s41571-023- 00766-x
17. Wu HJ, Wu E (2012) The role of gut microbiota in immune homeostasis and autoimmunity. Gut Microbes 3(1):4–14. https://doi. org/10.4161/gmic.19320
18. Ding X, Zhu T, Gu R, Li Y, Zhu Z (2025) Gut-microbiota-testis axis: the mechanism of gut microbiota regulating host male reproduction. Innov Life 3(4):100158. https:// doi.org/10.59717/j.xinn-life.2025.100158
19. Wang S, Zhou X, Ma Y, Zhang S, Gong X, Zhang B, Zhou J, Chen G-Q, Shi W, Yang Y (2024) Gut-to-brain neuromodulation by synthetic butyrate-producing commensal bacteria. Innov Life 2(3):100082. https://doi. org/10.59717/j.xinn-life.2024.100082
20. Gordon JH, Dubos R (1970) The anaerobic bacterial flora of the mouse cecum. J Exp Med 132(2):251–260. https://doi.org/10.1084/ jem.132.2.251

21. Harris MA, Reddy CA, Carter GR (1976) Anaerobic bacteria from the large intestine of mice. Appl Environ Microbiol 31(6): 907–912. https://doi.org/10.1128/aem.31.6.907-912.1976
22. Overacre-Delgoffe AE, Bumgarner HJ, Cillo AR, Burr AHP, Tometich JT, Bhattacharjee A, Bruno TC, Vignali DAA, Hand TW (2021) Microbiota-specific T follicular helper cells drive tertiary lymphoid structures and antitumor immunity against colorectal cancer. Immunity 54(12):2812–2824.e2814. https://doi.org/10.1016/j.immuni.2021.11.003
23. Wang Q, Hu T, Zhang Q, Zhang Y, Dong X, Jin Y, Li J, Guo Y, Guo F, Chen Z, Zhong P, Yang Y, Ma Y (2025) Fusobacterium nucleatum promotes colorectal cancer through neogenesis of tumor stem cells. J Clin Invest 135(3). https://doi.org/10.1172/jci181595
24. Hu M, Xu Y, Wang Y, Huang Z, Wang L, Zeng F, Qiu B, Liu Z, Yuan P, Wan Y, Ge S, Zhong D, Xiao S, Luo R, He J, Sun M, Zhuang X, Guo N, Cui C, Gao J, Zhou H, He X (2025) Gut microbial-derived N-acetylmuramic acid alleviates colorectal cancer via the AKT1 pathway. Gut 74(8):1230–1245. https://doi.org/10.1136/gutjnl-2024-332891
25. Liao Q, Zhou X, Wu L, Yang Y, Zhu X, Liao H, Zhang Y, Lian W, Zhang F, Wang H, Ding Y, Zhao L (2025) Gut microbial metabolite 4-hydroxybenzeneacetic acid drives colorectal cancer progression via accumulation of immunosuppressive PMN-MDSCs. J Clin Invest 135(11). https://doi.org/10.1172/jci181243
26. Bell HN, Rebernick RJ, Goyert J, Singhal R, Kuljanin M, Kerk SA, Huang W, Das NK, Andren A, Solanki S, Miller SL, Todd PK, Fearon ER, Lyssiotis CA, Gygi SP, Mancias JD, Shah YM (2022) Reuterin in the healthy gut microbiome suppresses colorectal cancer growth through altering redox balance. Cancer Cell 40(2):185–200.e186. https://doi.org/10.1016/j.ccell.2021.12.001
27. Arthur JC, Perez-Chanona E, Mühlbauer M, Tomkovich S, Uronis JM, Fan TJ, Campbell BJ, Abujamel T, Dogan B, Rogers AB, Rhodes JM, Stintzi A, Simpson KW, Hansen JJ, Keku TO, Fodor AA, Jobin C (2012) Intestinal inflammation targets cancer-inducing activity of the microbiota. Science 338(6103):120–123. https://doi.org/10.1126/science.1224820
28. Cougnoux A, Dalmasso G, Martinez R, Buc E, Delmas J, Gibold L, Sauvanet P, Darcha C, Déchelotte P, Bonnet M, Pezet D, Wodrich H, Darfeuille-Michaud A, Bonnet R (2014) Bacterial genotoxin colibactin promotes colon tumour growth by inducing a senescence-associated secretory phenotype. Gut 63(12):1932–1942. https://doi.org/10.1136/gutjnl-2013-305257
29. Cuevas-Ramos G, Petit CR, Marcq I, Boury M, Oswald E, Nougayrède JP (2010) Escherichia coli induces DNA damage in vivo and triggers genomic instability in mammalian cells. Proc Natl Acad Sci USA 107(25): 11537–11542. https://doi.org/10.1073/pnas.1001261107
30. Cao Y, Oh J, Xue M, Huh WJ, Wang J, Gonzalez-Hernandez JA, Rice TA, Martin AL, Song D, Crawford JM, Herzon SB, Palm NW (2022) Commensal microbiota from patients with inflammatory bowel disease produce genotoxic metabolites. Science 378(6618):eabm3233. https://doi.org/10.1126/science.abm3233
31. Clay SL, Fonseca-Pereira D, Garrett WS (2022) Colorectal cancer: the facts in the case of the microbiota. J Clin Invest 132(4). https://doi.org/10.1172/jci155101
32. Saftien A, Puschhof J, Elinav E (2023) Fungi and cancer. Gut 72(7):1410–1425. https://doi.org/10.1136/gutjnl-2022-327952
33. Sokol H, Leducq V, Aschard H, Pham HP, Jegou S, Landman C, Cohen D, Liguori G, Bourrier A, Nion-Larmurier I, Cosnes J, Seksik P, Langella P, Skurnik D, Richard ML, Beaugerie L (2017) Fungal microbiota dysbiosis in IBD. Gut 66(6):1039–1048. https://doi.org/10.1136/gutjnl-2015-310746
34. van Tilburg BE, Pettersen VK, Gutierrez MW, Laforest-Lapointe I, Jendzjowsky NG, Cavin JB, Vicentini FA, Keenan CM, Ramay HR, Samara J, MacNaughton WK, Wilson RJA, Kelly MM, McCoy KD, Sharkey KA, Arrieta MC (2020) Intestinal fungi are causally implicated in microbiome assembly and immune development in mice. Nat Commun 11(1): 2577. https://doi.org/10.1038/s41467-020-16431-1
35. van der Velden WJ, Plantinga TS, Feuth T, Donnelly JP, Netea MG, Blijlevens NM (2010) The incidence of acute graft-versus-host disease increases with Candida colonization depending the dectin-1 gene status. Clin Immunol 136(2):302–306. https://doi.org/10.1016/j.clim.2010.04.007
36. Jiang L, Stärkel P, Fan JG, Fouts DE, Bacher P, Schnabl B (2021) The gut mycobiome: a novel player in chronic liver diseases.

J Gastroenterol 56(1):1–11. https://doi.org/10.1007/s00535-020-01740-5

37. Uittamo J, Siikala E, Kaihovaara P, Salaspuro M, Rautemaa R (2009) Chronic candidosis and oral cancer in APECED-patients: production of carcinogenic acetaldehyde from glucose and ethanol by Candida albicans. Int J Cancer 124(3):754–756. https://doi.org/10.1002/ijc.23976
38. Tillonen J, Homann N, Rautio M, Jousimies-Somer H, Salaspuro M (1999) Role of yeasts in the salivary acetaldehyde production from ethanol among risk groups for ethanol-associated oral cavity cancer. Alcohol Clin Exp Res 23(8):1409–1415
39. Hsia CC, Sun TT, Wang YY, Anderson LM, Armstrong D, Good RA (1981) Enhancement of formation of the esophageal carcinogen benzylmethylnitrosamine from its precursors by Candida albicans. Proc Natl Acad Sci USA 78(3):1878–1881. https://doi.org/10.1073/pnas.78.3.1878
40. Leonardi I, Gao IH, Lin WY, Allen M, Li XV, Fiers WD, De Celie MB, Putzel GG, Yantiss RK, Johncilla M, Colak D, Iliev ID (2022) Mucosal fungi promote gut barrier function and social behavior via Type 17 immunity. Cell 185(5):831–846.e814. https://doi.org/10.1016/j.cell.2022.01.017
41. Li T, Coker OO, Sun Y, Li S, Liu C, Lin Y, Wong SH, Miao Y, Sung JJY, Yu J (2025) Multi-cohort analysis reveals altered archaea in colorectal cancer fecal samples across populations. Gastroenterology 168(3):525–538.e522. https://doi.org/10.1053/j.gastro.2024.10.023
42. Weinberger V, Darnhofer B, Thapa HB, Mertelj P, Stentz R, Jones E, Grabmann G, Mohammadzadeh R, Shinde T, Karner C, Ober J, Juodeikis R, Pernitsch D, Hingerl K, Zurabishvili T, Kumpitsch C, Kuehnast T, Rinner B, Strohmaier H, Kolb D, Gotts K, Weichhart T, Köcher T, Köfeler H, Carding SR, Schild S, Moissl-Eichinger C (2025) Proteomic and metabolomic profiling of extracellular vesicles produced by human gut archaea. Nat Commun 16(1):5094. https://doi.org/10.1038/s41467-025-60271-w
43. Tun HM, Peng Y, Massimino L, Sin ZY, Parigi TL, Facoetti A, Rahman S, Danese S, Ungaro F (2024) Gut virome in inflammatory bowel disease and beyond. Gut 73(2):350–360. https://doi.org/10.1136/gutjnl-2023-330001
44. Norman JM, Handley SA, Baldridge MT, Droit L, Liu CY, Keller BC, Kambal A, Monaco CL, Zhao G, Fleshner P, Stappenbeck TS, McGovern DP, Keshavarzian A, Mutlu EA, Sauk J, Gevers D, Xavier RJ, Wang D, Parkes M, Virgin HW (2015) Disease-specific alterations in the enteric virome in inflammatory bowel disease. Cell 160(3):447–460. https://doi.org/10.1016/j.cell.2015.01.002
45. Dong X, Pan P, Zheng DW, Bao P, Zeng X, Zhang XZ (2020) Bioinorganic hybrid bacteriophage for modulation of intestinal microbiota to remodel tumor-immune microenvironment against colorectal cancer. Sci Adv 6(20):eaba1590. https://doi.org/10.1126/sciadv.aba1590
46. Federici S, Kredo-Russo S, Valdés-Mas R, Kviatcovsky D, Weinstock E, Matiuhin Y, Silberberg Y, Atarashi K, Furuichi M, Oka A, Liu B, Fibelman M, Weiner IN, Khabra E, Cullin N, Ben-Yishai N, Inbar D, Ben-David H, Nicenboim J, Kowalsman N, Lieb W, Kario E, Cohen T, Geffen YF, Zelcbuch L, Cohen A, Rappo U, Gahali-Sass I, Golembo M, Lev V, Dori-Bachash M, Shapiro H, Moresi C, Cuevas-Sierra A, Mohapatra G, Kern L, Zheng D, Nobs SP, Suez J, Stettner N, Harmelin A, Zak N, Puttagunta S, Bassan M, Honda K, Sokol H, Bang C, Franke A, Schramm C, Maharshak N, Sartor RB, Sorek R, Elinav E (2022) Targeted suppression of human IBD-associated gut microbiota commensals by phage consortia for treatment of intestinal inflammation. Cell 185(16):2879–2898.e2824. https://doi.org/10.1016/j.cell.2022.07.003
47. Bruttin A, Brüssow H (2005) Human volunteers receiving Escherichia coli phage T4 orally: a safety test of phage therapy. Antimicrob Agents Chemother 49(7):2874–2878. https://doi.org/10.1128/aac.49.7.2874-2878.2005
48. Wong SH, Yu J (2019) Gut microbiota in colorectal cancer: mechanisms of action and clinical applications. Nat Rev Gastroenterol Hepatol 16(11):690–704. https://doi.org/10.1038/s41575-019-0209-8
49. McManus DP, Dunne DW, Sacko M, Utzinger J, Vennervald BJ, Zhou XN (2018) Schistosomiasis. Nat Rev Dis Primers 4(1):13. https://doi.org/10.1038/s41572-018-0013-8
50. Guérin A, Striepen B (2020) The biology of the intestinal intracellular parasite cryptosporidium. Cell Host Microbe 28(4):509–515. https://doi.org/10.1016/j.chom.2020.09.007
51. Adam RD (2021) Giardia duodenalis: biology and pathogenesis. Clin Microbiol Rev 34(4):e0002419. https://doi.org/10.1128/cmr.00024-19

52. Chen H, Tong T, Lu SY, Ji L, Xuan B, Zhao G, Yan Y, Song L, Zhao L, Xie Y, Leng X, Zhang X, Cui Y, Chen X, Xiong H, Yu T, Li X, Sun T, Wang Z, Chen J, Chen YX, Hong J, Fang JY (2023) Urea cycle activation triggered by host-microbiota maladaptation driving colorectal tumorigenesis. Cell Metab 35(4):651–666.e657. https://doi.org/10.1016/j.cmet.2023.03.003
53. Uebanso T, Shimohata T, Mawatari K, Takahashi A (2020) Functional roles of B-vitamins in the gut and gut microbiome. Mol Nutr Food Res 64(18):e2000426. https://doi.org/10.1002/mnfr.202000426
54. Li Q, Chan H, Liu WX, Liu CA, Zhou Y, Huang D, Wang X, Li X, Xie C, Liu WY, Wang XS, Ng SK, Gou H, Zhao LY, Fong W, Jiang L, Lin Y, Zhao G, Bai F, Liu X, Chen H, Zhang L, Wong SH, Chan MTV, Wu WKK, Yu J (2023) Carnobacterium maltaromaticum boosts intestinal vitamin D production to suppress colorectal cancer in female mice. Cancer Cell 41(8):1450–1465.e1458. https://doi.org/10.1016/j.ccell.2023.06.011
55. Jia D, Wang Q, Qi Y, Jiang Y, He J, Lin Y, Sun Y, Xu J, Chen W, Fan L, Yan R, Zhang W, Ren G, Xu C, Ge Q, Wang L, Liu W, Xu F, Wu P, Wang Y, Chen S, Wang L (2024) Microbial metabolite enhances immunotherapy efficacy by modulating T cell stemness in pan-cancer. Cell 187(7):1651–1665.e1621. https://doi.org/10.1016/j.cell.2024.02.022
56. Stoeva MK, Garcia-So J, Justice N, Myers J, Tyagi S, Nemchek M, McMurdie PJ, Kolterman O, Eid J (2021) Butyrate-producing human gut symbiont, Clostridium butyricum, and its role in health and disease. Gut Microbes 13(1):1–28. https://doi.org/10.1080/19490976.2021.1907272
57. Bai X, Wei H, Liu W, Coker OO, Gou H, Liu C, Zhao L, Li C, Zhou Y, Wang G, Kang W, Ng EK, Yu J (2022) Cigarette smoke promotes colorectal cancer through modulation of gut microbiota and related metabolites. Gut 71(12):2439–2450. https://doi.org/10.1136/gutjnl-2021-325021
58. Barcenilla A, Pryde SE, Martin JC, Duncan SH, Stewart CS, Henderson C, Flint HJ (2000) Phylogenetic relationships of butyrate-producing bacteria from the human gut. Appl Environ Microbiol 66(4):1654–1661. https://doi.org/10.1128/aem.66.4.1654-1661.2000
59. Campbell C, McKenney PT, Konstantinovsky D, Isaeva OI, Schizas M, Verter J, Mai C, Jin WB, Guo CJ, Violante S, Ramos RJ, Cross JR, Kadaveru K, Hambor J, Rudensky AY (2020) Bacterial metabolism of bile acids promotes generation of peripheral regulatory T cells. Nature 581(7809):475–479. https://doi.org/10.1038/s41586-020-2193-0
60. Wang G, Fan Y, Zhang G, Cai S, Ma Y, Yang L, Wang Y, Yu H, Qiao S, Zeng X (2024) Microbiota-derived indoles alleviate intestinal inflammation and modulate microbiome by microbial cross-feeding. Microbiome 12(1):59. https://doi.org/10.1186/s40168-024-01750-y
61. Li M, Ding Y, Wei J, Dong Y, Wang J, Dai X, Yan J, Chu F, Zhang K, Meng F, Ma J, Zhong W, Wang B, Gao Y, Yang R, Liu X, Su X, Cao H (2024) Gut microbiota metabolite indole-3-acetic acid maintains intestinal epithelial homeostasis through mucin sulfation. Gut Microbes 16(1):2377576. https://doi.org/10.1080/19490976.2024.2377576
62. Huang S, He C, Li J, Gao YZ, Wang Z, Wei Y (2023) Emerging paradigms in exploring the interactions among diet, probiotics, and cancer immunotherapeutic response. Innovation (Camb) 4(4):100456. https://doi.org/10.1016/j.xinn.2023.100456
63. Wilson CM, Loach D, Lawley B, Bell T, Sims IM, O'Toole PW, Zomer A, Tannock GW (2014) Lactobacillus reuteri 100-23 modulates urea hydrolysis in the murine stomach. Appl Environ Microbiol 80(19):6104–6113. https://doi.org/10.1128/aem.01876-14
64. Cui W, Hao M, Yang X, Yin C, Chu B (2025) Gut microbial metabolism in ferroptosis and colorectal cancer. Trends Cell Biol 35(4):341–351. https://doi.org/10.1016/j.tcb.2024.08.006
65. Cui W, Guo M, Liu D, Xiao P, Yang C, Huang H, Liang C, Yang Y, Fu X, Zhang Y, Liu J, Shi S, Cong J, Han Z, Xu Y, Du L, Yin C, Zhang Y, Sun J, Gu W, Chai R, Zhu S, Chu B (2024) Gut microbial metabolite facilitates colorectal cancer development via ferroptosis inhibition. Nat Cell Biol 26(1):124–137. https://doi.org/10.1038/s41556-023-01314-6
66. Chen C, Fang S, Wei H, He M, Fu H, Xiong X, Zhou Y, Wu J, Gao J, Yang H, Huang L (2021) Prevotella copri increases fat accumulation in pigs fed with formula diets. Microbiome 9(1):175. https://doi.org/10.1186/s40168-021-01110-0
67. Fleishman JS, Kumar S (2024) Bile acid metabolism and signaling in health and disease: molecular mechanisms and therapeutic targets. Signal Transduct Target Ther 9(1):

97. https://doi.org/10.1038/s41392-024-01811-6
68. Hylemon PB, Zhou H, Pandak WM, Ren S, Gil G, Dent P (2009) Bile acids as regulatory molecules. J Lipid Res 50(8):1509–1520. https://doi.org/10.1194/jlr.R900007-JLR200
69. Jones BV, Begley M, Hill C, Gahan CG, Marchesi JR (2008) Functional and comparative metagenomic analysis of bile salt hydrolase activity in the human gut microbiome. Proc Natl Acad Sci USA 105(36): 13580–13585. https://doi.org/10.1073/pnas.0804437105
70. Cong J, Liu P, Han Z, Ying W, Li C, Yang Y, Wang S, Yang J, Cao F, Shen J, Zeng Y, Bai Y, Zhou C, Ye L, Zhou R, Guo C, Cang C, Kasper DL, Song X, Dai L, Sun L, Pan W, Zhu S (2024) Bile acids modified by the intestinal microbiota promote colorectal cancer growth by suppressing CD8(+) T cell effector functions. Immunity 57(4):876–889.e811. https://doi.org/10.1016/j.immuni.2024.02.014
71. Wang H, Huang L, Deng J, Wang C, Jin S, Yang C, He X, Liu D, Wu Z, Yang G, Arjin C, Nyachoti CM, Li T, Yin Y, He L (2025) Astaxanthin in metabolic control: from regulatory mechanisms to therapeutic prospects. Innov Life 3(4):100161. https://doi.org/10.59717/j.xinn-life.2025.100161
72. Ocvirk S, O'Keefe SJD (2021) Dietary fat, bile acid metabolism and colorectal cancer. Semin Cancer Biol 73:347–355. https://doi.org/10.1016/j.semcancer.2020.10.003
73. Wang G, Qin S, Chen L, Geng H, Zheng Y, Xia C, Yao J, Deng L (2023) Butyrate dictates ferroptosis sensitivity through FFAR2-mTOR signaling. Cell Death Dis 14(4):292. https://doi.org/10.1038/s41419-023-05778-0
74. Kumar R, Herold JL, Schady D, Davis J, Kopetz S, Martinez-Moczygemba M, Murray BE, Han F, Li Y, Callaway E, Chapkin RS, Dashwood WM, Dashwood RH, Berry T, Mackenzie C, Xu Y (2017) Streptococcus gallolyticus subsp. gallolyticus promotes colorectal tumor development. PLoS Pathog 13(7): e1006440. https://doi.org/10.1371/journal.ppat.1006440
75. Taylor JC, Gao X, Xu J, Holder M, Petrosino J, Kumar R, Liu W, Höök M, Mackenzie C, Hillhouse A, Brashear W, Nunez MP, Xu Y (2021) A type VII secretion system of Streptococcus gallolyticus subsp. gallolyticus contributes to gut colonization and the development of colon tumors. PLoS Pathog 17(1):e1009182. https://doi.org/10.1371/journal.ppat.1009182
76. Aymeric L, Donnadieu F, Mulet C, du Merle L, Nigro G, Saffarian A, Bérard M, Poyart C, Robine S, Regnault B, Trieu-Cuot P, Sansonetti PJ, Dramsi S (2018) Colorectal cancer specific conditions promote streptococcus gallolyticus gut colonization. Proc Natl Acad Sci USA 115(2):E283–e291. https://doi.org/10.1073/pnas.1715112115
77. Lu Y, Li L, Yang H, Li B, Wang Z (2024) DNA bridging of FOXP3 ladder-like multimer: unveiling a novel transcriptional regulation paradigm. Innovation (Camb) 5(3): 100598. https://doi.org/10.1016/j.xinn.2024.100598
78. Yang J, Wei H, Zhou Y, Szeto CH, Li C, Lin Y, Coker OO, Lau HCH, Chan AWH, Sung JJY, Yu J (2022) High-fat diet promotes colorectal tumorigenesis through modulating gut microbiota and metabolites. Gastroenterology 162(1):135–149.e132. https://doi.org/10.1053/j.gastro.2021.08.041
79. Fu T, Huan T, Rahman G, Zhi H, Xu Z, Oh TG, Guo J, Coulter S, Tripathi A, Martino C, McCarville JL, Zhu Q, Cayabyab F, Low B, He M, Xing S, Vargas F, Yu RT, Atkins A, Liddle C, Ayres J, Raffatellu M, Dorrestein PC, Downes M, Knight R, Evans RM (2023) Paired microbiome and metabolome analyses associate bile acid changes with colorectal cancer progression. Cell Rep 42(8): 112997. https://doi.org/10.1016/j.celrep.2023.112997
80. Liu C, Lai P, Hu J, Yin S, Guo S, Liu D, Yu J, Liang D, Chen G, Xing M, Luo Z, He X, Gong J, Lan P, He Z (2025) Lactobacillus johnsonii synthesizes chenodeoxycholic acid to reduce susceptibility to high-fat diet-induced colorectal cancer. Cancer Res 85(23):4600–4615. https://doi.org/10.1158/0008-5472.Can-25-0879
81. Ji S, Liu Q, Zhang S, Chen Q, Wang C, Zhang W, Xiao C, Li Y, Nian C, Li J, Li J, Geng J, Hong L, Xie C, He Y, Chen X, Li X, Yin ZY, You H, Lin KH, Wu Q, Yu C, Johnson RL, Wang L, Chen L, Wang F, Zhou D (2019) FGF15 activates Hippo signaling to suppress bile acid metabolism and liver tumorigenesis. Dev Cell 48(4):460–474.e469. https://doi.org/10.1016/j.devcel.2018.12.021
82. Wang X, Du J, Li H, Cao Z, Cheng Z, Wang Z (2025) The Hippo signaling pathway modulates pancreatic tissue homeostasis. Cell Death Discov 11(1):343. https://doi.org/10.1038/s41420-025-02636-0
83. Alkhalaf LM, Ryan KS (2015) Biosynthetic manipulation of tryptophan in bacteria:

pathways and mechanisms. Chem Biol 22(3): 317–328. https://doi.org/10.1016/j.chembiol.2015.02.005

84. Agus A, Planchais J, Sokol H (2018) Gut microbiota regulation of tryptophan metabolism in health and disease. Cell Host Microbe 23(6):716–724. https://doi.org/10.1016/j.chom.2018.05.003
85. Jing W, Dong S, Xu Y, Liu J, Ren J, Liu X, Zhu M, Zhang M, Shi H, Li N, Xia P, Lu H, Wang S (2025) Gut microbiota-derived tryptophan metabolites regulated by Wuji Wan to attenuate colitis through AhR signaling activation. Acta Pharm Sin B 15(1):205–223. https://doi.org/10.1016/j.apsb.2024.11.009
86. Alexeev EE, Lanis JM, Kao DJ, Campbell EL, Kelly CJ, Battista KD, Gerich ME, Jenkins BR, Walk ST, Kominsky DJ, Colgan SP (2018) Microbiota-derived indole metabolites promote human and murine intestinal homeostasis through regulation of interleukin-10 receptor. Am J Pathol 188(5): 1183–1194. https://doi.org/10.1016/j.ajpath.2018.01.011
87. Hubbard TD, Murray IA, Perdew GH (2015) Indole and tryptophan metabolism: endogenous and dietary routes to Ah receptor activation. Drug Metab Dispos 43(10): 1522–1535. https://doi.org/10.1124/dmd.115.064246
88. Bender MJ, McPherson AC, Phelps CM, Pandey SP, Laughlin CR, Shapira JH, Medina Sanchez L, Rana M, Richie TG, Mims TS, Gocher-Demske AM, Cervantes-Barragan L, Mullett SJ, Gelhaus SL, Bruno TC, Cannon N, McCulloch JA, Vignali DAA, Hinterleitner R, Joglekar AV, Pierre JF, Lee STM, Davar D, Zarour HM, Meisel M (2023) Dietary tryptophan metabolite released by intratumoral Lactobacillus reuteri facilitates immune checkpoint inhibitor treatment. Cell 186(9):1846–1862.e1826. https://doi.org/10.1016/j.cell.2023.03.011
89. Ivashkiv LB (2018) IFNγ: signalling, epigenetics and roles in immunity, metabolism, disease and cancer immunotherapy. Nat Rev Immunol 18(9):545–558. https://doi.org/10.1038/s41577-018-0029-z
90. Wang L, Tu YX, Chen L, Yu KC, Wang HK, Yang SQ, Zhang Y, Zhang SJ, Song S, Xu HL, Yin ZC, Feng MQ, Yue JQ, Huang XH, Tang T, Wei SZ, Liang XJ, Chen ZX (2024) Black rice diet alleviates colorectal cancer development through modulating tryptophan metabolism and activating AHR pathway. iMeta 3(1):e165. https://doi.org/10.1002/imt2.165
91. Kimura I, Ichimura A, Ohue-Kitano R, Igarashi M (2020) Free fatty acid receptors in health and disease. Physiol Rev 100(1): 171–210. https://doi.org/10.1152/physrev.00041.2018
92. Xu C, Gu L, Hu L, Jiang C, Li Q, Sun L, Zhou H, Liu Y, Xue H, Li J, Zhang Z, Zhang X, Xu Q (2023) FADS1-arachidonic acid axis enhances arachidonic acid metabolism by altering intestinal microecology in colorectal cancer. Nat Commun 14(1):2042. https://doi.org/10.1038/s41467-023-37590-x
93. Okumura S, Konishi Y, Narukawa M, Sugiura Y, Yoshimoto S, Arai Y, Sato S, Yoshida Y, Tsuji S, Uemura K, Wakita M, Matsudaira T, Matsumoto T, Kawamoto S, Takahashi A, Itatani Y, Miki H, Takamatsu M, Obama K, Takeuchi K, Suematsu M, Ohtani N, Fukunaga Y, Ueno M, Sakai Y, Nagayama S, Hara E (2021) Gut bacteria identified in colorectal cancer patients promote tumourigenesis via butyrate secretion. Nat Commun 12(1): 5674. https://doi.org/10.1038/s41467-021-25965-x
94. He Y, Fu L, Li Y, Wang W, Gong M, Zhang J, Dong X, Huang J, Wang Q, Mackay CR, Fu YX, Chen Y, Guo X (2021) Gut microbial metabolites facilitate anticancer therapy efficacy by modulating cytotoxic CD8(+) T cell immunity. Cell Metab 33(5):988–1000.e1007. https://doi.org/10.1016/j.cmet.2021.03.002
95. Zhou J, Li L, Wu B, Feng Z, Lu Y, Wang Z (2024) MST1/2: important regulators of Hippo pathway in immune system associated diseases. Cancer Lett 587:216736. https://doi.org/10.1016/j.canlet.2024.216736
96. Tsenkova M, Brauer M, Pozdeev VI, Kasakin M, Busi SB, Schmoetten M, Cheung D, Meyers M, Rodriguez F, Gaigneaux A, Koncina E, Gilson C, Schlicker L, Herebian D, Schmitz M, de Nies L, Mayatepek E, Haan S, de Beaufort C, Cramer T, Meiser J, Linster CL, Wilmes P, Letellier E (2025) Ketogenic diet suppresses colorectal cancer through the gut microbiome long chain fatty acid stearate. Nat Commun 16(1):1792. https://doi.org/10.1038/s41467-025-56678-0
97. Zhang Y, Higgins CB, Tica S, Adams JA, Sun J, Kelly SC, Zong X, Dietzen DJ, Pietka T, Ballentine SJ, Shriver LP, Patti GJ, Cao Y, DeBosch BJ (2024) Hierarchical tricarboxylic acid cycle regulation by hepatocyte arginase 2 links the urea cycle to oxidative metabolism. Cell Metab 36(9):2069–2085.

e2068. https://doi.org/10.1016/j.cmet.2024.07.007

98. Lee JS, Adler L, Karathia H, Carmel N, Rabinovich S, Auslander N, Keshet R, Stettner N, Silberman A, Agemy L, Helbling D, Eilam R, Sun Q, Brandis A, Malitsky S, Itkin M, Weiss H, Pinto S, Kalaora S, Levy R, Barnea E, Admon A, Dimmock D, Stern-Ginossar N, Scherz A, Nagamani SCS, Unda M, Wilson DM 3rd, Elhasid R, Carracedo A, Samuels Y, Hannenhalli S, Ruppin E, Erez A (2018) Urea cycle dysregulation generates clinically relevant genomic and biochemical signatures. Cell 174(6):1559–1570.e1522. https://doi.org/10.1016/j.cell.2018.07.019
99. Keshet R, Szlosarek P, Carracedo A, Erez A (2018) Rewiring urea cycle metabolism in cancer to support anabolism. Nat Rev Cancer 18(10):634–645. https://doi.org/10.1038/s41568-018-0054-z
100. Li L, Mao Y, Zhao L, Li L, Wu J, Zhao M, Du W, Yu L, Jiang P (2019) p53 regulation of ammonia metabolism through urea cycle controls polyamine biosynthesis. Nature 567(7747):253–256. https://doi.org/10.1038/s41586-019-0996-7
101. Zhang T, Hu L, Tang JF, Xu H, Tian K, Wu MN, Huang SY, Du YM, Zhou P, Lu RJ, He S, Xu JM, Si JJ, Li J, Chen DL, Ran JH (2021) Metformin inhibits the urea cycle and reduces putrescine generation in colorectal cancer cell lines. Molecules 26(7). https://doi.org/10.3390/molecules26071990
102. Missiaen R, Anderson NM, Kim LC, Nance B, Burrows M, Skuli N, Carens M, Riscal R, Steensels A, Li F, Simon MC (2022) GCN2 inhibition sensitizes arginine-deprived hepatocellular carcinoma cells to senolytic treatment. Cell Metab 34(8):1151–1167.e1157. https://doi.org/10.1016/j.cmet.2022.06.010
103. Zhou M, Yuan M, Jin Y, Zhou Q, Yu Y, Li J, Dai X, Zheng X, Yu H, Wang W (2023) Vitamin B2-based ferroptosis promoter for sono-enhanced nanocatalytic therapy of triple-negative breast cancer. Adv Funct Mater 33(52):2303899. https://doi.org/10.1002/adfm.202303899
104. Li X, Yang J, Shi E, Lu Y, Song X, Luo H, Wang J, Liang C, Zhang J (2024) Riboflavin alleviates fluoride-induced ferroptosis by IL-17A-independent system Xc(-)/GPX4 pathway and iron metabolism in testicular Leydig cells. Environ Pollut 344:123332. https://doi.org/10.1016/j.envpol.2024.123332
105. Bonakdar M, Czuba LC, Han G, Zhong G, Luong H, Isoherranen N, Vaishnava S (2022) Gut commensals expand vitamin A metabolic capacity of the mammalian host. Cell Host Microbe 30(8):1084–1092.e1085. https://doi.org/10.1016/j.chom.2022.06.011
106. Ji L, Zhang H, Tian G, Xi S, Chu Y, Zhang Y, Liu J, Ning K, Yang J (2023) Tumor microenvironment interplay amid microbial community, host gene expression and pathological features elucidates cancer heterogeneity and prognosis risk. Innov Life 1(2):100028. https://doi.org/10.59717/j.xinn-life.2023.100028
107. Fujita K, Yamamoto H, Matsumoto T, Hirahashi M, Gushima M, Kishimoto J, Nishiyama K, Taguchi T, Yao T, Oda Y (2011) Sessile serrated adenoma with early neoplastic progression: a clinicopathologic and molecular study. Am J Surg Pathol 35(2):295–304. https://doi.org/10.1097/PAS.0b013e318205df36
108. Gao R, Wu C, Zhu Y, Kong C, Zhu Y, Gao Y, Zhang X, Yang R, Zhong H, Xiong X, Chen C, Xu Q, Qin H (2022) Integrated analysis of colorectal cancer reveals cross-cohort gut microbial signatures and associated serum metabolites. Gastroenterology 163(4):1024–1037.e1029. https://doi.org/10.1053/j.gastro.2022.06.069
109. Lee JWJ, Plichta DR, Asher S, Delsignore M, Jeong T, McGoldrick J, Staller K, Khalili H, Xavier RJ, Chung DC (2023) Association of distinct microbial signatures with premalignant colorectal adenomas. Cell Host Microbe 31(5):827–838.e823. https://doi.org/10.1016/j.chom.2023.04.007
110. Wu Y, Jiao N, Zhu R, Zhang Y, Wu D, Wang AJ, Fang S, Tao L, Li Y, Cheng S, He X, Lan P, Tian C, Liu NN, Zhu L (2021) Identification of microbial markers across populations in early detection of colorectal cancer. Nat Commun 12(1):3063. https://doi.org/10.1038/s41467-021-23265-y
111. Di Paola FJ, Alquati C, Conti G, Calafato G, Turroni S, D'Amico F, Ceccarelli C, Buttitta F, Bernardi A, Cuicchi D, Poggioli G, Turchetti D, Ferrari S, Cannizzaro R, Realdon S, Brigidi P, Ricciardiello L (2024) Interplay between WNT/PI3K-mTOR axis and the microbiota in APC-driven colorectal carcinogenesis: data from a pilot study and possible implications for CRC prevention. J Transl Med 22(1):631. https://doi.org/10.1186/s12967-024-05305-5
112. Kong C, Liang L, Liu G, Du L, Yang Y, Liu J, Shi D, Li X, Ma Y (2023) Integrated

metagenomic and metabolomic analysis reveals distinct gut-microbiome-derived phenotypes in early-onset colorectal cancer. Gut 72(6):1129–1142. https://doi.org/10.1136/gutjnl-2022-327156

113. Wang N, Fang JY (2023) Fusobacterium nucleatum, a key pathogenic factor and microbial biomarker for colorectal cancer. Trends Microbiol 31(2):159–172. https://doi.org/10.1016/j.tim.2022.08.010

114. Saito T, Karasawa H, Ouchi K, Ono T, Kajiwara T, Kohyama A, Ikeda-Ohtsubo W, Fukuda S, Fujishima F, Ozawa Y, Suzuki H, Watanabe K, Ishioka C, Kamei T, Ohnuma S, Abe T, Unno M (2025) Characteristics of gut microbiota in high-methylated colorectal cancer. Int J Clin Oncol 30(9):1805–1817. https://doi.org/10.1007/s10147-025-02812-3

115. Rubinstein MR, Wang X, Liu W, Hao Y, Cai G, Han YW (2013) Fusobacterium nucleatum promotes colorectal carcinogenesis by modulating E-cadherin/β-catenin signaling via its FadA adhesin. Cell Host Microbe 14(2):195–206. https://doi.org/10.1016/j.chom.2013.07.012

116. Yu J, Feng Q, Wong SH, Zhang D, Liang QY, Qin Y, Tang L, Zhao H, Stenvang J, Li Y, Wang X, Xu X, Chen N, Wu WK, Al-Aama J, Nielsen HJ, Kiilerich P, Jensen BA, Yau TO, Lan Z, Jia H, Li J, Xiao L, Lam TY, Ng SC, Cheng AS, Wong VW, Chan FK, Xu X, Yang H, Madsen L, Datz C, Tilg H, Wang J, Brünner N, Kristiansen K, Arumugam M, Sung JJ, Wang J (2017) Metagenomic analysis of faecal microbiome as a tool towards targeted non-invasive biomarkers for colorectal cancer. Gut 66(1):70–78. https://doi.org/10.1136/gutjnl-2015-309800

117. Thomas AM, Manghi P, Asnicar F, Pasolli E, Armanini F, Zolfo M, Beghini F, Manara S, Karcher N, Pozzi C, Gandini S, Serrano D, Tarallo S, Francavilla A, Gallo G, Trompetto M, Ferrero G, Mizutani S, Shiroma H, Shiba S, Shibata T, Yachida S, Yamada T, Wirbel J, Schrotz-King P, Ulrich CM, Brenner H, Arumugam M, Bork P, Zeller G, Cordero F, Dias-Neto E, Setubal JC, Tett A, Pardini B, Rescigno M, Waldron L, Naccarati A, Segata N (2019) Metagenomic analysis of colorectal cancer datasets identifies cross-cohort microbial diagnostic signatures and a link with choline degradation. Nat Med 25(4):667–678. https://doi.org/10.1038/s41591-019-0405-7

118. Li J, Chen Z, Wang Q, Du L, Yang Y, Guo F, Li X, Chao Y, Ma Y (2024) Microbial and metabolic profiles unveil mutualistic microbe-microbe interaction in obesity-related colorectal cancer. Cell Rep Med 5(3):101429. https://doi.org/10.1016/j.xcrm.2024.101429

119. Wirbel J, Pyl PT, Kartal E, Zych K, Kashani A, Milanese A, Fleck JS, Voigt AY, Palleja A, Ponnudurai R, Sunagawa S, Coelho LP, Schrotz-King P, Vogtmann E, Habermann N, Niméus E, Thomas AM, Manghi P, Gandini S, Serrano D, Mizutani S, Shiroma H, Shiba S, Shibata T, Yachida S, Yamada T, Waldron L, Naccarati A, Segata N, Sinha R, Ulrich CM, Brenner H, Arumugam M, Bork P, Zeller G (2019) Meta-analysis of fecal metagenomes reveals global microbial signatures that are specific for colorectal cancer. Nat Med 25(4):679–689. https://doi.org/10.1038/s41591-019-0406-6

120. Yang Y, Du L, Shi D, Kong C, Liu J, Liu G, Li X, Ma Y (2021) Dysbiosis of human gut microbiome in young-onset colorectal cancer. Nat Commun 12(1):6757. https://doi.org/10.1038/s41467-021-27112-y

121. Sun Y, Zhang X, Hang D, Lau HC, Du J, Liu C, Xie M, Pan Y, Wang L, Liang C, Zhou X, Chen D, Rong J, Zhao Z, Cheung AH, Wu Y, Gou H, Wong CC, Du L, Deng J, Hu Z, Shen H, Miao Y, Yu J (2024) Integrative plasma and fecal metabolomics identify functional metabolites in adenoma-colorectal cancer progression and as early diagnostic biomarkers. Cancer Cell 42(8):1386–1400.e1388. https://doi.org/10.1016/j.ccell.2024.07.005

122. Yue H, Shen P, Huang Q, Lin S, Shen J, Wang Y, Sun Z, Qi X, Jiang Y, Yuan C, Zou D, Zheng Y, Yang C (2025) Associations between dietary habits, Oral microbiota, and dental caries in adolescents. Phenomics. https://doi.org/10.1007/s43657-024-00208-3

123. Zhou SH, Du Y, Xue WQ, He MJ, Zhou T, Zhao ZY, Pei L, Chen YW, Xie JR, Huang CL, He YQ, Wang TM, Liao Y, Jia WH (2025) Oral microbiota signature predicts the prognosis of colorectal carcinoma. NPJ Biofilms Microbiomes 11(1):71. https://doi.org/10.1038/s41522-025-00702-0

124. Sadhana H, Santosh KA, Balaji PG, Yadav AK (2025) Microbiota-targeted therapies for colorectal cancer: current challenges and future directions. In: Mishra N, Garg A, Ashique S (eds) Role of gut microbiota and postbiotics for colorectal cancer: advancing therapeutic strategies. Springer Nature

Switzerland, Cham, pp 399–428. https://doi.org/10.1007/978-3-031-86065-2_17

125. O'Keefe SJ (2016) Diet, microorganisms and their metabolites, and colon cancer. Nat Rev Gastroenterol Hepatol 13(12):691–706. https://doi.org/10.1038/nrgastro.2016.165
126. Thanikachalam K, Khan G (2019) Colorectal cancer and nutrition. Nutrients 11(1). https://doi.org/10.3390/nu11010164
127. Wang W, Qu M, Zhang T, Wang Y, Zheng C, Lu F, Wang X, Liu F (2025) A microbiome war in our gut. Innov Life 3(1):100111. https://doi.org/10.59717/j.xinn-life.2024.100111
128. Xu H, Wang Y, Liu G, Zhu Z, Shahbazi MA, Reis RL, Kundu SC, Shi X, Zu M, Xiao B (2025) Nano-armed Limosilactobacillus reuteri for enhanced photo-immunotherapy and microbiota tryptophan metabolism against colorectal cancer. Adv Sci (Weinh) 12(7): e2410011. https://doi.org/10.1002/advs.202410011
129. Kaur P, Ghosh S, Bhowmick A, Gadhave K, Datta S, Ghosh A, Garg N, Mahajan RL, Basu B, Choudhury D (2022) Bacterioboat-A novel tool to increase the half-life period of the orally administered drug. Sci Adv 8(10): eabh1419. https://doi.org/10.1126/sciadv.abh1419
130. Zhang JG, Zhang XM, Wu X, Zhou CK, Liu ZZ, Luo XY, Zhang L, Chen W, Yang YJ (2025) Covalent organic frameworks-delivered Reuterin drives trained immunity in tumor-associated macrophages to enhance melanoma immunotherapy via glycerophospholipid metabolism. Adv Sci (Weinh) 12(36):e04784. https://doi.org/10.1002/advs.202504784
131. Hang D, Zeleznik OA, Lu J, Joshi AD, Wu K, Hu Z, Shen H, Clish CB, Liang L, Eliassen AH, Ogino S, Meyerhardt JA, Chan AT, Song M (2022) Plasma metabolomic profiles for colorectal cancer precursors in women. Eur J Epidemiol 37(4):413–422. https://doi.org/10.1007/s10654-021-00834-5
132. Baraibar I, Ros J, Saoudi N, Salvà F, García A, Castells MR, Tabernero J, Élez E (2023) Sex and gender perspectives in colorectal cancer. ESMO Open 8(2):101204. https://doi.org/10.1016/j.esmoop.2023.101204
133. Zhao F, Zai X (2025) Proceeding artificial intelligence technology towards vaccine designs. Innov Life 3(4):100162. https://doi.org/10.59717/j.xinn-life.2025.100162
134. Zhou L, Chen S, Liu J, Zhou Z, Yan Z, Li C, Zeng X, Tuan RS, Li ZA (2025) When artificial intelligence (AI) meets organoids and organs-on-chips (OoCs): game-changer for drug discovery and development? Innov Life 3(1):100115. https://doi.org/10.59717/j.xinn-life.2024.100115
135. Wu F, Zhang D, Ren T (2025) Intratumoral heterogeneity: targeting origin vs. variations and AI-driven non-invasive assessment. The innovation. Life 3(2):100123. https://doi.org/10.59717/j.xinn-life.2025.100123

Chapter 10

The Role of EBV Infection and Epigenetic Factors in Radioresistance of Colorectal Cancer

Yuanqi Li, Tao Huang, and You Zhou

Abstract

Despite great progress made in therapeutic strategies, the 5-year survival rate of colorectal cancer (CRC) remains low due to the high incidence rate and mortality. A pivotal factor in treatment failure is radioresistance—the ability of CRC cells to withstand high doses of ionizing radiation (IR). Understanding the cellular and molecular mechanisms that underline radiation resistance as well as characteristics of radioresistant cells is particularly warranted for exploiting novel treatment modalities. In the context of combating radioresistance in CRC, potential targets of interest include Epstein-Barr virus (EBV), microRNAs (miRNAs), long noncoding RNAs (lncRNAs), and their related signaling pathways amenable to pharmacological intervention. Additionally, radiotherapy reprograms the tumor microenvironment that has impacts on innate and adaptive immunity, such as recruitment of cancer-associated fibroblasts (CAFs), regulatory T cells (Tregs), myeloid-derived suppressor cells (MDSCs), and tumor-associated macrophages (TAMs). Targeting these immunosuppressive cells has the potential to enhance treatment responses in patients with radioresistant CRC. This review endeavors to depict a comprehensive landscape of factors influencing radiosensitivity of CRC, paving the way for advancements in future clinical applications.

Key words Radioresistance, Colorectal cancer, EBV, miRNA, lncRNA, Immunosuppression

1 Introduction

Advancements in elucidating the molecular basis of radioresistance pave the way for novel pharmacological interventions with the potential to be translated into clinic. This section attempts to summarize promising therapeutic candidates as well as their functional mechanisms, which consists of four primary research aspects: (1) Epstein-Barr virus (EBV) infection of epithelial cells causes DNA damage response (DDR) dysregulation that affects radiotherapy; (2) noncoding RNAs including microRNAs (miRNAs), long noncoding RNAs (lncRNAs) control radiosensitivity of malignant cells; (3) modulation of signaling pathways that involved in radioresistance; and (4) immunosuppressive components including cancer-associated fibroblasts (CAFs), regulatory T cells (Tregs),

Tao Huang (ed.), *Microbial 'Omics: Methods and Protocols*, Methods in Molecular Biology, vol. 3033,
https://doi.org/10.1007/978-1-0716-5264-0_10,

myeloid-derived suppressor cells (MDSCs), and tumor-associated macrophages (TAMs) endow certain influences on radiotherapy. Recognizing the nature of these factors is beneficial to augment additional members that improve radiotherapeutic outcomes of oncological patients, extending beyond colorectal cancer (CRC).

2 EBV Infection

As the first human oncovirus discovered, Epstein-Barr virus (EBV) is associated with a broad variety of cancers originating from epithelial cells, mesenchymal cells, and lymphocytes [1, 2]. EBV latent genes are essential for oncogenic transformation, accompanied by early and late lytic antigens to viral carcinogenesis [3–5]. The small minority of infected cells releases viral particles and then activates uninfected bystander neighbors, which also lead to more DNA mutations of highly proliferating cells. Furthermore, the release of a plenty of pro-inflammatory cytokines has been observed during EBV lytic course, increasing the risk of cancer onset [4]. Usually accompanied by immune dysfunction, chronic inflammation is predisposed to carcinogenesis especially in the case of microbial-causative cancers [6]. A typical example highlighting the interactions between chronic inflammation and cancers is colorectal cancer (CRC) in which patients suffer of inflammatory bowel diseases (IBD) associated with EBV infection [7–9].

Indeed, EBV infection of colonic epithelial cells has been reported to cause inflammation, DDR, and autophagy dysregulation by promoting secretion of cytokines and activation of related molecular pathways such as ERK1/2 [10] (Fig. 1). DDR is vital to maintain genome integrity whose elevated levels have been detected in cancer cells with permission in surviving and proliferating despite DNA damage induced by radiotherapy [11, 12]. DNA repair proteins include BRCA1/2, PARP, ATM, and CHK1/2 whose overexpression facilitates DNA damage repair and contributes to radioresistance [13–16]. Previous research has revealed that ERK1/2, activated by EBV, could accelerate CHK1 phosphorylation and degradation as well as impair DDR in EBV-infected colonic epithelial cells [10]. Therefore, digging out the mechanisms of DDR influenced by EBV infection and targeting EBV proteins may help scientists identify novel therapeutic drugs which potentially improve radiotherapy outcomes.

3 Noncoding RNAs

MiRNAs are composed of 18–24 nucleotides and endowed with ability to restrain their targets by complementing the 3′-untranslated region (3′-UTR) of target mRNAs, leading to

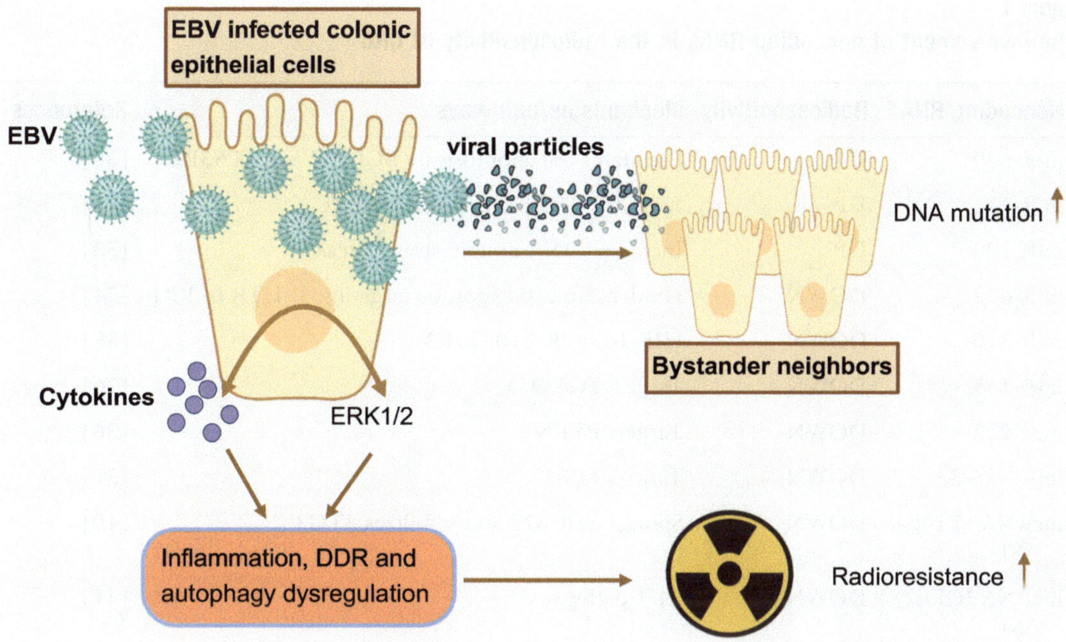

Fig. 1 Carcinogenic and radiotherapeutic effects of Epstein-Barr virus (EBV) on colonic epithelial cells. EBV-infected colonic epithelial cells induce inflammation, DDR, and autophagy dysregulation through cytokine secretion and ERK1/2 activation, resulting in radioresistance. Additionally, viral particles released by infected cells activate neighboring uninfected cells, promoting DNA mutations in highly proliferative cells

mRNA degradation or translation suppression [17]. Previous studies have demonstrated the crucial role of miRNAs in regulating multiple biological processes such as cellular growth, differentiation, death, and cancer-related inflammation [18–20]. Recently, accumulating evidence has identified scores of miRNAs as master regulators in modulating the sensitivity and resistance of CRC cells to radiotherapy, in which they were found to enhance or inhibit radiosensitivity [21, 22].

Post-radiotherapy, miR-630 facilitates cell apoptosis dependent on BCL2L2 and TP53RK, resulting in the enhancement of CRC sensitivity to radiotherapy. Also its expression is positively correlated with radiosensitivity of CRC cells [22]. Similarly, miR-124 confers CRC cells more sensitive to radiotherapy by directly retarding PRRX1 expression [23]. It has been demonstrated that radiation-induced DNA double-strand breaks (DSBs), the most important process involved in radiation-induced cell damage, could be increased by miR-100, and its overexpression contributes to radiosensitivity of CRC cells [24] (Table 1).

An example for miRNA with radioresistant role is miR-622, whose overexpression may render radioresistance of CRC cells [25]. Mechanically, miR-622 hinders Rb expression through directly targeting 3′-UTR of RB1. Prior research has also indicated

Table 1
The involvement of noncoding RNAs in the radiosensitivity of CRC

Noncoding RNA	Radiosensitivity	Mechanisms/pathways	References
miR-630	UP	Facilitates cell apoptosis by BCL2L2 and TP53RK	[30]
miR-124	UP	Retards PRRX1 expression	[31]
miR-100	UP	Increases DNA double-strand breaks	[33]
miR-622	DOWN	Hinders Rb expression by targeting 3′-UTR of RB1	[34]
miR-210	DOWN	HIF-1α/miR-210/Bcl-2	[35]
miR-155	DOWN	Targets FOXO3a	[36]
miR-222	DOWN	Targets PTEN	[36]
miR-338-3p	DOWN	Targets MSI1	[37]
lncRNA SP100-AS1	DOWN	Spongs miR-622 and stabilizes ATG3	[40]
lncRNA IGFL2-AS1	DOWN	AKT pathway	[41]
LINC00630	DOWN	Negatively regulates BEX1 methylation by binding to EZH2	[42]
lncRNA TINCR	DOWN	miR-137/TCF4 pathway	[43]
lncRNA MALAT1	DOWN	Glycosaminoglycan degradation pathway	[46]

that hypoxia-induced autophagy could reduce radiosensitivity of CRC cells by the HIF-1α/miR-210/Bcl-2 pathway [26]. Additionally, miR-155 and miR-222 are required for radioresistance of CRC cells by targeting FOXO3a and PTEN, respectively [27]. Another involved form of miRNAs in regulating radiosensitivity is its interaction with circular RNAs (circRNAs). A recent study has demonstrated that circ_0055625 could act as a sponge for miR-338-3p to regulate MSI1, contributing to impaired radiosensitivity of CRC cells [28] (Table 1). Therefore, miRNAs broadly participate in the regulation of radioresistance and could be used for predicting CRC response to pre- and postoperative radiotherapy.

LncRNAs, longer than 200 nucleotides, cannot produce functional proteins due to the absence of open reading frames (ORFs) [29]. They are vital in mediating a broad variety of cellular processes through their interactions with proteins and nucleic acids [30]. Recent evidence has also highlighted the significant role played by certain lncRNAs in controlling radioresistance in CRC. A recent study has identified a novel lncRNA SP100-AS1 which induces radioresistance of CRC via sponging miR-622 and stabilizing ATG3, suggesting its huge prospects as a therapeutic target for CRC radiotherapy [31]. The high expression levels of lncRNA

IGFL2-AS1 are associated with radioresistance of CRC cells. And further experiments reveal its downstream effector AKT pathway, attracting more attention on future clinical application [32]. Elevated LINC00630 has been reported to be associated with radioresistance and poor prognosis of CRC patients. Mechanically, this lncRNA negatively regulates BEX1 methylation by binding to EZH2 [33]. Knockdown of lncRNA TINCR has been shown to reduce radioresistance of CRC cells by modulating miR-137/TCF4 pathway [34]. As lncRNA MALAT1 has been involved in progression of multiple cancers, its key role in radioresistance also deserves noting [35, 36]. The high expression of MALAT1 is correlated with tumor stage and prognosis of CRC patients. The mechanism suggests that upregulation of MALAT1 promotes radioresistance by glycosaminoglycan degradation pathway [37] (Table 1). Although the majority of lncRNAs discovered act as radioresistance promoters, it should not be overlooked certain lncRNAs may act as radioresistance inhibitors which need further research.

Growing studies have emerged numerous benefits in the therapeutic applications of miRNAs and lncRNAs. First, their expression patterns presenting tissue-specific and cancer-specific make the research value of miRNAs and lncRNAs not only limited to malignant tumors but also for the management of nonneoplastic diseases. Second, increasing number of miRNAs and lncRNAs have been considered as biomarkers for diagnostic or prognostic or treatment efficacy evaluations [38–40]. The biggest advantage is their capability of selectively targeting certain genes or pathways which yields fewer adverse effects compared to traditional strategies [41]. Third, integration of miRNAs and lncRNAs with radiotherapy or chemotherapy or immunotherapy is feasible for contemporary therapeutic strategies to enhance therapeutic efficacy [42]. Nevertheless, the main limitation existed is our constrained understanding regarding their complicated interactions within cellular regulatory network [43]. Hence, it is imperative to conduct more in-depth research to escalate the development and approval of therapies based on miRNAs and lncRNAs.

4 Key Signaling Pathways Activated by IR

Similar to other DNA-damaging agents, IR leads to genomic instability and release of DNA or RNA into the cytosol of threatened cells, triggering a series of events including activation of key signaling pathways. These pathways not only alter cellular pathophysiological processes but also modulate their interactions with surrounding effector cells, thus influencing IR responses.

4.1 Transforming Growth Factor-β (TGF-β) Signaling

TGF-β, a multifunctional growth factor, plays crucial role in the regulation of cellular proliferation and maintenance of tissue homeostasis [44]. Studies have shown the dual functions of TGF-β in carcinogenesis, in which it exerts both tumor-suppressive and tumor-promoting role in established cancers [45–47]. Cells that mainly express TGF-β include endothelial, epithelial, hematopoietic, and connective tissue cell [48]. Classically, TGFβ-receptor complex phosphorylate and activate SMAD2/SMAD3/SMAD4 complexes subsequent by translocating into the nucleus to mediate the expression of target genes, resulting in tumor microenvironment (TME) reprogramming. Recent research has also shown elevated levels of TGF-β expression or activation within the TME, implying its close association with cancer radiotherapy [49–51]. Further mechanisms have indicated that TGF-β could augment phosphorylation of DNA damage repair proteins such as H2AX, ATM, and p53 and then decrease the sensitivity of cancer cells. Additionally, TGF-β can also induce epithelial-mesenchymal transition (EMT) in both SMAD-dependent and non-SMAD dependent ways upon ionizing radiation (IR), thereby promoting radioresistance [50, 52] (Fig. 2).

A recent study has shown that downregulation of TGF-β and its partner PODXL could block IR-induced migration and invasiveness of CRC cells, providing novel theoretical foundations and therapeutic targets for clinic [53]. Fibroblasts within the TME are commonly referred to as cancer-associated fibroblasts (CAFs), which are the most prevalent cell type in CRC [54]. Upregulation of TGF-β has been previously observed among cross talk between CAFs and CRC cells in irradiated nude mice [55]. Moreover, CAFs could promote stemness of CRC cells and hence increase radioresistance through activating TGF-β signaling [56].

4.2 NF-κB Signaling

The NF-κB pathway plays central role in a series of neoplastic processes including initiation, progression, and treatment resistance [57]. The mammalian NF-κB family of transcription factors includes RelA, c-Rel, RelB, NF-κB1, and NF-κB2. The canonical NF-κB pathway is most extensively studied and impinges upon RelA-p50 and c-Rel-p50 heterodimers [58]. The activation of the noncanonical NF-κB pathway relays on activation of RelB-p52 heterodimers [59]. Increasing evidence has demonstrated the involvement of NF-κB signaling in regulating radiosensitivity of CRC cells [60–63]. Previous studies have reported the suppression of NF-κB expression combination of chemotherapy could sensitize CRC cells to IR, suggesting an alternative to improve therapeutic efficacy [62, 63]. Furthermore, NF-κB signaling activated by IR also has fundamental influence on immune response. Noncanonical NF-κB in dendritic cells (DCs) antagonizes STING sensor-mediated DNA sensing in radiotherapy [64]. The activation of NF-κB in MDSCs boosts the migration, differentiation, and

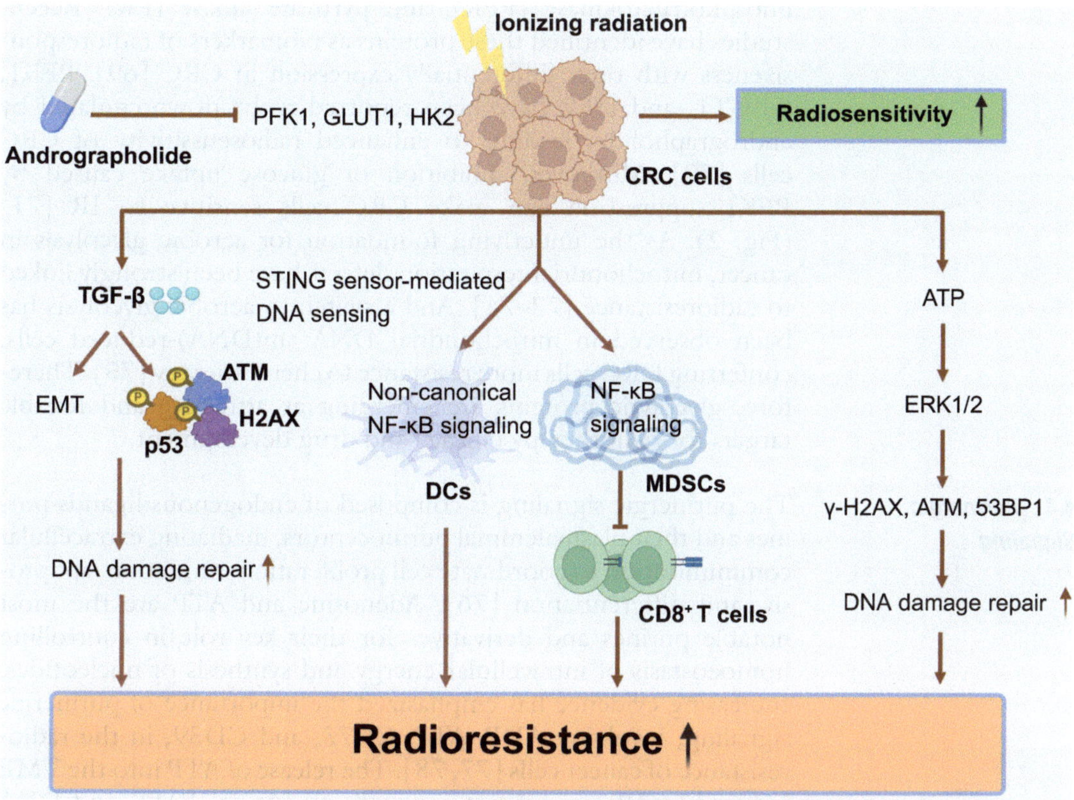

Fig. 2 Key signaling pathways activated by ionizing radiation (IR). IR triggers key signaling pathways. After IR, TGF-β enhances phosphorylation of DNA repair proteins (H2AX, ATM, p53) and induces EMT, promoting radioresistance. Further, noncanonical NF-κB in DCs antagonizes STING-mediated DNA sensing during radiotherapy. NF-κB activation in MDSCs impairs $CD8^+$ T-cell responses. Andrographolide downregulates PFK1, GLUT1, and HK2, increasing radiosensitivity in CRC cells. In addition, IR-induced ATP release in the TME activates its receptors, forming foci of γ-H2AX, ATM, and 53BP1, which mediate DNA repair in an ERK1/2-dependent manner. EMT, epithelial-mesenchymal transition; CRC colorectal cancer; DCs dendritic cells; MDSCs myeloid derived suppressive cells

suppressive role of tumor-infiltrating MDSCs, thus impeding $CD8^+$ T-cell response [65] (Fig. 2). Considering the great impacts of NF-κB signaling on both CRC cells and immune cells in response to IR, it is essential for understanding the dynamic regulation of this pathway within TME and developing targeted strategies.

4.3 Glycolysis

As a hallmark of cancer, cancer metabolism has gained continuous attention due to the influences of molecular metabolic changes on cellular growth, invasion, and metastasis, in which glycolysis is the vital component [66, 67]. Cancer cells exploit glycolysis to accelerate glucose absorption and produce lactic acid, contributing to survival of nutrient-deficient TME [68]. Glycolytic proteins contain glucose transporter (GLUT), hexokinase (HK), phosphofructokinase (PFK), and pyruvate kinase (PK). Recent studies have identified these proteins as biomarkers of radioresponsiveness with their differentially expression in CRC [69]. PFK1, GLUT1, and HK2 have been reported to be downregulated by andrographolide, leading to enhanced radiosenssitivity of CRC cells [70]. Moreover, inhibition of glucose uptake caused by PFK1 suppression can make CRC cells sensitive to IR [71] (Fig. 2). As the underlying foundation for aerobic glycolysis in cancer, mitochondrial respiration defects have been strongly linked to radioresistance [72–74]. And a switch to aerobic glycolysis has been observed in mitochondrial DNA (mtDNA)-reduced cells, conferring CRC cells more resistance to chemotherapy [75]. Therefore, glycolytic proteins are emerging as attractive and feasible targets for radiotherapy efficacy and drug development.

4.4 Purinergic Signaling

The purinergic signaling is composed of endogenous ligands purines and their plasmalemmal purinoceptors, mediating extracellular communication to coordinate cell proliferation, migration, apoptosis, and differentiation [76]. Adenosine and ATP are the most notable purines and derivatives for their key role in controlling homoeostasis of intracellular energy and synthesis of nucleotides. Increasing evidence has emphasized the importance of purinergic signaling, involving ADP, ATP, CD73, and CD39, in the radioresistance of cancer cells [77, 78]. The release of ATP into the TME induced by IR can promote growth, migration, EMT, and DNA damage repair of cancer cells. Mechanically, ATP can bind to its receptors to form an autocrine-paracrine signaling loop. Also, ATP can activate its receptors to form foci consisted of γ-H2AX, ATM, and 53BP1, mediating DNA damage repair in an ERK1/2-dependent manner [79, 80] (Fig. 2). Furthermore, purinergic signaling has also been considered as a vital determinant of anti-tumor immune responses after chemotherapy in CRC [81]. Meanwhile, purinergic receptor exhibits correlations with regulatory T-cell infiltration in post-radiochemotherapy cancer patients [82]. Due to preliminary steps made in elucidating the key role of purinergic signaling in CRC radioresistance, current evidence has given us sufficient confidence in targeting this pathway for potential drug exploitation.

5 Radioimmunotherapy

In addition to directly killing cancer cells, radiotherapy also remodels TME and regulates immune responses to indirectly destroy cancer cells. However, many treatments fail due to IR-induced suppression. Specifically, IR induces the recruitment of immunosuppressive cells, including cancer-associated fibroblasts (CAFs), regulatory T cells (Tregs), myeloid-derived suppressor cells

(MDSCs), and tumor-associated macrophages (TAMs), into the TME. Targeting these cells will hold the key to improving radiotherapy.

5.1 Cancer-Associated Fibroblasts (CAFs)

CAFs are the most prevalent cell type among stromal cells in the TME of CRC, which exert their functions by releasing cytokines, growth factors, and components of the extracellular matrix [54]. They participate in numerous aspects of cell behavior such as proliferation, apoptosis, migration, invasion, differentiation, as well as other physiological processes vital for the proper function of organisms [83]. Radiotherapy-activated CAFs induce DNA damage, cell cycle arrest, and secretion of IGF1, subsequently by survival and metabolic switch of CRC cells, which worsens CRC progression [84]. Moreover, CAF-derived exosomes decrease the sensitivity of CRC cells to IR through regulation of CLCA4-dependent PI3K/Akt pathway and promotion of CRC stem cells phenotype [56, 85] (Fig. 3). Recent studies have also underscored their associations with therapeutic responses including radioresistance and chemotherapy resistance, thereby complicating treatment outcomes [86]. CAFs have been reported to be correlated with poor prognosis in CRC patients after chemoradiotherapy [87].

In addition to affect cancer cells, CAFs can also interact with immune cells within the TME. For example, CAFs express high levels of ICAM-1 to augment monocytes infiltration into cancer tissues [88]. CAFs secrete exosomes to promote the differentiation of monocytes into M2 macrophages, thus increasing angiogenesis and metastasis of CRC cells [89]. CAFs produce a large amount of lactate to inhibit the maturation of tumor-associated dendritic cells. CAFs can also suppress migration of cytotoxic T lymphocytes and weaken antitumor activity. CAFs expressing CD70 facilitate immune evasion through adding the number and survival rate of T cells [90]. Consequently, a deep insight into the mechanisms of specific CAF subpopulations is instrumental for effective targeting.

5.2 Regulatory T Cells (Tregs)

As a unique subset of $CD4^+$ T cells, Tregs are characterized by expression of FOXP3 and CD25. Tregs can exert immunosuppressive effects on cancer progression through inhibiting the activation of diverse immune cells [91, 92]. Recently, a large body of evidence has demonstrated the abundant Tregs within the TME could be associated with poor prognosis and favorable tumor stage of CRC patients [93–98]. Following IR, increased IL-35^+ Tregs significantly enhance immune evasion by accelerating T-cell exhaustion, limiting antitumor immunity [99] (Fig. 3). Additionally, Tregs proliferate most rapidly among T-cell subpopulations independent of TGF-β and IL-33 signaling [100]. These data support the notion that radiotherapy combined with Tregs-targeting agents

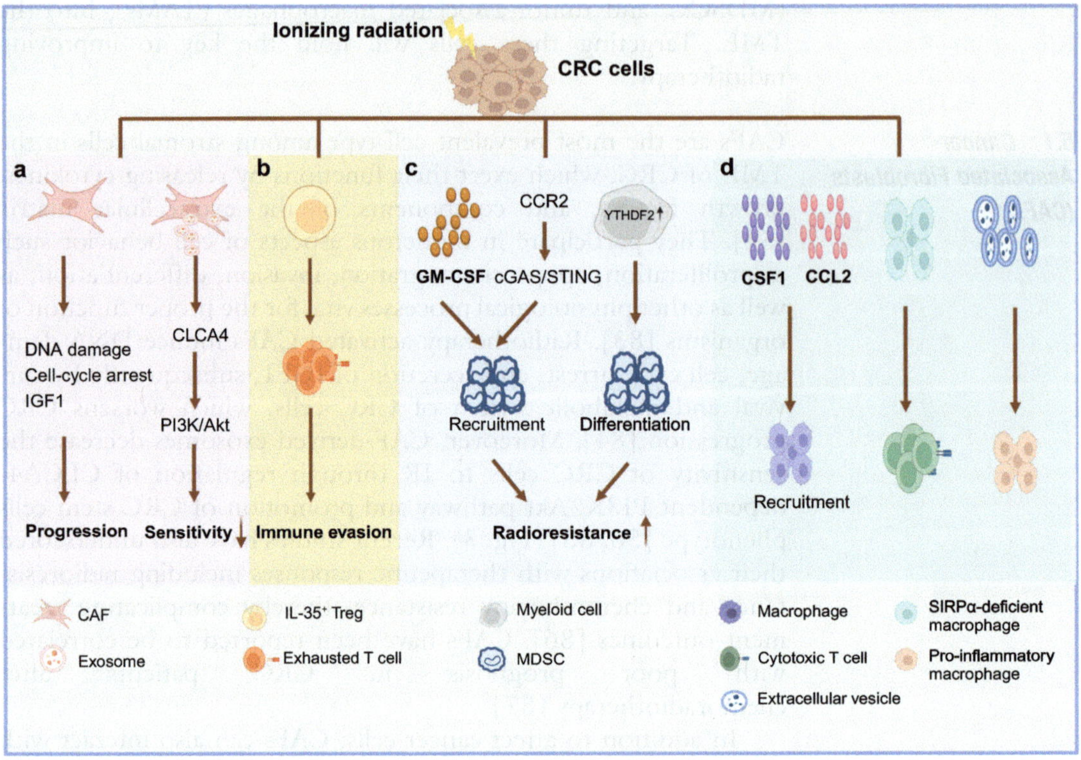

Fig. 3 Impact of IR on immunosuppressive cells. (**a**) Radiotherapy-activated CAFs induce DNA damage, cell cycle arrest, and secretion of IGF1, which facilitates CRC progression. Additionally, CAF-derived exosomes decrease the sensitivity of CRC cells to IR through regulation of CLCA4-dependent PI3K/Akt pathway. (**b**) Following IR, increased IL-35$^+$ Tregs significantly enhance immune evasion by accelerating T-cell exhaustion. (**c**) In the context of radiation, MDSCs are recruited by GM-CSF stimulation, and the cGAS/STING pathway further enhances immunosuppression in tumors by recruiting MDSCs in a CCR2-dependent manner, leading to CRC radioresistance. Following IR, YTHDF2 overexpression in myeloid cells promotes CRC radioresistance by altering MDSC differentiation. (**d**) IR-induced macrophage recruitment is triggered by CSF1 and CCL2. Following IR, SIRPα-deficient macrophages activate tumor antigen-specific cytotoxic T cells, while short-term radiotherapy promotes pro-inflammatory macrophages through extracellular vesicle. CAFs cancer-associated fibroblasts; Tregs, regulatory T cells; MDSCs myeloid-derived suppressive cells; TAMs tumor-associated macrophages

would maximize antitumor efficacy, establishing a preclinical rationale to further improve CRC patient survival.

5.3 Myeloid-Derived Suppressor Cells (MDSCs)

Derived from common myeloid progenitor cells in the bone marrow, MDSCs are aggregates of heterogeneous cells and perform immunosuppressive function in the TME through inhibiting T-cell responses, promoting Tregs generation, and blocking lymphocytes homing [101–103]. In the context of radiation, MDSCs are recruited upon stimulation of GM-CSF and respond to cancer-related stress, acting as extrinsic radioresistance [104, 105]. Prior investigations have demonstrated the cGAS/STING pathway could enhance immunosuppression in tumors through recruiting MDSCs in a CCR2-dependent manner, resulting in CRC radioresistance [106]. Following IR, overexpression of YTHDF2 in myeloid cells impedes antitumor immunity and enhances CRC radioresistance by altering MDSC differentiation and inhibiting MDSC infiltration and suppressive function [65] (Fig. 3). Recent studies have revealed the radioprotective properties of MDSCs in CRC by arginase-induced L-arginine depletion [107]. The number of MDSCs is reduced by intratumoral injection of IL-2 in murine models, resulting in enhanced local and abscopal effects of radiotherapy [108]. Moreover, a systemic elevation of MDSCs induced by local radiation can increase the potential for metastatic dissemination in distal sites [109]. Thus, MDSCs are promising targets to improve radiotherapy as well as antitumor response.

5.4 Tumor-Associated Macrophages (TAMs)

According to different functions, TAMs can be divided into antitumorigenic M1 type and pro-tumorigenic M2 type [110]. In general, recruitment of macrophage upon IR is initiated by CSF1 and CCL2. Subsequently, the polarization direction of TAMs is strongly associated with TME [111]. It was suggested that M1-like phenotype can be induced by IR, and M2-like macrophages display high radioresistance [112]. However, the concept of M1/M2 polarization is gradually abandoned, and diverse TAM phenotypes have been identified [113]. A recent study in a treatment-resistant CRC model has demonstrated that intratumoral SIRPα-deficient macrophages activate tumor antigen-specific cytotoxic T cells under radiotherapy [114]. Several research have also indicated the relationship between radiosensitivity of CRC and autophagy of TAMs [115, 116]. Additionally, short-term radiotherapy facilitates pro-inflammatory macrophages via extracellular vesicles in human rectal cancer, providing a novel strategy for clinical application [117] (Fig. 3). More research is needed to comprehensively clarify the relevance of TAMs to the radiotherapy efficacy.

6 Concluding Remarks

In conclusion, the integration of EBV, miRNAs, lncRNAs, and their related signaling pathways presents comprehensive strategies to combat radioresistance, heralding a novel era in personalized treatment of CRC.

Ethics Approval and Consent to Participate Animal or human subjects are not used in this article.

CRediT Authorship Contribution Statement *Yuanqi Li*: writing—original draft, writing—review and editing. *Tao Huang*:

funding acquisition, writing—review and editing. *You Zhou*: funding acquisition, writing—review and editing.

Conflict of Interest The authors declare that they have no conflicts of interest with the contents of this article.

Funding

This project was supported by grants from National Natural Science Foundation of China (31701111), Strategic Priority Research Program of Chinese Academy of Sciences (XDB38050200, XDB38040202, XDA26040304), National Key R&D Program of China (2022YFF1203202), Natural Science Foundation of Jiangsu Province (BK20241777), Natural Science Foundation of Universities in Jiangsu Province (23KJB310024), Self-supporting Program of Guangzhou Laboratory (SRPG22–007), Major Project of Guangzhou National Laboratory (GZNL2024A01003), and Top Talent of Changzhou "The 14th Five-Year Plan" High-Level Health Talents Training Project (2022260).

References

1. Ko YH (2015) EBV and human cancer. Exp Mol Med 47(1):e130. https://doi.org/10.1038/emm.2014.109
2. El-Sharkawy A, Al Zaidan L, Malki A (2018) Epstein-Barr virus-associated malignancies: roles of viral oncoproteins in carcinogenesis. Front Oncol 8:265. https://doi.org/10.3389/Fonc.2018.00265
3. Gaur N, Gandhi J, Robertson ES, Verma SC, Kaul R (2015) Epstein-Barr virus latent antigens EBNA3C and EBNA1 modulate epithelial to mesenchymal transition of cancer cells associated with tumor metastasis. Tumour Biol 36(4):3051–3060. https://doi.org/10.1007/s13277-014-2941-6
4. Rosemarie Q, Sugden B (2020) Epstein-Barr virus: how its lytic phase contributes to oncogenesis. Microorganisms 8(11):1824. https://doi.org/10.3390/microorganisms8111824
5. Fang CY, Huang SY, Wu CC, Hsu HY, Chou SP, Tsai CH, Chang Y, Takada K, Chen JY (2012) The synergistic effect of chemical carcinogens enhances Epstein-Barr virus reactivation and tumor progression of nasopharyngeal carcinoma cells. PLoS One 7(9):e44810. https://doi.org/10.1371/journal.pone.0044810
6. Greten FR, Grivennikov SI (2019) Inflammation and cancer: triggers, mechanisms, and consequences. Immunity 51(1):27–41. https://doi.org/10.1016/j.immuni.2019.06.025
7. Xu WJ, Jiang XY, Chen JJ, Mao QQ, Zhao XG, Sun X, Zhong L, Rong L (2020) Chronic active Epstein-Barr virus infection involving gastrointestinal tract mimicking inflammatory bowel disease. BMC Gastroenterol 20(1):257. https://doi.org/10.1186/s12876-020-01395-9
8. Tan G, Huang CY, Chen JY, Zhi FC (2020) HMGB1 released from GSDME-mediated pyroptotic epithelial cells participates in the tumorigenesis of colitis-associated colorectal cancer through the ERK1/2 pathway. J Hematol Oncol 13(1):149. https://doi.org/10.1186/s13045-020-00985-0
9. Ma WL, Feng L, Zhang SS, Zhang HJ, Zhang X, Qi XK, Zhang YC, Feng QS, Xiang T, Zeng YX (2018) Induction of chemokine (C-C motif) ligand 5 by Epstein-Barr virus infection enhances tumor angiogenesis in nasopharyngeal carcinoma. Cancer Sci 109(5):1710–1722. https://doi.org/10.1111/cas.13584

10. Santarelli R, Evangelista L, Pompili C, Lo Presti S, Rossi A, Arena A, Gaeta A, Gonnella R, Gilardini Montani MS, Cirone M (2023) EBV infection of primary colonic epithelial cells causes inflammation, DDR and autophagy dysregulation, effects that may predispose to IBD and carcinogenesis. Virus Res 338:199236. https://doi.org/10.1016/j.virusres.2023.199236
11. Brandsma I, Fleuren EDG, Williamsona CT, Lord CJ (2017) Directing the use of DDR kinase inhibitors in cancer treatment. Expert Opin Investig Drugs 26(12):1341–1355. https://doi.org/10.1080/13543784.2017.1389895
12. Nickoloff JA, Boss MK, Allen CP, LaRue SM (2017) Translational research in radiation-induced DNA damage signaling and repair. Transl Cancer Res 6:S875–S891. https://doi.org/10.21037/tcr.2017.06.02
13. Dilmac S, Ozpolat B (2023) Mechanisms of PARP-inhibitor-resistance in BRCA-mutated breast cancer and new therapeutic approaches. Cancer 15(14):3642. https://doi.org/10.3390/Cancers15143642
14. Okabe S, Tanaka Y, Moriyama M, Gotoh A (2023) WEE1 and PARP-1 play critical roles in myelodysplastic syndrome and acute myeloid leukemia treatment. Cancer Cell Int 23(1):128. https://doi.org/10.1186/s12935-023-02961-3
15. Obata H, Ogawa M, Zalutsky MR (2023) DNA repair inhibitors: potential targets and partners for targeted radionuclide therapy. Pharmaceutics 15(7):1926. https://doi.org/10.3390/pharmaceutics15071926
16. Tang Q, Wang X, Wang HX, Zhong L, Zou DL (2023) Advances in ATM, ATR, WEE1, and CHK1/2 inhibitors in the treatment of PARP inhibitor-resistant ovarian cancer. Cancer Biol Med 20(12). https://doi.org/10.20892/j.issn.2095-3941.2023.0260
17. Amirkhah R, Schmitz U, Linnebacher M, Wolkenhauer O, Farazmand A (2015) MicroRNA-mRNA interactions in colorectal cancer and their role in tumor progression. Genes Chromosomes Cancer 54(3):129–141. https://doi.org/10.1002/gcc.22231
18. Chatterjee B, Sarkar M, Bose S, Alam MT, Chaudhary AA, Dixit AK, Tripathi PP, Srivastava AK (2024) MicroRNAs: key modulators of inflammation-associated diseases. Semin Cell Dev Biol 154(Pt C):364–373. https://doi.org/10.1016/j.semcdb.2023.01.009
19. Lv LL, Feng Y, Wu M, Wang B, Li ZL, Zhong X, Wu WJ, Chen J, Ni HF, Tang TT, Tang RN, Lan HY, Liu BC (2020) Exosomal miRNA-19b-3p of tubular epithelial cells promotes M1 macrophage activation in kidney injury. Cell Death Differ 27(1):210–226. https://doi.org/10.1038/s41418-019-0349-y
20. Chakraborty C, Sharma AR, Sharma G, Lee SS (2020) The interplay among miRNAs, major cytokines, and cancer-related inflammation. Mol Ther-Nucl Acids 20:606–620. https://doi.org/10.1016/j.omtn.2020.04.002
21. Wu QB, Sheng X, Zhang N, Yang MW, Wang F (2018) Role of microRNAs in the resistance of colorectal cancer to chemoradiotherapy. Mol Clin Oncol 8(4):523–527. https://doi.org/10.3892/mco.2018.1578
22. Zhu YQ, Peng QL, Lin YX, Zou L, Shen PP, Chen FF, Min M, Shen L, Chen JJ, Shen BR (2017) Identification of biomarker microRNAs for predicting the response colorectal cancer to neoadjuvant chemoradiotherapy based on microRNA regulatory network. Oncotarget 8(2):2233–2248. https://doi.org/10.18632/oncotarget.13659
23. Zhang YQ, Zheng L, Huang J, Gao F, Lin XS, He L, Li D, Li ZJ, Ding Y, Chen LH (2014) MiR-124 radiosensitizes human colorectal cancer cells by targeting PRRX1. PLoS One 9(4):e93917. https://doi.org/10.1371/journal.pone.0093917
24. Yang XD, Xu XH, Zhang SY, Wu Y, Xing CG, Ru G, Xu HT, Cao JP (2015) Role of miR-100 in the radioresistance of colorectal cancer cells. Am J Cancer Res 5(2):545–559
25. Ma WH, Yu J, Qi XL, Liang L, Zhang Y, Ding Y, Lin XS, Li GX, Ding YQ (2015) Radiation-induced microRNA-622 causes radioresistance in colorectal cancer cells by down-regulating Rb. Oncotarget 6(18):15984–15994. https://doi.org/10.18632/oncotarget.3762
26. Sun Y, Xing X, Liu Q, Wang Z, Xin YH, Zhang P, Hu CS, Liu Y (2015) Hypoxia-induced autophagy reduces radiosensitivity by the HIF-1α/miR-210/Bcl-2 pathway in colon cancer cells. Int J Oncol 46(2):750–756. https://doi.org/10.3892/ijo.2014.2745
27. Khoshinani HM, Afshar S, Pashaki AS, Mahdavinezhad A, Nikzad S, Najafi R, Amini R, Gholami MH, Khoshghadam A, Saidijam M (2017) Involvement of miR-155/FOXO3a and miR-222/PTEN in acquired radioresistance of colorectal cancer cell line. Jpn J Radiol 35(11):664–672. https://doi.org/10.1007/s11604-017-0679-y
28. Gao C, Zhang Y, Tian Y, Han C, Wang L, Ding B, Tian H, Zhou C, Ju Y, Peng A, Yu Q

(2021) Circ_0055625 knockdown inhibits tumorigenesis and improves radiosensitivity by regulating miR-338-3p/MSI1 axis in colon cancer. World J Surg Oncol 19(1): 131. https://doi.org/10.1186/s12957-021-02234-1

29. Herman AB, Tsitsipatis D, Gorospe M (2022) Integrated lncRNA function upon genomic and epigenomic regulation. Mol Cell 82(12): 2252–2266. https://doi.org/10.1016/j.molcel.2022.05.027

30. Grammatikakis I, Lal A (2022) Significance of lncRNA abundance to function. Mamm Genome 33(2):271–280. https://doi.org/10.1007/s00335-021-09901-4

31. Zhou Y, Shao YJ, Hu WW, Zhang JP, Shi YF, Kong XY, Jiang JT (2022) A novel long noncoding RNA SP100-AS1 induces radioresistance of colorectal cancer via sponging miR-622 and stabilizing ATG3. Cell Death Differ 30(1):111–124. https://doi.org/10.1038/s41418-022-01049-1

32. Lee J, Kim DY, Kim Y, Shin US, Kim KS, Kim EJ (2023) IGFL2-AS1, a long non-coding RNA, is associated with radioresistance in colorectal cancer. Int J Mol Sci 24(2). https://doi.org/10.3390/ijms24020978

33. Liu F, Huang W, Hong J, Cai C, Zhang W, Zhang J, Kang Z (2020) Long noncoding RNA LINC00630 promotes radio-resistance by regulating BEX1 gene methylation in colorectal cancer cells. IUBMB Life 72(7): 1404–1414. https://doi.org/10.1002/iub.2263

34. Kang Z, Jifu E, Guo K, Ma X, Zhang Y, Yu E (2019) Knockdown of long non-coding RNA TINCR decreases radioresistance in colorectal cancer cells. Pathol Res Pract 215(11): 152622. https://doi.org/10.1016/j.prp.2019.152622

35. Bernard D, Prasanth KV, Tripathi V, Colasse S, Nakamura T, Xuan ZY, Zhang MQ, Sedel F, Jourdren L, Coulpier F, Triller A, Spector DL, Bessis A (2010) A long nuclear-retained non-coding RNA regulates synaptogenesis by modulating gene expression. EMBO J 29(18):3082–3093. https://doi.org/10.1038/emboj.2010.199

36. Schmidt LH, Spieker T, Koschmieder S, Humberg J, Jungen D, Bulk E, Hascher A, Wittmer D, Marra A, Hillejan L, Wiebe K, Berdel WE, Wiewrodt R, Muller-Tidow C (2011) The long noncoding MALAT-1 RNA indicates a poor prognosis in non-small cell lung cancer and induces migration and tumor growth. J Thorac Oncol 6(12):1984–1992. https://doi.org/10.1097/JTO.0b013e3182307eac

37. Shen W, Yu Q, Pu Y, Xing C (2022) Upregulation of long noncoding RNA MALAT1 in colorectal cancer promotes Radioresistance and aggressive malignance. Int J Gener Med 15:8365–8380. https://doi.org/10.2147/IJGM.S393270

38. Bhat AA, Afzal M, Moglad E, Thapa R, Ali H, Almalki WH, Kazmi I, Alzarea SI, Gupta G, Subramaniyan V (2024) lncRNAs as prognostic markers and therapeutic targets in cuproptosis-mediated cancer. Clin Exp Med 24(1):226. https://doi.org/10.1007/s10238-024-01491-0

39. Sarfi M, Abbastabar M, Khalili E (2019) Long noncoding RNAs biomarker-based cancer assessment. J Cell Physiol 234(10): 16971–16986. https://doi.org/10.1002/jcp.28417

40. Lv DZ, Xu K, Jin XY, Li JY, Shi YC, Zhang MY, Jin XY, Li YS, Xu J, Li X (2020) LncSpA: LncRNA spatial atlas of expression across nrmal and cancer tissues. Cancer Res 80(10): 2067–2071. https://doi.org/10.1158/0008-5472.CAN-19-2687

41. Gloss BS, Dinger ME (2016) the specificity of long noncoding RNA expression. Bba-Gene Regul Mech 1859(1):16–22. https://doi.org/10.1016/j.bbagrm.2015.08.005

42. Yan YY, Kumar AB, Finnes H, Markovic SN, Park S, Dronca RS, Dong HD (2018) Combining immune checkpoint inhibitors with conventional cancer therapy. Front Immunol 9:1739. https://doi.org/10.3389/Fimmu.2018.01739

43. Zhao SS, Zhang X, Chen S, Zhang S (2021) Long noncoding RNAs: fine-tuners hidden in the cancer signaling network. Cell Death Discov 7(1):283. https://doi.org/10.1038/s41420-021-00678-8

44. Bakalenko N, Kuznetsova E, Malashicheva A (2024) The complex interplay of TGF-β and notch signaling in the pathogenesis of fibrosis. Int J Mol Sci 25(19):10803. https://doi.org/10.3390/Ijms251910803

45. Akhurst RJ, Hata A (2012) Targeting the TGFβ signalling pathway in disease. Nat Rev Drug Discov 11(10):790–811. https://doi.org/10.1038/nrd3810

46. Javelaud D, Alexaki VI, Dennler S, Mohammad KS, Guise TA, Mauviel A (2011) TGF-β/SMAD/GLI2 signaling axis in cancer progression and metastasis. Cancer Res 71(17):5606–5610. https://doi.org/10.1158/0008-5472.CAN-11-1194

47. Johansson J, Berg T, Kurzejamska E, Pang MF, Fuxe J (2013) MiR-155-mediated loss of C/EBPβ shifts the TGF-β response from

growth inhibition to epithelial-mesenchymal transition, invasion and metastasis in breast cancer. Oncogene 32:5614
48. Achyut BR, Yang L (2011) Transforming growth factor-β in the gastrointestinal and hepatic tumor microenvironment. Gastroenterology 141(4):1167–1178. https://doi.org/10.1053/j.gastro.2011.07.048
49. Wang J, Xu Z, Wang Z, Du G, Lun L (2021) TGF-beta signaling in cancer radiotherapy. Cytokine 148:155709. https://doi.org/10.1016/j.cyto.2021.155709
50. Farhood B, Khodamoradi E, Hoseini-Ghahfarokhi M, Motevaseli E, Mirtavoos-Mahyari H, Eleojo Musa A, Najafi M (2020) TGF-beta in radiotherapy: mechanisms of tumor resistance and normal tissues injury. Pharmacol Res 155:104745. https://doi.org/10.1016/j.phrs.2020.104745
51. Wang L, Lynch C, Pitroda SP, Piffko A, Yang K, Huser AK, Liang HL, Weichselbaum RR (2024) Radiotherapy and immunology. J Exp Med 221(7). https://doi.org/10.1084/jem.20232101
52. Yadav P, Shankar BS (2019) Radio resistance in breast cancer cells is mediated through TGF-β signalling, hybrid epithelial-mesenchymal phenotype and cancer stem cells. Biomed Pharmacother 111:119–130
53. Lee H, Kong JS, Lee SS, Kim A (2021) Radiation-induced overexpression of TGFbeta and PODXL contributes to colorectal cancer cell Radioresistance through enhanced motility. Cells-Basel 10(8). https://doi.org/10.3390/cells10082087
54. Zhang YL, Lv N, Li MS, Liu M, Wu CL (2023) Cancer-associated fibroblasts: tumor defenders in radiation therapy. Cell Death Dis 14(8):541. https://doi.org/10.1038/s41419-023-06060-z
55. Chen X, Liu J, Zhang Q, Liu B, Cheng Y, Zhang Y, Sun Y, Ge H, Liu Y (2020) Exosome-mediated transfer of miR-93-5p from cancer-associated fibroblasts confer radioresistance in colorectal cancer cells by downregulating FOXA1 and upregulating TGFB3. J Exper Clin Cancer Res 39(1):65. https://doi.org/10.1186/s13046-019-1507-2
56. Liu L, Zhang Z, Zhou L, Hu L, Yin C, Qing D, Huang S, Cai X, Chen Y (2020) Cancer associated fibroblasts-derived exosomes contribute to radioresistance through promoting colorectal cancer stem cells phenotype. Exp Cell Res 391(2):111956. https://doi.org/10.1016/j.yexcr.2020.111956
57. Karin M (2006) Nuclear factor-κB in cancer development and progression. Nature 441(7092):431–436. https://doi.org/10.1038/nature04870
58. Sun SC (2012) The noncanonical NF-κB pathway. Immunol Rev 246:125–140. https://doi.org/10.1111/j.1600-065X.2011.01088.x
59. Vallabhapurapu S, Karin M (2009) Regulation and function of NF-κB transcription factors in the immune system. Annu Rev Immunol 27:693–733. https://doi.org/10.1146/annurev.immunol.021908.132641
60. Huang WL, Wu SF, Xu ST, Ma YC, Wang R, Jin SJ, Zhou S (2020) Allicin enhances the radiosensitivity of colorectal cancer cells via inhibition of NF-κB signaling pathway. J Food Sci 85(6):1924–1931. https://doi.org/10.1111/1750-3841.15156
61. Yu L, Guo Q, Luo Z, Wang Y, Weng J, Chen Y, Liang W, Li Y, Zhang Y, Chen K, Chen Z, Ding Y, Zhang Y (2022) TXN inhibitor impedes radioresistance of colorectal cancer cells with decreased ALDH1L2 expression via TXN/NF-kappaB signaling pathway. Br J Cancer 127(4):637–648. https://doi.org/10.1038/s41416-022-01835-1
62. Lee SC, Shen CY, Wang WH, Lee YP, Liang KW, Chou YH, Tyan YS, Hwang JJ (2023) Synergistic effect of ginsenoside Rh2 combines with ionizing radiation on CT26/luc colon carcinoma cells and tumor-bearing animal model. Pharmaceuticals 16(9). https://doi.org/10.3390/ph16091188
63. Kuo YC, Lin WC, Chiang IT, Chang YF, Chen CW, Su SH, Chen CL, Hwang JJ (2012) Sorafenib sensitizes human colorectal carcinoma to radiation via suppression of NF-kappaB expression in vitro and in vivo. Biomed Pharmacother 66(1):12–20. https://doi.org/10.1016/j.biopha.2011.09.011
64. Hou Y, Liang H, Rao E, Zheng W, Huang X, Deng L, Zhang Y, Yu X, Xu M, Mauceri H, Arina A, Weichselbaum RR, Fu YX (2018) Non-canonical NF-kappaB antagonizes STING sensor-mediated DNA sensing in radiotherapy. Immunity 49(3):490–503 e494. https://doi.org/10.1016/j.immuni.2018.07.008
65. Wang L, Dou X, Chen S, Yu X, Huang X, Zhang L, Chen Y, Wang J, Yang K, Bugno J, Pitroda S, Ding X, Piffko A, Si W, Chen C, Jiang H, Zhou B, Chmura SJ, Luo C, Liang HL, He C, Weichselbaum RR (2023) YTHDF2 inhibition potentiates radiotherapy antitumor efficacy. Cancer Cell 41(7):

1294–1308. e1298. https://doi.org/10.1016/j.ccell.2023.04.019

66. Pavlova NN, Zhu J, Thompson CB (2022) The hallmarks of cancer metabolism: still emerging. Cell Metab 34(3):355–377. https://doi.org/10.1016/j.cmet.2022.01.007
67. Li Y, Huang H, Wu S, Zhou Y, Huang T, Jiang J (2023) The role of RNA m(6)A modification in cancer glycolytic reprogramming. Curr Gene Ther 23(1):51–59. https://doi.org/10.2174/1566523222666220830150446
68. Lunt SY, Vander Heiden MG (2011) Aerobic glycolysis: meeting the metabolic requirements of cell proliferation. Annu Rev Cell Dev Biol 27:441–464. https://doi.org/10.1146/annurev-cellbio-092910-154237
69. Islam Khan MZ, Tam SY, Azam Z, Law HKW (2022) Proteomic profiling of metabolic proteins as potential biomarkers of radioresponsiveness for colorectal cancer. J Proteome 262:104600. https://doi.org/10.1016/j.jprot.2022.104600
70. Li X, Tian R, Liu L, Wang L, He D, Cao K, Ma JK, Huang C (2020) Andrographolide enhanced radiosensitivity by downregulating glycolysis via the inhibition of the PI3K-Akt-mTOR signaling pathway in HCT116 colorectal cancer cells. J Int Med Res 48(8):300060520946169. https://doi.org/10.1177/0300060520946169
71. Tian RF, Li XF, Xu C, Wu H, Liu L, Wang LH, He D, Cao K, Cao PG, Ma JK, Huang CH (2020) SiRNA targeting PFK1 inhibits proliferation and migration and enhances radiosensitivity by suppressing glycolysis in colorectal cancer. Am J Transl Res 12(9):4923–4940
72. Shi Y, Wang Y, Jiang H, Sun X, Xu H, Wei X, Wei Y, Xiao G, Song Z, Zhou F (2021) Mitochondrial dysfunction induces radioresistance in colorectal cancer by activating [Ca(2+)](m)-PDP1-PDH-histone acetylation retrograde signaling. Cell Death Dis 12(9):837. https://doi.org/10.1038/s41419-021-03984-2
73. Sun C, Liu XX, Wang B, Wang ZH, Liu Y, Di CX, Si J, Li HY, Wu QF, Xu D, Li J, Li G, Wang YP, Wang F, Zhang H (2019) Endocytosis-mediated mitochondrial transplantation: transferring normal human astrocytic mitochondria into glioma cells rescues aerobic respiration and enhances radiosensitivity. Theranostics 9(12):3595–3607. https://doi.org/10.7150/thno.33100
74. Wei YH, Chen LL, Xu H, Xie CH, Zhou YF, Zhou FX (2018) Mitochondrial dysfunctions regulated Radioresistance through mitochondria-to-nucleus retrograde signaling pathway of NF-κB/PI3K/AKT2/mTOR. Radiat Res 190(2):204–215. https://doi.org/10.1667/Rr15021.1
75. Mou JJ, Peng J, Shi YY, Li N, Wang Y, Ke Y, Zhou YF, Zhou FX (2018) Mitochondrial DNA content reduction induces aerobic glycolysis and reversible resistance to drug-induced apoptosis in SW480 colorectal cancer cells. Biomed Pharmacother 103:729–737. https://doi.org/10.1016/j.biopha.2018.04.099
76. Huang Z, Xie N, Illes P, Di Virgilio F, Ulrich H, Semyanov A, Verkhratsky A, Sperlagh B, Yu SG, Huang C, Tang Y (2021) From purines to purinergic signalling: molecular functions and human diseases. Signal Transduct Target Ther 6(1):162. https://doi.org/10.1038/s41392-021-00553-z
77. Zanoni M, Sarti AC, Zamagni A, Cortesi M, Pignatta S, Arienti C, Tebaldi M, Sarnelli A, Romeo A, Bartolini D, Tosatto L, Adinolfi E, Tesei A, Di Virgilio F (2022) Irradiation causes senescence, ATP release, and P2X7 receptor isoform switch in glioblastoma. Cell Death Dis 13(1):80. https://doi.org/10.1038/s41419-022-04526-0
78. Kojima S, Nakayama K, Ishida H (2004) Low dose γ-rays activate immune functions via induction of glutathione and delay tumor growth. J Radiat Res 45(1):33–39. https://doi.org/10.1269/Jrr.45.33
79. Ide S, Nishimaki N, Tsukimoto M, Kojima S (2014) Purine receptor P2Y6 mediates cellular response to γ-ray-induced DNA damage. J Toxicol Sci 39(1):15–23
80. Nishimaki N, Tsukimoto M, Kitami A, Kojima S (2012) Autocrine regulation of γ-irradiation-induced DNA damage response via extracellular nucleotides-mediated activation of P2Y6 and P2Y12 receptors. DNA Repair 11:657
81. Huang KC, Chiang SF, Lin PC, Hong WZ, Yang PC, Chang HP, Peng SL, Chen TW, Ke TW, Liang JA, Chen WT, Chao KSC (2024) TNFalpha modulates PANX1 activation to promote ATP release and enhance P2RX7-mediated antitumor immune responses after chemotherapy in colorectal cancer. Cell Death Dis 15(1):24. https://doi.org/10.1038/s41419-023-06408-5
82. Vacchelli E, Semeraro M, Enot DP, Chaba K, Kroemer G (2015) Negative prognostic impact of regulatory T cell infiltration in surgically resected esophageal cancer post-radiochemotherapy. Oncotarget 6(25):20840–20850

83. Chen X, Song E (2019) Turning foes to friends: targeting cancer-associated fibroblasts. Nat Publ Group (2)
84. Tommelein J, De Vlieghere E, Verset L, Melsens E, Leenders J, Descamps B, Debucquoy A, Vanhove C, Pauwels P, Gespach CP, Vral A, De Boeck A, Haustermans K, de Tullio P, Ceelen W, Demetter P, Boterberg T, Bracke M, De Wever O (2018) Radiotherapy-activated cancer-associated fibroblasts promote tumor progression through paracrine IGF1R activation. Cancer Res 78(3):659–670. https://doi.org/10.1158/0008-5472.CAN-17-0524
85. Chen XJ, Liu YQ, Zhang QL, Liu BX, Cheng Y, Zhang YL, Sun YN, Liu JQ (2021) Exosomal miR-590-3p derived from cancer-associated fibroblasts confers radioresistance in colorectal cancer. Mol Ther Nucl Acids 24:113–126. https://doi.org/10.1016/j.omtn.2020.11.003
86. Yan L, Zheng J, Wang Q, Hao H (2023) Role of cancer-associated fibroblasts in colorectal cancer and their potential as therapeutic targets. Biochem Biophys Res Commun 681:127–135. https://doi.org/10.1016/j.bbrc.2023.09.065
87. Saigusa S, Toiyama Y, Tanaka K, Yokoe T, Okugawa Y, Fujikawa H, Matsusita K, Kawamura M, Inoue Y, Miki C, Kusunoki M (2011) Cancer-associated fibroblasts correlate with poor prognosis in rectal cancer after chemoradiotherapy. Int J Oncol 38(3):655–663. https://doi.org/10.3892/ijo.2011.906
88. Schellerer VS, Langheinrich M, Hohenberger W, Croner RS, Merkel S, Rau TT, Sturzl M, Naschberger E (2014) Tumor-associated fibroblasts isolated from colorectal cancer tissues exhibit increased ICAM-1 expression and affinity for monocytes. Oncol Rep 31(1):255–261. https://doi.org/10.3892/or.2013.2860
89. Wang M, Su Z, Barnie PA (2020) Crosstalk among colon cancer-derived exosomes, fibroblast-derived exosomes, and macrophage phenotypes in colon cancer metastasis. Int Immunopharmacol 81:106298
90. Jacobs J, Deschoolmeester V, Zwaenepoel K, Flieswasser T, Deben C, Van den Bossche J, Hermans C, Rolfo C, Peeters M, De Wever O, Lardon F, Siozopoulou V, Smits E, Pauwels P (2018) Unveiling a CD70-positive subset of cancer-associated fibroblasts marked by pro-migratory activity and thriving regulatory T cell accumulation. Onco Targets Ther 7(7):e1440167. https://doi.org/10.1080/2162402X.2018.1440167
91. Wolf D, Sopper S, Pircher A, Gastl G, Wolf AM (2015) Treg(s) in cancer: friends or foe? J Cell Physiol 230(11):2598–2605. https://doi.org/10.1002/jcp.25016
92. Sakaguchi S, Miyara M, Costantino CM, Hafler DA (2010) FOXP3+ regulatory T cells in the human immune system. Nat Rev Immunol 10(7):490–500. https://doi.org/10.1038/nri2785
93. Togashi Y, Shitara K, Nishikawa H (2019) Regulatory T cells in cancer immunosuppression - implications for anticancer therapy. Nat Rev Clin Oncol 16(6):356–371. https://doi.org/10.1038/s41571-019-0175-7
94. Wing JB, Tanaka A, Sakaguchi S (2019) Human FOXP3(+) regulatory T cell heterogeneity and function in autoimmunity and cancer. Immunity 50(2):302–316. https://doi.org/10.1016/j.immuni.2019.01.020
95. Napolitano M, D'Alterio C, Cardone E, Trotta AM, Pecori B, Rega D, Pace U, Scala D, Scognamiglio G, Tatangelo F, Cacciapuoti C, Pacelli R, Delrio P, Scala S (2015) Peripheral myeloid-derived suppressor and T regulatory PD-1 positive cells predict response to neoadjuvant short-course radiotherapy in rectal cancer patients. Oncotarget 6(10):8261–8270. https://doi.org/10.18632/oncotarget.3014
96. Schmidt MA, Förtsch C, Schmidt M, Rau TT, Fietkau R, Distel LV (2012) Circulating regulatory T cells of cancer patients receiving radiochemotherapy may be useful to individualize cancer treatment. Radiother Oncol 104(1):131–138
97. Herrera FG, Bourhis J, Coukos G (2017) Radiotherapy combination opportunities leveraging immunity for the next oncology practice. CA Cancer J Clin 67(1):65–85. https://doi.org/10.3322/caac.21358
98. Persa E, Balogh A, Safrany G, Lumniczky K (2015) The effect of ionizing radiation on regulatory T cells in health and disease. Cancer Lett 368(2):252–261. https://doi.org/10.1016/j.canlet.2015.03.003
99. Turnis ME, Sawant DV, Szymczak-Workman AL, Andrews LP, Delgoffe GM, Yano H, Beres AJ, Vogel P, Workman CJ, Vignali DA (2016) Interleukin-35 limits anti-tumor immunity. Immunity 44(2):316–329. https://doi.org/10.1016/j.immuni.2016.01.013
100. Muroyama Y, Nirschl TR, Kochel CM, Lopez-Bujanda Z, Theodros D, Mao W, Carrera-Haro MA, Ghasemzadeh A, Marciscano AE, Velarde E, Tam AJ, Thoburn CJ, Uddin M, Meeker AK, Anders RA, Pardoll DM, Drake CG (2017) Stereotactic

radiotherapy increases functionally suppressive regulatory T cells in the tumor microenvironment. Cancer Immunol Res 5(11): 992–1004. https://doi.org/10.1158/2326-6066.CIR-17-0040

101. Yin Z, Li C, Wang J, Xue L (2019) Myeloid-derived suppressor cells: roles in the tumor microenvironment and tumor radiotherapy. Int J Cancer 144(5):933–946. https://doi.org/10.1002/ijc.31744
102. Zhao Y, Wu T, Shao S, Shi B, Zhao Y (2016) Phenotype, development, and biological function of myeloid-derived suppressor cells. Onco Targets Ther 5(2):e1004983. https://doi.org/10.1080/2162402X.2015.1004983
103. Liu Y, Wei J, Guo G, Zhou J (2015) Norepinephrine-induced myeloid-derived suppressor cells block T-cell responses via generation of reactive oxygen species. Immunopharmacol Immunotoxicol 37(4): 359–365. https://doi.org/10.3109/08923973.2015.1059442
104. Chajon E, Castelli J, Marsiglia H, De Crevoisier R (2017) The synergistic effect of radiotherapy and immunotherapy: a promising but not simple partnership. Crit Rev Oncol Hematol 111:124–132. https://doi.org/10.1016/j.critrevonc.2017.01.017
105. Wang L, Katipally RR, Liang HL, Yang K, Pitroda SP, He C, Weichselbaum RR (2023) RNA m(6)A methylation and MDSCs: roles and therapeutic implications for radiotherapy. Med 4(12):863–874. https://doi.org/10.1016/j.medj.2023.09.001
106. Liang H, Deng L, Hou Y, Meng X, Huang X, Rao E, Zheng W, Mauceri H, Mack M, Xu M, Fu YX, Weichselbaum RR (2017) Host STING-dependent MDSC mobilization drives extrinsic radiation resistance. Nat Commun 8(1):1736. https://doi.org/10.1038/s41467-017-01566-5
107. Leonard W, Dufait I, Schwarze JK, Law K, Engels B, Jiang H, Van den Berge D, Gevaert T, Storme G, Verovski V, Breckpot K, De Ridder M (2016) Myeloid-derived suppressor cells reveal radioprotective properties through arginase-induced l-arginine depletion. Radiother Oncol 119(2): 291–299. https://doi.org/10.1016/j.radonc.2016.01.014
108. Yasuda K, Nirei T, Tsuno NH, Nagawa H, Kitayama J (2011) Intratumoral injection of interleukin-2 augments the local and abscopal effects of radiotherapy in murine rectal cancer. Cancer Sci 102(7):1257–1263. https://doi.org/10.1111/j.1349-7006.2011.01940.x
109. Hou Y, Yang K, Wang L, Wang J, Huang X, Piffko A, Luo SZ, Yu X, Rao E, Martinez C, Bugno J, Mack M, Vokes EE, Pitroda SP, Chmura SJ, Weichselbaum RR, Liang HL (2024) Radiotherapy enhances metastasis through immune suppression by inducing PD-L1 and MDSC in distal sites. Clin Cancer Res 30(9):1945–1958. https://doi.org/10.1158/1078-0432.CCR-23-3206
110. Sica A, Larghi P, Mancino A, Rubino L, Porta C, Totaro MG, Rimoldi M, Biswas SK, Allavena P, Mantovani A (2008) Macrophage polarization in tumour progression. Semin Cancer Biol 18(5):349–355. https://doi.org/10.1016/j.semcancer.2008.03.004
111. Mosser DM, Edwards JP (2008) Exploring the full spectrum of macrophage activation. Nat Rev Immunol 8(12):958–969. https://doi.org/10.1038/nri2448
112. Leblond MM, Pérès EA, Helaine C, Gérault AN, Valable MBS (2017) M2 macrophages are more resistant than M1 macrophages following radiation therapy in the context of glioblastoma. Oncotarget 8(42): 72597–72612
113. Ma RY, Black A, Qian BZ (2022) Macrophage diversity in cancer revisited in the era of single-cell omics. Trends Immunol 43(7): 546–563. https://doi.org/10.1016/j.it.2022.04.008
114. Bian Z, Shi L, Kidder K, Zen K, Garnett-Benson C, Liu Y (2021) Intratumoral SIRPalpha-deficient macrophages activate tumor antigen-specific cytotoxic T cells under radiotherapy. Nat Commun 12(1): 3229. https://doi.org/10.1038/s41467-021-23442-z
115. Shao LN, Zhu BS, Xing CG, Yang XD, Young W, Cao JP (2016) Effects of autophagy regulation of tumor-associated macrophages on radiosensitivity of colorectal cancer cells. Mol Med Rep 13(3): 2661–2670. https://doi.org/10.3892/mmr.2016.4820
116. Shao L, Wu Y, Cao J, Zhong F, Yang X, Xing C (2024) Activation of M2 macrophage autophagy by rapamycin increases the radiosensitivity of colorectal cancer xenografts. J Cancer Res Ther 20(2):695–705. https://doi.org/10.4103/jcrt.jcrt_215_23
117. Stary V, Wolf B, Unterleuthner D, List J, Talic M, Laengle J, Beer A, Strobl J, Stary G, Dolznig H, Bergmann M (2020) Short-course radiotherapy promotes pro-inflammatory macrophages via extracellular vesicles in human rectal cancer. J Immunother Cancer 8(2). https://doi.org/10.1136/jitc-2020-000667

Chapter 11

Infection-Associated Microecology and Hepatocellular Carcinoma

Qian Song

Abstract

Hepatocellular carcinoma (HCC) is the second leading cause of cancer-related mortality worldwide and remains one of the few malignancies with steadily increasing incidence and death rates over recent years. Globally, major etiological drivers of HCC include chronic hepatitis B virus (HBV) and hepatitis C virus (HCV) infections, consumption of aflatoxin-contaminated food, heavy alcohol intake, obesity, cigarette smoking, and type 2 diabetes mellitus. In China, however, the exceptionally high burden of HCC is primarily attributable to chronic HBV infection and aflatoxin exposure, with approximately 80% of Chinese HCC patients testing positive for HBV. The natural history of the disease generally progresses from chronic hepatitis to liver cirrhosis and ultimately to HCC. Throughout this trajectory, the infection-associated microecological environment exerts a crucial influence on hepatocarcinogenesis. This chapter will examine the role and mechanistic underpinnings of infection-related microecology—particularly the gut microbiome—in the development of HCC and highlight the clinical relevance of infection-associated microbial components in liver cancer treatment.

Key words Infection-associated microecology, Hepatocellular carcinoma, HBV, MASLD

1 Infection-Associated Microecology and the Development of Hepatocellular Carcinoma

The liver maintains an intimate anatomical and functional relationship with the intestine through the portal venous system and the enterohepatic circulation. The portal vein delivers gut-derived nutrients and potentially harmful substances to the liver, while the liver, through the secretion of bile, regulates intestinal function and shapes microbial homeostasis [1, 2]. Together, these bidirectional interactions constitute the "gut–liver axis."

Perturbations in the gut microbiota not only disrupt local intestinal physiology but also exert profound downstream effects on the liver via this axis. Under physiological conditions, the liver efficiently clears translocated bacteria, fungi, and their associated toxic metabolites. However, when liver function becomes

Tao Huang (ed.), *Microbial 'Omics: Methods and Protocols*, Methods in Molecular Biology, vol. 3033, https://doi.org/10.1007/978-1-0716-5264-0_11, compromised—such as in acute or chronic hepatitis, alcohol-associated liver disease, or metabolic dysfunction-associated steatotic liver disease (MASLD)—this homeostasis is disrupted, resulting in varying degrees of microbial dysbiosis. Beneficial butyrate-producing bacteria (e.g., *Ruminococcus*, *Oscillibacter*, *Faecalibacterium*, *Clostridium*, *Eubacterium*) are markedly diminished, whereas potential pathogenic species capable of producing lipopolysaccharide (LPS), including *Klebsiella* and *Haemophilus*, proliferate significantly. Additional depletion of other probiotics, such as *Bifidobacterium* and *Akkermansia*, is also commonly observed.

This dysbiotic shift facilitates hepatocarcinogenesis through multiple mechanisms. Certain pathogenic species, such as *Enterococcus faecalis*, secrete gelatinase E, which degrades tight junction proteins [3] and heightens intestinal permeability. As a consequence, bacteria and their metabolites (e.g., LPS) translocate into the liver via the portal circulation, where they induce chronic inflammation and ultimately promote hepatic fibrosis and the progression to hepatocellular carcinoma [3, 4].

MASLD, the most prevalent chronic liver disease globally—has emerged as a major etiological driver of HCC. The gut microecosystem plays a pivotal regulatory role in the transition from MASLD to HCC. Patients with MASLD demonstrate significantly increased intestinal permeability [5], accompanied by microbial dysbiosis characterized by the overgrowth of Enterobacteriaceae (such as *Escherichia coli*) and the depletion of beneficial taxa including *Faecalibacterium prausnitzii* and related commensals [6]. As the disease progresses to cirrhosis, dysbiosis becomes further exacerbated, with the enrichment of pathogenic species such as *Clostridium bolteae*, *Streptococcus parasanguinis*, and *Klebsiella pneumoniae*, while beneficial genera such as *Blautia* and *Roseburia* continue to decline [7]. Notably, animal studies have demonstrated that dietary manipulation of gut microbiota in MASLD mice can directly influence the progression from MASLD to HCC [8], providing causal evidence for the role of the gut microecosystem in hepatocarcinogenesis.

Similar patterns of microbial dysbiosis and gut–liver axis dysfunction are observed in virus-associated HCC. Patients with cirrhosis resulting from chronic viral hepatitis (e.g., HBV infection) display marked reductions in *Bifidobacterium*, *Lachnospiraceae*, and *Ruminococcaceae*, leading to compromised intestinal barrier integrity and facilitating infection and inflammation [9]. Importantly, individuals with high HBV viral loads exhibit more severe dysbiosis, and this vicious cycle of "virus–microbiota–liver injury" substantially increases HCC risk among those with elevated viral titers [10].

Furthermore, HCV infection accelerates liver disease progression by inducing gut microbial imbalance. HCV-related dysbiosis is characterized by the abnormal expansion of *Clostridium* sensu

Table 1
Infection-associated microecological alterations across etiologies of chronic liver disease

Liver disease type	Characteristic microbial alterations
MASLD	↑ *Enterobacteriaceae* ↓*Faecalibacterium prausnitzii* ↓SCFA-producing bacteria
HBV	↓ *Bifidobacterium*, ↓ *Lachnospiraceae*, ↓ *Ruminococcaceae*
HCV	↑*Clostridium* sensu stricto
Liver cirrhosis	↓ *Faecalibacterium*, ↓ *Lachnospiraceae*, ↓ *Ruminococcaceae*, ↑ *Streptococcus*, ↑ *Veillonella*

stricto, which disrupts immune homeostasis and amplifies hepatic inflammatory responses [11].

Cirrhosis—the final common pathway of multiple chronic liver diseases—is marked by distinct microbial alterations. Beneficial anti-inflammatory and barrier-protective taxa, including *Faecalibacterium*, *Lachnospiraceae*, and *Ruminococcaceae*, show sustained depletion, while potentially pathogenic genera such as *Streptococcus* and *Veillonella* are significantly enriched [12]. This characteristic microbial signature correlates strongly with cirrhosis severity and promotes hepatocarcinogenesis by perpetuating hepatic inflammation and driving aberrant tissue repair (Table 1).

2 Mechanisms Underlying Infection-Associated Microecology in the Progression of Hepatocellular Carcinoma

The gut microecosystem plays a pivotal role in hepatocarcinogenesis through its intimate interplay with the gut–liver axis. Across the full continuum of chronic liver diseases—including viral hepatitis, alcoholic liver disease (ALD), and MASLD—structural and functional abnormalities of the gut microbiota continuously drive and accelerate disease evolution, from early inflammation and fibrogenesis to cirrhosis and ultimately malignant transformation. The core initiating event is the disruption of intestinal barrier integrity—commonly described as "leaky gut"—which enables pathological translocation of gut microbes and their metabolites. This translocation induces systemic inflammation, immune dysregulation, metabolic derangement, and direct genotoxic injury, collectively constituting a multifaceted cascade that promotes the development of HCC.

Leaky gut serves as the critical mechanistic bridge linking microbial dysbiosis to hepatic pathology. In chronic liver disease, impaired intestinal epithelial tight junctions permit the excessive entry of pathogen-associated molecular patterns (PAMPs), such as lipopolysaccharide (LPS), into the portal circulation. LPS activates Toll-like receptor 4 (TLR4) on hepatic immune cells—particularly Kupffer cells—triggering sustained NF-κB activation and downstream pro-inflammatory signaling. This fosters a chronic hepatic inflammatory milieu that drives fibrosis and cirrhosis, thereby establishing the pathological substrate for hepatocarcinogenesis. Notably, TLR4 activation by LPS enhances invasiveness and epithelial–mesenchymal transition (EMT) in HCC cell lines. Clinically, elevated serum LPS levels are consistently detected in patients with cirrhosis and HCC, and high circulating anti-LPS antibody titers are independently associated with increased HCC risk. Beyond LPS, increased intestinal permeability facilitates the translocation of bacterial DNA (a TLR9 agonist) and flagellin (a TLR5 ligand), amplifying inflammatory responses. Together, these mediators form a tightly interconnected, TLR-driven inflammatory network that fuels hepatocarcinogenesis [13]. Following intestinal barrier disruption, *Candidatus Saccharibacter* can colonize the liver and selectively activate TLR4–NF-κB signaling in senescent hematopoietic stem cells. This activation promotes NLRP3 inflammasome assembly and increases the production of IL-1β and IL-6, thereby accelerating HCC progression [14]. These findings suggest that microbial dysbiosis contributes to hepatocarcinogenesis not only by promoting leaky gut but also by enriching pro-metastatic or pro-inflammatory bacterial taxa that intensify pathogenic signaling cascades.

The progression of ALD and MASLD further reinforces this mechanistic paradigm. In ALD, ethanol and its metabolite acetaldehyde directly impair intestinal tight junctions, markedly increasing bacterial translocation. In MASLD, intestinal permeability is likewise elevated. Under dysbiotic conditions, dietary choline is converted to trimethylamine (TMA), reducing host phosphatidylcholine synthesis, impairing very-low-density lipoprotein secretion, and diminishing hepatic lipid export—mechanisms that directly initiate and exacerbate fatty liver disease [13].

3 Diagnosis and Treatment Strategies Targeting Infection-Associated Microecology in Hepatocellular Carcinoma

Given the central role of the gut microecosystem in hepatocarcinogenesis, therapeutic targeting of the gut microbiota–liver axis has emerged as a highly promising strategy for both prevention and disease modification. This axis is closely intertwined with numerous complications of chronic liver disease, making it an appealing multitarget pathway capable of delivering "one intervention–multiple benefits." Beyond reducing the risk of HCC development, interventions aimed at this axis can simultaneously mitigate hepatic fibrosis, alleviate portal hypertension, decrease the incidence of spontaneous bacterial peritonitis, and improve manifestations of hepatic encephalopathy.

Among the growing array of therapeutic approaches, antibiotics represent a potent means of interrupting the tumor-promoting gut–liver axis. By reducing the total bacterial load—particularly pathogenic taxa with high translocation capacity, such as *Enterobacteriaceae*—antibiotics effectively suppress the hepatic influx of pro-inflammatory microbial signals (e.g., LPS). Selective antibiotics such as vancomycin, which targets *Clostridium* species, can inhibit defined metabolic pathways and markedly reduce the production of procarcinogenic metabolites, including secondary bile acids such as deoxycholic acid (DCA).

Probiotics provide a microbial "ecological restoration strategy," restoring ecological balance disrupted in chronic liver disease. Supplementation with defined beneficial strains can strengthen intestinal barrier integrity, reduce pathogenic colonization, and reestablish immune–metabolic homeostasis. Clinical trials using *Lactobacillus* and *Bifidobacterium* species have reported varying degrees of benefit in MASLD, ALD, and hepatic encephalopathy [13]. Preclinical studies further demonstrate that extracellular polysaccharides or postbiotics derived from *Lactobacillus* species ameliorate MASLD and metabolic inflammation.

A more comprehensive ecological reconstruction is offered by fecal microbiota transplantation (FMT). By transferring a fully functional microbial ecosystem from healthy donors, FMT enables deep microbial reconstitution, restores impaired gut barrier integrity, and enhances the synthesis of protective metabolites such as short-chain fatty acids (SCFAs). Additionally, hepatic 16S rRNA abundance may serve as a biomarker of gut barrier failure and systemic microbial translocation, with potential utility in predicting immunotherapy response and identifying patients who may benefit from microbiota-modulating interventions [15].

Despite these promising avenues, substantial challenges hinder clinical translation. Broad-spectrum antibiotics can destabilize microbial homeostasis and promote antimicrobial resistance; in advanced cirrhosis, severe intestinal permeability limits the colonization and efficacy of probiotics; and the long-term safety, stability, and donor–recipient compatibility of FMT remain uncertain. Mechanistic complexity, interindividual heterogeneity, limited high-quality clinical evidence, and concerns regarding long-term safety continue to impede widespread adoption.

Future research must prioritize the identification of key molecular nodes mediating host–microbiota interactions, develop personalized ecologically informed interventions, and conduct rigorously designed, large-scale randomized controlled trials to validate clinical benefit. In summary, therapeutic strategies targeting the gut–liver axis provides a multifaceted approach to HCC prevention and management by interrupting core pathways of carcinogenesis while simultaneously improving the hepatic microenvironment. Although substantial translational hurdles remain, interventions aimed at modulating infection-associated microecology hold remarkable therapeutic versatility and cross-disease potential, marking a promising frontier in the translational medicine landscape for HCC prevention and treatment.

Conflicts of Interest

The authors have no conflicts of interest to declare that are relevant to the content of this chapter.

References

1. Murray CJ, Vos T, Lozano R, Naghavi M, Flaxman AD, Michaud C, Ezzati M, Shibuya K, Salomon JA, Abdalla S (2012) Disability-adjusted life years (DALYs) for 291 diseases and injuries in 21 regions, 1990–2010: a systematic analysis for the Global Burden of Disease Study 2010. Lancet 380(9859): 2197–2223
2. Llovet JM, Burroughs A, Bruix J (2003) Hepatocellular carcinoma. Lancet 362(9399): 1907–1917
3. Schnabl B, Brenner DA (2014) Interactions between the intestinal microbiome and liver diseases. Gastroenterology 146(6):1513–1524
4. Dapito DH, Mencin A, Gwak G-Y, Pradere J-P, Jang M-K, Mederacke I, Caviglia JM, Khiabanian H, Adeyemi A, Bataller R (2012) Promotion of hepatocellular carcinoma by the intestinal microbiota and TLR4. Cancer Cell 21(4):504–516
5. Miele L, Valenza V, La Torre G, Montalto M, Cammarota G, Ricci R, Masciana R, Forgione A, Gabrieli ML, Perotti G, Vecchio FM, Rapaccini G, Gasbarrini G, Day CP, Grieco A (2009) Increased intestinal permeability and tight junction alterations in nonalcoholic fatty liver disease. Hepatology 49(6): 1877–1887. https://doi.org/10.1002/hep.22848
6. de Vos WM, Tilg H, Van Hul M, Cani PD (2022) Gut microbiome and health: mechanistic insights. Gut 71(5):1020–1032. https://doi.org/10.1136/gutjnl-2021-326789
7. Behary J, Amorim N, Jiang XT, Raposo A, Gong L, McGovern E, Ibrahim R, Chu F, Stephens C, Jebeili H, Fragomeli V, Koay YC, Jackson M, O'Sullivan J, Weltman M, McCaughan G, El-Omar E, Zekry A (2021) Gut microbiota impact on the peripheral immune response in non-alcoholic fatty liver disease related hepatocellular carcinoma. Nat Commun 12(1):187. https://doi.org/10.1038/s41467-020-20422-7
8. Zhang X, Coker OO, Chu ES, Fu K, Lau HC, Wang Y-X, Chan AW, Wei H, Yang X, Sung JJ (2021) Dietary cholesterol drives fatty liver-associated liver cancer by modulating gut microbiota and metabolites. Gut 70(4): 761–774
9. Shu W, Shanjian C, Jinpiao L, Qishui O (2022) Gut microbiota dysbiosis in patients with hepatitis B virus-related cirrhosis. Ann Hepatol 27(2):100676. https://doi.org/10.1016/j.aohep.2022.100676
10. Chen C-J, Yang H-I, Su J, Jen C-L, You S-L, Lu S-N, Huang G-T, Iloeje UH, Group R-HS (2006) Risk of hepatocellular carcinoma across a biological gradient of serum hepatitis B virus DNA level. JAMA 295(1):65–73
11. Ali RO, Quinn GM, Umarova R, Haddad JA, Zhang GY, Townsend EC, Scheuing L, Hill KL, Gewirtz M, Rampertaap S, Rosenzweig SD, Remaley AT, Han JM, Periwal V, Cai H, Walter PJ, Koh C, Levy EB, Kleiner DE, Etzion O, Heller T (2023) Longitudinal multi-omics analyses of the gut-liver axis reveals metabolic dysregulation in hepatitis C infection and cirrhosis. Nat Microbiol 8(1): 12–27. https://doi.org/10.1038/s41564-022-01273-y
12. Qin N, Yang F, Li A, Prifti E, Chen Y, Shao L, Guo J, Le Chatelier E, Yao J, Wu L, Zhou J,

Ni S, Liu L, Pons N, Batto JM, Kennedy SP, Leonard P, Yuan C, Ding W, Chen Y, Hu X, Zheng B, Qian G, Xu W, Ehrlich SD, Zheng S, Li L (2014) Alterations of the human gut microbiome in liver cirrhosis. Nature 513(7516):59–64. https://doi.org/10.1038/nature13568

13. Yu LX, Schwabe RF (2017) The gut microbiome and liver cancer: mechanisms and clinical translation. Nat Rev Gastroenterol Hepatol 14(9):527–539. https://doi.org/10.1038/nrgastro.2017.72
14. Lau HC, Zhang X, Yu J (2025) Gut microbiome in metabolic dysfunction-associated steatotic liver disease and associated hepatocellular carcinoma. Nat Rev Gastroenterol Hepatol. https://doi.org/10.1038/s41575-025-01089-1
15. Schneider KM, Mohs A, Gui W, Galvez EJC, Candels LS, Hoenicke L, Muthukumarasamy U, Holland CH, Elfers C, Kilic K, Schneider CV, Schierwagen R, Strnad P, Wirtz TH, Marschall HU, Latz E, Lelouvier B, Saez-Rodriguez J, de Vos W, Strowig T, Trebicka J, Trautwein C (2022) Imbalanced gut microbiota fuels hepatocellular carcinoma development by shaping the hepatic inflammatory microenvironment. Nat Commun 13(1):3964. https://doi.org/10.1038/s41467-022-31312-5

Chapter 12

Microbial Omics Analysis for Multispecies Symbioses in Staple and Traditional Fermented Foods

Fan Wang and Haibo Zhang

Abstract

Fermented foods represent a cornerstone of global culinary traditions, underpinned by intricate microbial communities that drive flavor development and preservation. This chapter examines the microbial ecology and functional interactions across representative staple and traditional fermented foods, including fermented vegetables (sauerkraut, kimchi), fermented condiments (soy sauce, vinegar, miso, doubanjiang), and fermented alcoholic beverages (Chinese baijiu, wine, sake, Huangjiu). Recent meta-omics advances have revealed dynamic community succession, cross-feeding networks, quorum-sensing mechanisms, and key microorganisms contributing to flavor metabolites and health-promoting compounds. Furthermore, synthetic microbial communities and starter engineering are discussed as strategies to standardize fermentation, enhance quality, and develop novel functional products.

Key words Microbial community, Fermented vegetables, Fermented condiments, Fermented alcoholic beverages, Flavor metabolism

1 Introduction

Fermented foods constitute a diverse group of traditional products that have sustained human societies for millennia. These foods owe their distinct characteristics to the metabolic activities of complex microbial communities, which drive fermentation processes while conferring health benefits and preservation properties. Microbial communities in food fermentation processes exhibit spatial and temporal heterogeneity, driven by localized pH, temperature and moisture gradients, and resource availability. These conditions foster multispecies interactions, including syntrophy (metabolic cross-feeding), competition, and mutualism [1]. For instance, in tempeh fermentation, *Rhizopus oligosporus* secretes proteases and lipases to hydrolyze soy proteins and fats, while nitrogen-fixing bacteria (*Bacillus* spp.) enhance amino acid availability [2]. Such symbiotic

Tao Huang (ed.), *Microbial 'Omics: Methods and Protocols*, Methods in Molecular Biology, vol. 3033,
https://doi.org/10.1007/978-1-0716-5264-0_12,

relationships are critical to the efficiency and functionality of food fermentation processes.

Traditional culture-dependent methods, though foundational, struggle to capture the full diversity and functional complexity of microbial communities in staple and traditional fermented foods. Many keystone species remain unculturable, and these methods often overlook transient interactions and metabolic redundancies. Advances in high-throughput microbial omics, as well as the rapid iteration in data analysis software [3–6], have revolutionized our ability to dissect these symbioses, offering unprecedented resolution into microbial behavior and community dynamics [7–9]. Metagenomics provides a census of microbial populations and their functional gene repertoire [10–12]. By sequencing DNA directly from fermented samples, researchers can identify species, strains, and genes involved in substrate degradation, stress tolerance, and secondary metabolite production [13]. Moreover, microbial omics analysis allows us to observe shifts in the microbial community through various stages of fermentation and to explore metabolic networks within complex communities, provides insights into how microbial interactions can affect metabolite production and degradation, and expands the applications of these functional microbes [14–16].

One of the most significant implications of microbial omics in food fermentation processes is its potential to enhance the scalability and sustainability of fermentation, and omics-driven SSF innovations are paving the way for precision fermentation [17]. By understanding the microbial interactions and metabolic potentials, researchers can engineer optimal fermentation conditions tailored to the chosen substrates [18]. This is also beneficial for exploring functional microorganisms and genes, designing microorganisms or synthetic microbiomes and promoting their wide applications [19–21]. Ultimately, integrating microbial omics with fermentation processes contributes to a more informed and efficient production of fermented foods, which aligns with food safety regulations and consumer preferences for high-quality, health-promoting products [22]. This chapter integrates recent metagenomic, transcriptomic, and metabolomic studies to elucidate the ecological principles governing fermented food microbiomes. By bridging fundamental microbiology and applied food science, we aim to inspire innovations in starter culture design, process optimization, and functional food development.

2 Fermented Vegetables

2.1 Sauerkraut

Sauerkraut, one of the world's oldest and most widespread fermented cabbage products, showcases the complexity and richness of microbial communities in food preservation and flavor

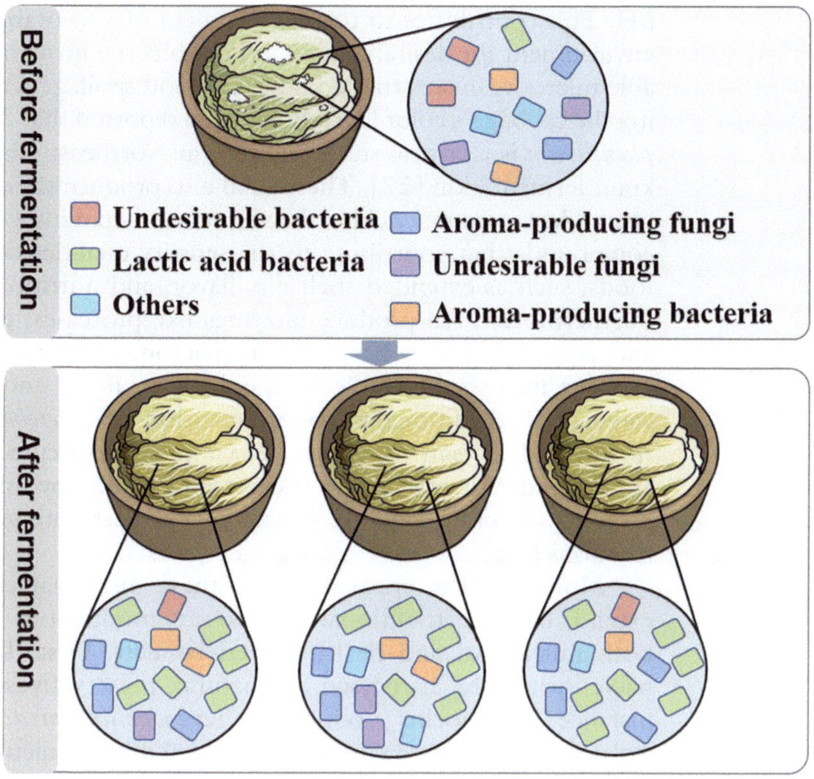

Fig. 1 The spontaneous fermentation of sauerkraut

enhancement. The primary and functional microbial composition of sauerkraut fermentation is generally dominated by lactic acid bacteria (LAB), such as *Leuconostoc*, *Lactobacillus*, *Weissella*, and *Pediococcus*, which undergo dynamic succession. Initially, undesirable bacteria including *Pseudomonas*, *Halomonas*, *Pectobacterium*, and *Clostridium* may be more abundant, as they are presented in untreated raw cabbages. *Clostridium* has been reported to lead to severe infections when the normal balance of gut microbiota is disrupted, which is considered to be undesirable [23]. As fermentation starts, *Lactococcus* and *Weissella* grow rapidly and, together with the bacteriostatic effect of salt, replace undesirable bacteria. Subsequently, the community succession continues with oxygen consumption, substrate changes, and acidity increase. Despite distribution differences, LAB strains show similar fluctuation patterns throughout sauerkraut fermentation (Fig. 1) [24]. Dominant fungal communities, though less studied, include *Tausonia*, *Wickerhamomyces*, *Dipodascus*, *Clavispora*, and *Candida*, which may contribute to secondary metabolite production [25].

For the sauerkraut fermentation, LAB are primarily responsible for conversion of sugars in cabbage to lactic acid and carbon dioxide, quickly lowering the environmental oxygen concentration and pH. This contributes to the development of a favorable anaerobic environment for desirable taxa and inhibits the growth of undesirable microorganisms that could cause food spoilage while preserving the cabbage's color [26]. It has been reported that *Lactobacillus plantarum* is a critical starter culture in Northeast Chinese sauerkraut fermentation [27]. These LAB also produce various metabolites like organic acids, bacteriocins, vitamins, and flavor compounds that contribute to the sensory qualities of fermented foods, such as extended shelf life, flavor, and nutritional content. Furthermore, LAB produce enzymes like proteases, glycosidases, and amylases, which break down macromolecular substances in cabbage into smaller molecules, enhancing flavor and nutritional value [28]. Additionally, co-fermentation of *Lactobacillus plantarum* and *Zygosaccharomyces rouxii* can help degrade biogenic amines and reduced nitrites (decreased by 76%) by inhibiting the growth of gram-negative bacteria or performing enzymatic reactions [29].

Moreover, the applications of these microbial communities extend beyond traditional sauerkraut production, influencing both nutritional and health-related benefits. First, LAB-derived bacteriocins are harnessed as natural preservatives [30]. For instance, bacteriocins produced by *Lactobacillus paracasei* HD1-7 isolated from Chinese sauerkraut juice exhibit antimicrobial activity against *Bacillus subtilis* [31]. Then, research has suggested that the live microorganisms present in sauerkraut may aid in digestion, enhance immune responses, and exert anti-inflammatory effects [32–35]. Other applications involve the use of starter cultures (e.g., *Lactobacillus plantarum*) to standardize fermentation processes and improve fermentation efficiency and product quality, ensuring consistent flavor, acidity, and shelf-life [36, 37]. Additionally, synthetic microbial communities mimicking sauerkraut ecosystems are engineered to notably reduce fermentation duration [38] and to enhance bacteriocins [39]. Therefore, sauerkraut serves as an excellent model for exploring the diverse and complex world of microbial communities in solid-state fermentation (SSF), offering valuable insights into microbial interactions, flavor development, and health applications in fermented foods. Further research on the interplay between the diverse microbial functions, such as high-salt environment growth, acid production, exopolysaccharide synthesis, and volatile compound generation, would be helpful.

2.2 Kimchi

Kimchi, a traditional fermented vegetable dish, is primarily composed of cabbage, radishes, garlic, and various spices, embodying a rich tapestry of flavors that arise from complex microbial interactions during fermentation. The main microbial community of kimchi is predominantly characterized by LAB, especially species of *Leuconostoc*, *Lactobacillus*, *Weissella*, and *Lactococcus*. Furthermore, a small number of yeasts also exist in kimchi, such as *Pichia*,

Kazachstania, *Saccharomyces*, and *Rhodotorula* [40, 41]. Studies have shown that the microbial composition of kimchi varies significantly depending on several environmental factors, including temperature, duration of fermentation, and the specific ingredients used. For example, lower fermentation temperatures and longer fermentation times typically promote *Leuconostoc* species, while warmer temperatures tend to favor the growth of *Lactobacillus* species [42]. The pH, salt concentration, and even the microbiome of the raw materials have also been shown to exert significant influence on the development of the microbial community in kimchi. For example, studies have shown that at low salt concentrations, *Leuconostoc gelidum* is the dominant microorganism, while at high salt concentrations, *Aerococcus funiformis* becomes the primary species [43]. Seasonal harvesting of kimchi cabbage further influences microbial profiles: *Latilactobacillus* prevails in autumn-harvested batches, while *Weissella* dominates in spring, summer, and winter samples [44]. Raw ingredients, such as garlic and red pepper, also introduce niche-specific microbes (e.g., *Leuconostoc* from cabbage, *Weissella* from garlic), which determine starter culture efficacy and fermentation outcomes [45].

Advanced microbial omics analyses, including metagenomics and metabolomics, have begun to elucidate the relationship between specific microbial taxa and flavor compounds. *Leuconostoc* species are critical for the initial stages of fermentation due to their ability to produce carbon dioxide and lactic acid, leading to the characteristic tanginess and effervescence of kimchi. *Lactobacillus* species dominate in the later stages of fermentation and contribute to the final acidity and flavor profile. *Leuconostoc mesenteroides* synthesizes dextran, contributing to the overall viscosity of the fermentation milieu. Meanwhile, the organic acids produced by these LAB could create an unfavorable environment for pathogenic bacteria [46]. Metabolomic profiling identifies ethyl esters (e.g., ethyl lactate) and amino acid derivatives (e.g., glutamic acid) as key flavor compounds, modulated by microbial enzymes (such as proteases and glycosidases) during fermentation. For example, *Lactobacillus plantarum* produces proteases that degrade proteins into amino acids, imparting a savory taste to kimchi [47]. Meanwhile, yeasts produce fruity aromatic compounds that mitigate kimchi's acidic and moldy flavors, positively influencing its sensory qualities. However, high yeast loads, particularly those forming white colonies on kimchi's surface during late fermentation, can negatively impact product quality [48]. Furthermore, *Rhodotorula* has the ability to synthesize triglycerides, phospholipids, sterol esters, polyol esters of fatty acids, β-carotene, γ-carotene, torulene, and torularhodin [49].

In recent years, the application of microbial communities in kimchi production has gained renewed attention not only for its implications in flavor enhancement but also for its potential health benefits. Starter cultures, such as *Leuconostoc mesenteroides*, are engineered to standardize fermentation and enhance functional properties, including anti-inflammatory metabolites (e.g., short-chain fatty acids) [50]. And culture starters elevate lactic acid and leucine levels, improving kimchi's nutritional profile [46]. The probiotic properties of certain LAB strains present in kimchi, such as *Lactobacillus kimchii* and *Lactobacillus plantarum*, also have been widely studied for their positive effects on gut health, immune function, cholesterol levels, anticancer activity, and antimicrobial properties [51–55]. Moreover, the fermentation of plant-based substrates, like those used in kimchi, provides a sustainable model for producing functional foods rich in bioactive compounds [52]. Additionally, kimchi can be used to mine functional microbial resources. For example, *Bacillus methylotrophicus* isolated from kimchi can produce gibberellins and support plant growth and enhances the nutrition of lettuce [56]. Antifungal agents produced by LAB from kimchi can inhibit spoilage molds in food preservation [57]. The ongoing research exploring these microbial applications and their relationships within multispecies symbioses would continue to transform the understanding of kimchi fermentation, promising innovation in both culinary arts and nutrition science.

3 Fermented Condiments

3.1 Soy Sauce

Soy sauce, a traditional fermented condiment originating from East Asia, exemplifies the complexity and dynamism of microbial communities in food fermentation processes. The microbial community composition of soy sauce is predominantly characterized by a consortium of bacteria, yeasts, and molds. Bacterial genera such as *Weissella*, *Staphylococcus*, *Tetragenococcus*, *Bacillus*, *Lactobacillus*, *Streptococcus*, *Enterococcus*, *Kurthia*, and *Klebsiella*, mold genera like *Aspergillus*, as well as yeasts like *Zygosaccharomyces*, *Candida*, and *Debaryomyces*, are the main components of soy sauce microbial communities across fermentation stages [58]. Environmental factors, including temperature, humidity, and substrate composition, as well as fermentation containers, significantly influence the microbial community's dynamics [59–61]. For instance, higher humidity levels promote the growth of molds, while temperature fluctuations can affect the metabolic activities of different microbial taxa, ultimately shaping the fermentation characteristics and flavor profile of soy sauce. Additionally, the use of specific raw materials, such as the proportion of soybeans to wheat, can alter the microbial community, leading to diverse soy sauce varieties characterized by unique flavors and aromas. Furthermore, high salt content inhibits the growth of most microorganisms but promotes the proliferation of halotolerant species like *Zygosaccharomyces rouxii* [62].

Interactions among these microbial communities are central to the fermentation process, contributing to the production of complex flavors and aromas synonymous with soy sauce. During the initial fermentation stages, *Aspergillus oryzae* plays a critical role in the hydrolysis of soy proteins and carbohydrates, producing essential amino acids, sugars, and bioactive compounds, which serve as substrates for subsequent fermentation by yeasts and bacteria. Then, the production of alcohol, organic acids, and volatile compounds by yeasts and LAB not only enhances flavor complexity but also creates an environment that inhibits the growth of spoilage organisms [58]. *Zygosaccharomyces rouxii* was a highly salt-tolerant yeast, playing an important role in soy sauce fermentation. The formation of desirable esters, phenols, and other aromatic compounds during fermentation can also be attributed to the complex biochemical interactions among different microorganisms, each contributing unique metabolites that define the final sensory attributes of soy sauce.

Advances in microbial omics analysis have facilitated the characterization of these communities at a deeper level, enabling the identification of specific strains with desirable traits, such as enhanced flavor production or improved fermentation efficiency of soy sauce production. For instance, *Aspergillus oryzae* JQ09, *Wickerhamomyces anomalus* HJ07, *Zygosaccharomyces rouxii* JZ11, *Staphylococcus carnosus* QJ26, *Weissella paramesenteroides* ZJ19, and *Tetragenococcus halophilus* GY03 were used to rebuild a synthetic microbial community, enabling salt-reduced soy sauce fermentation [63]. *Weissella paramesenteroides* JL-5 and *Bacillus amyloliquefaciens* JDF1 have been used to produce salt-reduced soy sauce with enhanced aroma and saltiness [64]. The integration of microbial omics tools in soy sauce production underscores the potential of leveraging natural microbial diversity to innovate traditional fermentation processes, ensuring the preservation of cultural heritage while meeting contemporary consumer demands for flavor quality and nutritional value.

3.2 Vinegar

Vinegar production relies on complex microbial communities whose dynamics and metabolic activities are shaped by raw material composition, fermentation parameters, and inoculation strategies. Traditional vinegar fermentation involves four successive stages: *koji* preparation, saccharification of starch and alcohol fermentation, acetic-acid fermentation, and maturation (Fig. 2). Metagenomic studies reveal that molds like *Aspergillus niger*, harboring starch-degrading ability, hydrolyze the carbohydrates into sugars, while yeasts such as *Saccharomyces*, *Candida*, *Hanseniaspora*, *Pichia*, and *Zygosaccharomyces rouxii* dominate the initial phase, converting sugars into ethanol. Subsequently, AAB (*Acetobacter*, *Gluconobacter*, *Gluconacetobacter*), LAB (*Lactobacillus*, *Pediococcus*), and *Bacillus* are the main genera in the acetic-acid

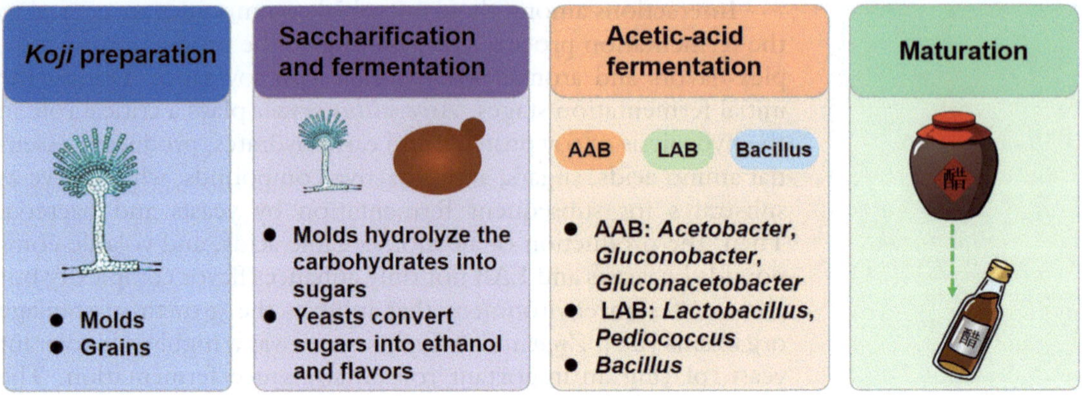

Fig. 2 The vinegar production process and the microorganisms involved

fermentation of vinegars (Fig. 2) [65, 66]. Various factors influence this microbial composition, including the substrate used for fermentation, environmental conditions like temperature and pH, and the presence of other microbial species [67].

Interactions between microbial communities in vinegar not only influence fermentation outcomes but also contribute significantly to flavor development. The complex interplay between AAB and yeasts during fermentation results in a dynamic ecosystem where various biochemical pathways are activated. For instance, yeasts facilitate the initial conversion of sugars into alcohol and glycerol, which serve as substrates for AAB and contribute to mouthfeel. Conversely, AAB may release siderophores that promote yeast iron uptake, enhancing ethanol production [68]. Competitions for resources, such as between *Acetobacter* and *Komagataeibacter* species, can influence acetic acid yields [69]. Flavor profiles arise from microbial cross talk: yeasts synthesize esters (e.g., ethyl acetate) and higher alcohols (e.g., isoamyl alcohol), while AAB generate acetic acid and other organic acids (e.g., gluconic acid). Secondary metabolites like polyphenols and bioactive compounds (e.g., gluconacetobacter cellulose in apple vinegar) further enrich sensory and health attributes [68]. In addition, *Lactobacillus* spp. in rice vinegar fermentation produce lactic acid and diacetyl, enhancing umami notes and aroma complexity [70]. And *Aspergillus niger* has been reported to produce L-malic acid [71]. Thus, the synergistic and antagonistic relationships between different microorganisms can shape the overall microbiota, further influencing flavor complexity and depth.

The applications of microbial communities in vinegar production extend beyond traditional fermentation practices. Starter cultures derived from omics-guided strain selection enhance fermentation reproducibility and shelf stability. Functional vinegar products, enriched with probiotic yeasts or bacteriocins from AAB, are emerging in the health sector [72]. In industrial vinegar production, omics tools identify metabolic bottlenecks, enabling strain engineering for higher acetic acid titers. Traditional vinegar systems, with their multispecies symbioses, also serve as models for studying microbial succession and stress adaptation, offering insights applicable to environmental microbiology and synthetic biology.

3.3 Miso

Miso is a traditional Japanese fermented soybean paste made through SSF processes. The primary microbial community that characterizes miso fermentation is predominantly composed of bacteria (*Bacillus*, *Enterococcus*, *Staphylococcus*, and various LAB) and yeast (*Zygosaccharomyces rouxii*) with the species *Aspergillus oryzae* playing a critical role as the starter culture. The microbial community composition in miso can be influenced by several factors, including the type of raw materials (e.g., soybeans, rice, salt), fermentation temperature, duration, and environmental conditions such as humidity and oxygen levels. The substrate's nutrient availability also plays a pivotal role, as different carbon and nitrogen sources can select for specific microbial taxa. Consequently, variations in these factors can lead to distinct miso varieties—characterized by flavor profiles, umami, and aromas—due to changes in the prevailing microbial populations [73, 74].

The interactions between the microbial communities in miso fermentation are intricate, often described through synergistic, antagonistic, or commensal relationships. For instance, *Aspergillus oryzae* initiates the fermentation process by breaking down polysaccharides into sugars, which are subsequently utilized by LAB and yeasts, creating a complex network of cross-feeding interactions. *A. oryzae* also initiates proteolysis, releasing amino acids (e.g., glutamate) and peptides that enhance umami taste, while other microorganisms such as *Lactobacillus* species metabolize glucose to lactic acid, acidifying the environment and triggering yeast fermentation of residual sugars into ethanol and esters (e.g., ethyl acetate), which impart fruity and alcoholic notes [75]. And the LAB-derived organic acids also suppress pathogen growth, ensuring food safety [76]. Furthermore, omics integration elucidates how these microbial interactions are mediated by quorum sensing and nutrient competition. This symbiotic relationship not only ensures the stability of the microbial community but also enhances the development of flavor and aroma compounds often associated with miso.

The applications of microbial communities and specific microorganisms in the production and enhancement of miso are continually expanding. As with other fermented foods, by understanding the metabolic pathways and interactions among microbial populations, producers can select for beneficial strains to optimize traditional fermentation processes, improve flavor profiles, and enhance nutritional values. Furthermore, the manipulation of microbial consortia can also pave the way for the creation of novel miso products tailored for specific health benefits or culinary applications. For example, the incorporation of probiotics in miso through selected LAB strains could yield health-promoting properties while appealing to a broader consumer base [73]. The study of microbial communities and their functions through omics approaches provides a pathway toward innovating traditional practices in miso production, highlighting the significance of microbial symbiosis in achieving desirable sensory and nutritional outcomes in SSF processes.

3.4 Doubanjiang

Doubanjiang is a traditional Chinese fermented soybean paste with a unique spicy flavor, produced through spontaneous SSF. The primary microbial community composition of Doubanjiang primarily consists of bacteria (*Staphylococcus*, *Bacillus*, *Weissella*, *Enterococcus*, *Pediococcus*, *Tetragenococcus*, *Lactobacillus*), yeasts (*Candida*, *Millerozyma*, *Zygosaccharomyces*, *Wickerhamomyces*, *Meyerozyma*), and certain fungi (*Aspergillus*, *Tulasnella*) [77, 78]. Various factors substantially affect the microbial community composition [77, 79, 80]. For instance, raw materials provide the nutrients and substrates for microbial growth, resulting in varying microbial communities. In addition, salt concentration affects microbial growth and metabolic activities. *Lactobacillus* emerged as the predominant bacteria in low-salt samples, while *Staphylococcus* is the dominant in the high-salt samples. Seasonal temperature variations lead to seasonal distribution patterns of microbial communities in Doubanjiang. Fermentation equipment and environmental conditions, such as hygiene levels and aeration, also affect microbial community composition and succession.

Interactions between microbial communities in Doubanjiang fermentation are multilayered and dynamic. Synergistic interactions, such as those between bacteria and yeasts, can result in enhanced biochemical transformations that lead to the synthesis of volatile compounds, contributing significantly to the flavor and aromas. For example, *Bacillus* exhibits strong protease activity, capable of breaking down soybean proteins into peptides and amino acids, providing nitrogen sources for other microorganisms and contributing to the umami flavor of Doubanjiang [81]. *Lactobacillus* produces lactic acid through fermentation, lowering the pH of the environment, creating favorable conditions for the growth of other acid-tolerant microbes, inhibiting the growth of spoilage organisms, and enhancing the acidity and freshness of Doubanjiang [82]. *Aspergillus* secretes a wide range of enzymes, such as proteases, amylases, and lipases, which break down proteins, starches, and fats in soybeans into smaller molecules, enriching the flavor compounds of Doubanjiang; *Saccharomycetales* promote the formation of jiang flavor via synergistic effect with *Lactobacillales* [77]. *Zygosaccharomyces rouxii* produces enzymes like esterases and proteases, generating various volatile compounds such as alcohols, esters, aldehydes, and ketones, which impart the distinctive aroma of Doubanjiang [83]. Additionally, *Pseudomonas* is highly correlated with volatile compounds like 3-methyl butanal and 2-methyl butanal; *Bacillus* and *Lactobacillus* also positively influence certain flavor compounds, such as Asp, Leu, β-ionone, and 2-acetylpyrrole [84].

The microbial diversity present in Doubanjiang can be exploited to develop starter cultures aimed at standardizing production while enhancing flavor and safety. For instance, inoculating *Bacillus* and *Aspergillus* strains with strong protease activity can enhance protein breakdown, increase amino acid content, and amplify the umami flavor [85]. Recent advances in microbial omics analysis, such as metagenomics and metabolomics, provide insight into the complex yet delicate balance of microbial interactions, paving the way for optimizing fermentation processes. This knowledge can lead to innovative applications in developing new products that harness the rich diversity of microbial species found in Doubanjiang, such as functional foods and nutraceuticals targeting specific health benefits. Moreover, the insights gained from this microbial analysis can also be applied in biotechnological advancements, including sustainable fermentation practices that maximize resource efficiency while minimizing waste [86].

4 Fermented Alcoholic Beverages

4.1 Chinese Baijiu

Chinese Baijiu (Chinese distilled spirit or Chinese liquor) is produced through a complex SSF process with *Jiuqu* as starter (Fig. 3). Different types of *Jiuqu* have different dominant molds, bacteria, and yeast, which predominately determine the microbial consortium and its enzymes for liquor fermentation as well as contribute to the flavor formulation [87]. In addition, the microbial community composition of Chinese Baijiu is also influenced by various factors including the type of raw materials used (such as sorghum with or without rice, wheat), the fermentation conditions (temperature, humidity, and duration), and the geographic region in which the production occurs [88]. Additionally, the age and composition of *Zaopei* (fermented grains) and pit mud determine microbial succession; for example, *Lactobacillus* dominates earlier in aged pit mud fermentations, accelerating ethanol production, whereas new pit mud fosters prolonged *Lactobacillus* activity, increasing acidity [89].

The interactions among microbial communities within the fermentation environment are intricate and multifaceted. In Baijiu fermentation systems, mutualistic relationships are particularly important. For instance, *Lactobacillus* produces organic acids that lower the pH, creating favorable conditions for *Saccharomyces*

Fig. 3 Differences raw materials and source of microbial community among Chinese baijiu, wine, sake, and Huangjiu

cerevisiae fermentation. In return, ethanol produced by *Saccharomyces cerevisiae* suppresses the growth of other microorganisms, providing a competitive advantage to *Lactobacillus* [90]. Such synergistic interactions promote the stability and efficiency of the fermentation process. This dynamic can lead to a stratified microbial ecosystem where community members engage in competitive and cooperative behaviors, facilitating complex biochemical pathways that yield a rich array of volatile compounds associated with desirable flavors and aromas [91]. Quorum sensing (QS) is another significant interaction mechanism. Microbes release chemical signal molecules to regulate microbial gene expression and behavior, such as biofilm formation and secretion of lipopeptides, thereby influencing microbial community structure and function of Chinese Baijiu [92, 93]. Moreover, the combined action of these microorganisms results in the formation of various flavor compounds in Chinese Baijiu, such as alcohols, aldehydes, carboxylic acids, esters, ketones, and phenols, giving Baijiu its distinctive aroma and flavor profile [89]. *Pichia anomala* generates acetic acid, octanoic acid, lactic acid, phenylethyl alcohol, 2,3-butanediol, ethyl myristate, and 2-furanethanol; *Saccharomyces cerevisiae* releases monoterpenes from glycoside precursors through β-glucosidase secretion; and *Lactobacillus* promotes the Maillard reaction, producing lactic acid-related organic acids and diacetyl. Thus, different types of Chinese Baijiu exhibit distinct flavor characteristics due to variations in microbial community composition and metabolic pathways [94].

The knowledge gained from studying microbial communities and their interactions in Chinese Baijiu production can be translated into practical applications for enhancing liquor quality and consistency. For instance, the inoculation of *Bacillus* into *Daqu* can alter its microbial community structure, enriching functional microorganisms and improving flavor attributes [95]. Additionally, the insights gained can guide fermentation technology improvements, such as bioprocess optimization and the development of bioreactors tailored to support specific microbial interactions [90]. Furthermore, the study of microbial communities in Chinese Baijiu provides insights into the ecological mechanisms of microbial symbiosis and interactions, offering valuable references for microbial community ecology research. In turn, it helps uncover the mysteries of Chinese Baijiu brewing, advancing the theoretical and technological development of traditional fermentation industries.

4.2 Wine

Wine production is a complex process involving the participation of multiple microbial communities. From the vineyard to the winery, microbes play a decisive role in shaping the characteristics of wine (Fig. 3). The main players in the microbial community of wine are *Saccharomyces cerevisiae*, which dominates alcoholic fermentation, alongside wild yeasts such as *Hanseniaspora*, *Kluyveromyces*, and non-*Saccharomyces* species, which contribute to aroma and flavor profiles during fermentation [96]. LAB, predominantly *Oenococcus oeni*, are responsible for malolactic fermentation. While AAB like *Acetobacter* and *Gluconobacter* can influence wine quality during fermentation and storage [97]. Factors affecting these communities include grape variety, terroir, fermentation temperature, and the winemaker's choices regarding inoculation practices and sanitation. For instance, research has shown that terroir not only encompasses the geographical and climatic factors but also includes the microbiota present in the vineyard, which can transition into the winery ecosystem, leading to variable wine profiles from year to year [98, 99]. In addition, spontaneous fermentation involves a greater diversity of fungal communities but fewer bacterial communities compared to inoculated fermentation [97].

Inter-species interactions within microbial communities are also critical to understanding how flavor compounds are produced during fermentation. Multi-omics analyses reveal that the interaction between yeasts and bacteria contributes over 60% of geographic flavor markers [100]. For example, *Saccharomyces* is the primary driver of alcoholic fermentation and is closely associated with the biosynthesis of major metabolites such as alcohols, acids, and esters in wine. LAB, on the other hand, can convert malic acid into lactic acid during malolactic fermentation, reducing wine acidity and enhancing its flavor complexity. Some LAB, such as *Lactobacillus*, can release free aromatic compounds by secreting glycosidases and alter the concentration of ester compounds in wine through esterase secretion, thereby influencing its overall flavor. *Bacillus velezensis* BI1 has been reported to increase total esters by 20% while suppressing higher alcohols, improving sensory profiles [101]. Interactions between yeasts also lead to the synergistic production of various compounds. For example, non-*Saccharomyces* yeasts can coexist with *Saccharomyces* and hydrolyze polysaccharides and produce compounds like glycerol and higher alcohols, which contribute to the mouthfeel and aroma of wine [100].

With advancements in microbial omics technology, winemakers can now harness specific strains or synthetic microbial communities for controlled fermentation processes, leading to improved consistency and quality in wine production [102, 103]. Additionally, the application of microbial inoculants, including carefully selected non-*Saccharomyces* yeasts or LAB, can lead to tailored flavor enhancements and greater stability in wines [104, 105]. This strategic orchestration of microbial communities, informed by omics analyses, opens new avenues for innovation in the wine industry, enabling the production of distinctive wines that resonate with consumers' preferences while maintaining the integrity of the traditional winemaking crafts [106]. Furthermore, microbial engineering focuses on enhancing functional components like γ-aminobutyric acid (GABA) and melatonin, aligning with health-conscious consumer trends [107]. By embracing the symbiotic relationships inherent in multispecies interactions during fermentation, researchers and winemakers can unlock the full potential of their products, cultivating wines that are not only rich in flavor but are also a testament to the intricate biological processes at play.

4.3 Sake

Sake, a traditional Japanese alcoholic beverage, is a product of a unique fermentation process that involves the intricate interplay of various microorganisms, predominantly yeasts and molds. The primary microbial community composition of sake is dominated by *Saccharomyces cerevisiae* and *Aspergillus oryzae* (Fig. 3) [108]. LAB also present in sake with a relatively small quantity [109]. Several factors influence these microbial communities, including the specific rice varieties used, environmental conditions such as temperature and humidity, and the presence of other microbial species during fermentation. The rice used for sake brewing, particularly sake-specific rice varieties like Yamada Nishiki, have unique characteristics such as low protein content, large grain size, and the presence of a *shinpaku* (starchy core) in the center of the grain, which affect the growth and metabolic activities of microorganisms during brewing. Additionally, the inoculum process, where specific yeast strains or molds are introduced, can dictate the dominance of certain microorganisms, thus influencing the overall sensory profile of the final product.

Interactions between microbial communities in sake brewing are complex and multifaceted, where cooperation and competition among species drive fermentation outcomes. *Aspergillus oryzae* not only initiates starch breakdown through the production of amylase enzymes but also creates an environment favorable for *Saccharomyces cerevisiae* growth by producing organic acids, which inhibit the growth of unwanted microorganisms [108]. In traditional kimoto brewing, LAB produce lactic acid, which inhibits the growth of other microorganisms while interacting with sake yeast to generate compounds that contribute to the flavor of sake. However, ethanol-tolerant LAB and interactions between yeast and LAB have been reported to cause spoilage of sake [109]. The relationship between microorganisms and flavor compounds is particularly noteworthy—*Saccharomyces cerevisiae* can produce varying concentrations of esters, higher alcohols, and other volatile compounds, contributing to the complex aroma and taste profile that sake aficionados appreciate. For example, isoamyl acetate and ethyl caproate impart a fruity aroma reminiscent of apples or bananas, while organic acids like lactic acid, malic acid, and succinic acid influence the taste of sake [110].

Traditionally, natural brewing practices are susceptible to environmental factors and may result in inconsistent product quality. Modern sake brewing increasingly employs pure cultures of sake yeast to enhance the stability and consistency of the product. Advances in microbial omics analyses allow for detailed characterization of microbial populations and their metabolic outputs, which can be harnessed to optimize brewing processes. For instance, understanding the genetic makeup of *Aspergillus oryzae* can lead to the development of specialized strains that enhance saccharification efficiency, thereby improving yield [111]. Additionally, insights derived from microbial community analysis enable brewers to implement biocontrol strategies that leverage beneficial microbes to suppress spoilage organisms, thus ensuring consistency and quality in sake production [112]. As such, the integration of microbial omics into sake brewing not only enhances traditional practices but also paves the way for novel applications that can revolutionize the industry, making it a vibrant field of study at the intersection of microbiology, food science, and sensory analysis.

4.4 Huangjiu (Chinese Rice Wine)

Huangjiu, a traditional Chinese rice wine, exemplifies the intricate dynamics of microbial communities and their fermentation processes, predominantly employing SSF techniques. The microbial community composition of Huangjiu depends on the used starter (Fig. 3) and is predominantly comprised of yeasts, molds, and bacteria, with species such as *Saccharomyces*, *Aspergillus*, *Rhizopus*, *Mucor*, *Bacillus*, and *Lactobacillus* spp. being integral to its production [113]. Factors influencing this microbial composition include the microbial diversity in raw materials, the conditions of fermentation (e.g., temperature, humidity), and the geographical origin of the rice and starter cultures utilized in the fermentation process.

For instance, glutinous rice, with its high amylopectin content, is favored for traditional aromatic Huangjiu production [114]. Additionally, fermentation techniques and processing procedures, such as fermentation duration and saccharification methods, also play a role in shaping microbial communities.

During Huangjiu fermentation, microorganisms exhibit complex symbiotic and synergistic relationships. For example, *Aspergillus* breaks down starch into sugars, providing substrates for *Saccharomyces* fermentation. Meanwhile, *Saccharomyces* metabolizes sugars to produce ethanol and carbon dioxide, creating an anaerobic environment that promotes the growth of LAB like *Lactobacillus.* LAB, in turn, produces organic acids that influence the pH of the fermentation system, further affecting the metabolic activities of other microorganisms [115]. The relationship between microorganisms and flavor in Huangjiu is profoundly shaped by these interactions. Microbial metabolism produces various flavor compounds, such as esters, alcohols, acids, and aldehydes, which together shape the unique flavor of Huangjiu. For example, *Saccharomyces* produces ethanol and esters like ethyl acetate, which impart a rich aroma to Huangjiu. LAB generate organic acids such as lactic acid and acetic acid, adding a unique sourness. The interaction between yeast and LAB can produce compounds like ethyl lactate and ethyl acetate, further enriching the flavor profile of Huangjiu. Furthermore, microorganisms like *Rhizopus* and *Mucor* secrete proteases and lipases, breaking down proteins and lipids in the raw materials into amino acids, peptides, fatty acids, and other substances. These compounds not only serve as precursors for flavor formation but also directly contribute to the flavor of Huangjiu.

The application of microbial communities in Huangjiu production primarily focuses on starter cultivation and microbial strain selection. Traditional starters often contain complex microbial communities with unstable fermentation performance. By using microbial separation and cultivation techniques, specific strains can be isolated from starters and then mixed or cocultivated to create artificial starters. This approach enhances the saccharification power and fermentation capacity of starters, improving the quality and consistency of Huangjiu. Inoculation strategies using functional strains, such as *Bacillus subtilis* spores, enhance flavor consistency by upregulating 13 key aroma compounds while suppressing undesirable metabolites [116]. *Saccharomyces cerevisiae* SC-6 was screened and used to reduce its bitterness [117]. Additionally, a synthetic microbial community not only replicated the major flavor profiles of traditional pour-over rice wine but also further proved that the core species directly determine the main flavor of pour-over rice wine [118].

5 Summary and Perspectives

Traditional fermented foods represent intricate ecosystems in which microbial symbiosis drives product stability, flavor complexity, and functional attributes. Across diverse fermentation systems including fermented vegetables, condiments, and alcoholic beverages, common patterns of microbial succession, interspecies cooperation, and metabolic complementarity emerge. Lactic acid bacteria, yeasts, molds, and various functional bacterial taxa often coexist and interact through nutrient cross-feeding, quorum sensing, and niche partitioning. These interactions not only stabilize fermentation environments but also shape key sensory compounds such as organic acids, esters, alcohols, and volatile flavor molecules. Despite differences in raw materials, salinity, environmental conditions, and cultural practices, a striking commonality across traditional fermentation systems is the presence of multispecies consortia whose synergies are essential for product quality, safety, and nutritional outcomes.

Furthermore, multi-omics technologies will continue to revolutionize our understanding of traditional fermentation. High-resolution metagenomics, metatranscriptomics, metabolomics, and single-cell approaches are enabling deeper insights into microbial dynamics, functional gene expression, metabolic pathways, and interspecies signaling processes. These tools will support the precise characterization of keystone taxa, the decoding of flavor metabolism networks, and the identification of microbial markers linked to fermentation performance or regional characteristics. Integration of machine learning with omics datasets will further enhance predictive modeling of fermentation processes, enabling real-time monitoring and control. Synthetic microbial community engineering, guided by omics data, also presents emerging opportunities to rationally design fermentation consortia for improved flavor, reduced salt content, higher fermentation efficiency, and enhanced functional metabolite production while preserving the cultural heritage of traditional foods.

Notably, the vast microbial diversity embedded within traditional fermentation ecosystems provides an invaluable reservoir for microbial resource mining. Many strains demonstrate desirable traits such as salt tolerance, acid resistance, production of bioactive compounds, enzymatic activities, or specific aroma-generating pathways. These native microbes represent promising candidates for developing next-generation starter cultures, functional probiotics, biopreservatives, and biotechnological enzymes with industrial relevance. Furthermore, exploring microbial resources from traditional foods can support innovation in plant-based foods, low-salt fermentation, sustainable bioprocessing, and health-oriented functional food development. Expanding culture collections, applying high-throughput isolation strategies, and coupling microbial screening with phenotypic and genomic data will greatly accelerate the discovery and application of these microbial resources.

Conflict of Interest

The authors have no conflicts of interest to declare that are relevant to the content of this chapter.

References

1. Liu L, Chen G, Liu J, Bao W, Li X, Yang K, Shi S, Zhao B, Wang Q, Cao X, Wang J, Men X, Wang F, Zhang G, Zhang L, Zhang H (2023) Sequential production of secondary metabolites by one operon affects interspecies interactions in *Enterobacter* sp. CGMCC 5087. Innov Life 1(2):100023. https://doi.org/10.59717/j.xinn-life.2023.100023
2. Yarlina VP, Djali M, Andoyo R, Nurmilah S, Lani MN (2024) Metagenomic insights into enhancing protein content and digestibility in Jack Bean (Canavalia ensiformis) tempeh: unraveling microbial dynamics during fermentation. Appl Food Res 4(2):100588. https://doi.org/10.1016/j.afres.2024.100588
3. Yu G (2024) Thirteen years of clusterProfiler. Innovation 5(6):100722. https://doi.org/10.1016/j.xinn.2024.100722
4. Wu T, Hu E, Xu S, Chen M, Guo P, Dai Z, Feng T, Zhou L, Tang W, Zhan L, Fu X, Liu S, Bo X, Yu G (2021) clusterProfiler 4.0: a universal enrichment tool for interpreting omics data. Innovation 2(3):100141. https://doi.org/10.1016/j.xinn.2021.100141
5. Dong P, Chen Y, Wei Y, Zhao X, Wang T, Jiang S, Xu J, Ren T, Li M, Zhang L (2025) Dix-seq: an integrated pipeline for fast amplicon data analysis. Innov Life 3(1):100120. https://doi.org/10.59717/j.xinn-life.2024.100120
6. Xu S, Zhan L, Tang W, Wang Q, Dai Z, Zhou L, Feng T, Chen M, Wu T, Hu E, Yu G (2023) MicrobiotaProcess: a comprehensive R package for deep mining microbiome. Innovation 4(2):100388. https://doi.org/10.1016/j.xinn.2023.100388
7. Wen L, Li G, Huang T, Geng W, Pei H, Yang J, Zhu M, Zhang P, Hou R, Tian G, Su W, Chen J, Zhang D, Zhu P, Zhang W, Zhang X, Zhang N, Zhao Y, Cao X, Peng G, Ren X, Jiang N, Tian C, Chen Z-J (2022) Single-cell technologies: from research to application. Innovation 3(6):100342. https://doi.org/10.1016/j.xinn.2022.100342
8. Nair S, Zhang Z, Wang X, Zhang B, Jiao N, Zhang Y (2025) Engineering microbiomes to enhance macroalgal health, biomass yield, and carbon sequestration. Green Carbon 3(1):63–73. https://doi.org/10.1016/j.greenca.2024.11.001
9. Huang Y, Wen L, Zhang L, Xu J, Wang W, Hu H, Xu P, Li Z, Tang H (2023) Community-integrated multi-omics facilitates the isolation of an organohalide dehalogenation microorganism. Innovation 4(1):100355. https://doi.org/10.1016/j.xinn.2022.100355
10. Chen S, Xu J, Cao J, Bahadur A, Wu M, Wang Z, Chen J, Wang J, Shi Y (2024) Depth-related microbial communities and functional genes in alpine permafrost. Innov Life 2(3):100081. https://doi.org/10.59717/j.xinn-life.2024.100081
11. Jin W-Y, Guo J-X, Zhang M, Teng L-Z, Chao Y, Sansonetti PJ, Gao Y-Z (2023) Absolute quantification of the microbiota spatial distribution in the murine large intestine. Innov Life 1(2):100030. https://doi.org/10.59717/j.xinn-life.2023.100030
12. Nagy LG (2023) Mushroom functional genomics springs up. Innov Life 1(1):100005. https://doi.org/10.59717/j.xinn-life.2023.100005
13. Luo M, Yang W, Bai L, Zhang L, Huang J-W, Cao Y, Xie Y, Tong L, Zhang H, Yu L, Zhou L-W, Shi Y, Yu P, Wang Z, Yuan Z, Zhang P, Zhang Y, Ju F, Zhang H, Wang F, Cui Y, Zhang J, Jia G, Wan D, Ruan C, Zeng Y, Wu P, Gao Z, Zhao W, Xu Y, Yu G, Tian C, Jin LN, Dai J, Xia B, Sun B, Chen F, Gao Y-Z, Wang H, Wang B, Zhang D, Cao X, Wang H, Huang T (2024) Artificial intelligence for life sciences: a comprehensive guide and future

trends. Innov Life 2(4):100105. https://doi.org/10.59717/j.xinn-life.2024.100105

14. Banerjee S, Schlaeppi K, van der Heijden MGA (2018) Keystone taxa as drivers of microbiome structure and functioning. Nat Rev Microbiol 16(9):567–576. https://doi.org/10.1038/s41579-018-0024-1
15. Jing X, Gong Y, Diao Z, Ma Y, Meng Y, Chen J, Ren Y, Liang Y, Li Y, Sun W, Zhang J, Ji Y, Cong Z, Li S, Ma B, Cui Z, Ma L, Xu J (2025) Phylogeny-metabolism dual-directed single-cell genomics for dissecting and mining ecosystem function by FISH-scRACS-seq. Innovation 6(3):100759. https://doi.org/10.1016/j.xinn.2024.100759
16. Bahadur A, Sajjad W, Banerjee A, Ilahi N, SZU D, Chen T, Zhang W, Chen S (2024) Could microbes inhabiting extreme desert environments be a gateway to life on the Martian surface? Innov Life 2(4):100091. https://doi.org/10.59717/j.xinn-life.2024.100091
17. Niyigaba T, Küçükgöz K, Kołożyn-Krajewska D, Królikowski T, Trząskowska M (2025) Advances in fermentation technology: a focus on health and safety. Appl Sci 15(6):3001. https://doi.org/10.3390/app15063001
18. Shi H, An F, Lin H, Li M, Wu J, Wu R (2022) Advances in fermented foods revealed by multi-omics: a new direction toward precisely clarifying the roles of microorganisms. Front Microbiol 13:1044820. https://doi.org/10.3389/fmicb.2022.1044820
19. Zheng S, Li R, Huang Y, Yang M, Chen W, Mo S, Qi L, Wang W, Wan D, Yin Y, Peng J (2025) Gut microbiome of black soldier fly larvae for efficient use and purification of organic waste: an environmentally friendly development concept. Innov Life 3:100134. https://doi.org/10.59717/j.xinn-life.2025.100134
20. Zha Y, Chen C, Jiao Q, Zeng X, Cui X, Ning K (2024) Comprehensive profiling of antibiotic resistance genes in diverse environments and novel function discovery. Innov Life 2(1):100054. https://doi.org/10.59717/j.xinn-life.2024.100054
21. Hu Y, Liu D, Jin X, Feng Y, Guo Y (2023) Synthetic microbiome for a sustainable poultry industry. Innovation 4(1):100357. https://doi.org/10.1016/j.xinn.2022.100357
22. Mukherjee A, Gómez-Sala B, O'Connor EM, Kenny JG, Cotter PD (2022) Global regulatory frameworks for fermented foods: a review. Front Nutr 9:902642. https://doi.org/10.3389/fnut.2022.902642
23. Wang L, Lee C, Gálvez JAV, Kelly CP, Lin Q, Yao J, Chen X, Cao Y (2025) Exploring the epidemiology, pathogenesis, and immunotherapeutic advancements in *Clostridium* difficile infection. Innov Life 3(1):100107. https://doi.org/10.59717/j.xinn-life.2024.100107
24. Xu J, Peng S, Xiong Y, Zheng Z, Liu M, Xu J, Chen W, Liu M, Kong J, Wang C, Wang Y, Rao L, Zhao L, Liao X (2024) A review on fermented vegetables: microbial community and potential upgrading strategy via inoculated fermentation. Compr Rev Food Sci Food Saf 23(3):e13362. https://doi.org/10.1111/1541-4337.13362
25. Satora P, Strnad S (2024) The influence of fermentation vessels on yeast microbiota and main parameters of sauerkraut. Appl Sci 14(1): 236. https://doi.org/10.3390/app14010236
26. Ameen SM, Caruso G (2017) Lactic acid and lactic acid bacteria: current use and perspectives in the food and beverage industry. In: Ameen SM, Caruso G (eds) Lactic acid in the food industry. Springer International Publishing, Cham, pp 33–44. https://doi.org/10.1007/978-3-319-58146-0_5
27. Yang X, Hu W, Xiu Z, Jiang A, Yang X, Sarengaowa, Ji Y, Guan Y, Feng K (2020) Comparison of northeast sauerkraut fermentation between single lactic acid bacteria strains and traditional fermentation. Food Res Int 137:109553. https://doi.org/10.1016/j.foodres.2020.109553
28. Peñas E, Martinez-Villaluenga C, Frias J, Sánchez-Martínez MJ, Pérez-Corona MT, Madrid Y, Cámara C, Vidal-Valverde C (2012) Se improves indole glucosinolate hydrolysis products content, se-methylselenocysteine content, antioxidant capacity and potential anti-inflammatory properties of sauerkraut. Food Chem 132(2):907–914. https://doi.org/10.1016/j.foodchem.2011.11.064
29. Wu C, Zheng J, Huang J, Zhou R (2014) Reduced nitrite and biogenic amine concentrations and improved flavor components of Chinese sauerkraut via co-culture of *Lactobacillus plantarum* and *Zygosaccharomyces rouxii*. Ann Microbiol 64(2):847–857. https://doi.org/10.1007/s13213-013-0724-8
30. Thierry A, Baty C, Marche L, Chuat V, Picard O, Lortal S, Valence F (2023) Lactofermentation of vegetables: an ancient method of preservation matching new trends.

Trends Food Sci Technol 139:104112. https://doi.org/10.1016/j.tifs.2023.07.009
31. Ge J, Sun Y, Xin X, Wang Y, Ping W (2016) Purification and partial characterization of a novel bacteriocin synthesized by *Lactobacillus paracasei* HD1-7 isolated from Chinese sauerkraut juice. Sci Rep 6(1):19366. https://doi.org/10.1038/srep19366
32. Shahbazi R, Sharifzad F, Bagheri R, Alsadi N, Yasavoli-Sharahi H, Matar C (2021) Anti-inflammatory and immunomodulatory properties of fermented plant foods. Nutrients 13(5):1516. https://doi.org/10.3390/nu13051516
33. Wei L, Marco Maria L (2025) The fermented cabbage metabolome and its protection against cytokine-induced intestinal barrier disruption of Caco-2 monolayers. Appl Environ Microbiol 91(5):e02234–02224. https://doi.org/10.1128/aem.02234-24
34. Wu J, Wu Z, Dong S, Wang Q, Zhong Q (2024) Simulated gastrointestinal digestion and fecal fermentation characteristics of exopolysaccharides synthesized by *Schleiferilactobacillus harbinensis* Z171. J Agric Food Chem 72(36):19748–19765. https://doi.org/10.1021/acs.jafc.4c02803
35. Diez-Ozaeta I, Astiazaran OJ (2022) Fermented foods: an update on evidence-based health benefits and future perspectives. Food Res Int 156:111133. https://doi.org/10.1016/j.foodres.2022.111133
36. Zhao Y, Zhao Z, Gao Y, Yang G, Liu X, Huang R, Liang W, Li S (2023) Assessment of autochthonous lactic acid bacteria as starter culture for improving traditional Chinese Dongbei Suancai fermentation. LWT Food Sci Technol 178:114615. https://doi.org/10.1016/j.lwt.2023.114615
37. Hu W, Yang X, Ji Y, Guan Y (2021) Effect of starter cultures mixed with different autochthonous lactic acid bacteria on microbial, metabolome and sensory properties of Chinese northeast sauerkraut. Food Res Int 148:110605. https://doi.org/10.1016/j.foodres.2021.110605
38. Nikoloudaki O, Aheto F, Di Cagno R, Gobbetti M (2024) Synthetic microbial communities: a gateway to understanding resistance, resilience, and functionality in spontaneously fermented food microbiomes. Food Res Int 192:114780. https://doi.org/10.1016/j.foodres.2024.114780
39. Ye Z, Jiang B, Gao D, Ping W, Ge J (2021) Bacillus spp. increase the Paracin 1.7 titer of L. paracasei HD1.7 in sauerkraut juice: emphasis on the influence of inoculation conditions on the symbiotic relationship. LWT Food Sci Technol 146:111443. https://doi.org/10.1016/j.lwt.2021.111443
40. Lee SH, Whon TW, Roh SW, Jeon CO (2020) Unraveling microbial fermentation features in kimchi: from classical to meta-omics approaches. Appl Microbiol Biotechnol 104(18):7731–7744. https://doi.org/10.1007/s00253-020-10804-8
41. Lee M, Song JH, Jung MY, Lee SH, Chang JY (2017) Large-scale targeted metagenomics analysis of bacterial ecological changes in 88 kimchi samples during fermentation. Food Microbiol 66:173–183. https://doi.org/10.1016/j.fm.2017.05.002
42. Jung S, Hwang IM, Lee J-H (2024) Temperature impact on microbial and metabolic profiles in kimchi fermentation. Heliyon 10(5):e27174. https://doi.org/10.1016/j.heliyon.2024.e27174
43. Lee M-A, Choi Y-J, Lee H, Hwang S, Lee HJ, Park SJ, Chung YB, Yun Y-R, Park S-H, Min S, Kwon L-S, Seo H-Y (2021) Influence of salinity on the microbial community composition and metabolite profile in kimchi. Fermentation-Basel 7(4):308. https://doi.org/10.3390/fermentation7040308
44. Bae S, Choi Y-J, Park S-E, Kim E-J, Lee MJ, Chung YB, Park SH, Min SG, Ku K-M, Seo H-Y, Son HS (2024) Effects of seasonal harvest of kimchi cabbage on microbial and metabolic profiles of kimchi. Food Res Int 188:114476. https://doi.org/10.1016/j.foodres.2024.114476
45. Song HS, Whon TW, Kim J, Lee SH, Kim JY, Kim YB, Choi H-J, Rhee J-K, Roh SW (2020) Microbial niches in raw ingredients determine microbial community assembly during kimchi fermentation. Food Chem 318:126481. https://doi.org/10.1016/j.foodchem.2020.126481
46. Park S-E, Seo S-H, Kim E-J, Byun S, Na C-S, Son H-S (2019) Changes of microbial community and metabolite in kimchi inoculated with different microbial community starters. Food Chem 274:558–565. https://doi.org/10.1016/j.foodchem.2018.09.032
47. Park S-E, Yoo S-A, Seo S-H, Lee K-I, Na C-S, Son H-S (2016) GC-MS based metabolomics approach of Kimchi for the understanding of *Lactobacillus plantarum* fermentation characteristics. LWT Food Sci Technol 68:313–321. https://doi.org/10.1016/j.lwt.2015.12.046
48. Kim M-J, Lee H-W, Kim JY, Kang SE, Roh SW, Hong SW, Yoo SR, Kim T-W (2020) Impact of fermentation conditions on the diversity of white colony-forming yeast and analysis of metabolite changes by white colony-forming yeast in kimchi. Food Res

Int 136:109315. https://doi.org/10.1016/j.foodres.2020.109315

49. Zhang T-T, Wu A-H, Aslam M, Song J-Z, Chi Z-M, Liu G-L (2024) Developing *Rhodotorula* as microbial cell factories for the production of lipids and carotenoids. Green Carbon 2(4):409–420. https://doi.org/10.1016/j.greenca.2024.09.004
50. Lee KW, Kim GS, Baek AH, Hwang HS, Kwon DY, Kim SG, Lee SY (2020) Isolation and characterization of kimchi starters *Leuconostoc mesenteroides* PBio03 and *Leuconostoc mesenteroides* PBio104 for manufacture of commercial kimchi. J Microbiol Biotechnol 30(7):1060–1066. https://doi.org/10.4014/jmb.2001.01011
51. Ampemohotti T, Aida G, Christopher P, Charles B, Hao Van TT (2025) Fermented vegetables: their microbiology and impact on gut microbiota and overall health benefits. Food Rev Int 41:1–24. https://doi.org/10.1080/87559129.2025.2486274
52. Abbaspour N (2024) Fermentation's pivotal role in shaping the future of plant-based foods: an integrative review of fermentation processes and their impact on sensory and health benefits. Applied Food Research 4(2):100468. https://doi.org/10.1016/j.afres.2024.100468
53. Maoloni A, Ferrocino I, Milanovic V, Cocolin L, Corvaglia MR, Ottaviani D, Bartolini C, Talevi G, Belleggia L, Cardinali F, Marabini R, Aquilanti L, Osimani A (2020) The microbial diversity of non-Korean kimchi as revealed by viable counting and metataxonomic sequencing. Foods 9(11):1568. https://doi.org/10.3390/foods9111568
54. Liu R, Kim AH, Kwak M-K, Kang S-O (2017) Proline-based cyclic dipeptides from Korean fermented vegetable kimchi and from *Leuconostoc mesenteroides* LBP-K06 have activities against multidrug-resistant bacteria. Front Microbiol 8:761. https://doi.org/10.3389/fmicb.2017.00761
55. Huang S, He C, Li J, Gao Y-Z, Wang Z, Wei Y (2023) Emerging paradigms in exploring the interactions among diet, probiotics, and cancer immunotherapeutic response. Innovation 4(4):100456. https://doi.org/10.1016/j.xinn.2023.100456
56. Radhakrishnan R, Lee I-J (2016) Gibberellins producing *Bacillus methylotrophicus* KE2 supports plant growth and enhances nutritional metabolites and food values of lettuce. Plant Physiol Biochem 109:181–189. https://doi.org/10.1016/j.plaphy.2016.09.018
57. Bianchini A (2015) Lactic acid bacteria as antifungal agents. In: Holzapfel W (ed) Advances in fermented foods and beverages. Woodhead Publishing, pp 333–353. https://doi.org/10.1016/B978-1-78242-015-6.00014-1
58. Elhalis H, Chin XH, Chow Y (2024) Soybean fermentation: microbial ecology and starter culture technology. Crit Rev Food Sci Nutr 64(21):7648–7670. https://doi.org/10.1080/10408398.2023.2188951
59. Feng Y, Xie Z, Huang M, Tong X, Hou S, Tin H, Zhao M (2024) Decoding temperature-driven microbial community changes and flavor regulation mechanism during winter fermentation of soy sauce. Food Res Int 177:113756. https://doi.org/10.1016/j.foodres.2023.113756
60. Feng Y, Zeng J, Lei H, Zhao M (2024) Effect of fermentation containers on the taste characteristics and microbiota succession of soy sauce. Food Chem 448:139066. https://doi.org/10.1016/j.foodchem.2024.139066
61. Nguyen NTH, Huang MB, Liu FY, Huang WL, Tran H-T, Hsu TW, Huang C-L, Chiang T-Y (2023) Deciphering microbial community dynamics along the fermentation course of soy sauce under different temperatures using metagenomic analysis. Biosci Microbiota Food Health 42(2):104–113. https://doi.org/10.12938/bmfh.2022-01
62. Pei R, Lv G, Guo B, Li Y, Ai M, He B, Wan R (2022) Physiological and transcriptomic analyses revealed the change of main flavor substance of *Zygosaccharomyces rouxii* under salt treatment. Front Nutr 9:990380. https://doi.org/10.3389/fnut.2022.990380
63. Liu H, Zhou W, Lu J, Wu D, Ge F (2025) Construction of a synthetic microbial community and its application in salt-reduced soy sauce fermentation. Food Microbiol 128:104738. https://doi.org/10.1016/j.fm.2025.104738
64. Hu G, Wang Y, Chen J, Du G, Fang F (2024) Synergistic fermentation with functional bacteria for production of salt-reduced soy sauce with enhanced aroma and saltiness. Food Biosci 57:103459. https://doi.org/10.1016/j.fbio.2023.103459
65. Wu J, Li Q, Hu K, Li J, Durán-Guerrero E, Liu S, Guo M, Liu A (2024) Microbial characterization of Sichuan Baoning vinegar: lactic acid bacteria, acetic acid bacteria and yeasts. Arch Microbiol 206(2):59. https://doi.org/10.1007/s00203-023-03784-6
66. Lu Z-M, Wang Z-M, Zhang X-J, Mao J, Shi J-S, Xu Z-H (2018) Microbial ecology of cereal vinegar fermentation: insights for

driving the ecosystem function. Curr Opin Biotechnol 49:88–93. https://doi.org/10.1016/j.copbio.2017.07.006

67. Xu L, Yang T, Zhang X-J, Chai L-J, Li X, Shi J-S, Li B, Huang WE, Wang Y, Lu Z-M, Xu Z-H (2024) Predicting the multispecies solid-state vinegar fermentation process using single-cell Raman spectroscopy combined with machine learning. LWT Food Sci Technol 192:115708. https://doi.org/10.1016/j.lwt.2023.115708
68. Ge Y, Wu Y, Aihaiti A, Wang L, Wang Y, Xing J, Zhu M, Hong J (2025) The metabolic pathways of yeast and acetic acid bacteria during fruit vinegar fermentation and their influence on flavor development. Microorganisms 13(3):477. https://doi.org/10.3390/microorganisms13030477
69. Hua S, Wang Y, Wang L, Zhou Q, Li Z, Liu P, Wang K, Zhu Y, Han D, Yu Y (2024) Regulatory mechanisms of acetic acid, ethanol and high temperature tolerances of acetic acid bacteria during vinegar production. Microb Cell Factories 23(1):324. https://doi.org/10.1186/s12934-024-02602-y
70. Wang Y, Wu J, Lv M, Shao Z, Hungwe M, Wang J, Bai X, Xie J, Wang Y, Geng W (2021) Metabolism characteristics of lactic acid bacteria and the expanding applications in food industry. Front Bioeng Biotechnol 9:612285. https://doi.org/10.3389/fbioe.2021.612285
71. Xi Y, Fan F, Zhang X (2023) Microbial L-malic acid production: history, current progress, and perspectives. Green Carbon 1(2):118–132. https://doi.org/10.1016/j.greenca.2023.10.005
72. Huang L, Wang Y, Zhong K, Jiang Z, Jia H, Chen S, Zhao Z, Chen X (2024) In vitro characterization and identification of potential probiotic yeasts isolated from Zaopocu, a traditional fermented dregs vinegar from Hainan Island. Probiotics Antimicrob Proteins. https://doi.org/10.1007/s12602-024-10347-8
73. Saeed F, Afzaal M, Shah YA, Khan MH, Hussain M, Ikram A, Ateeq H, Noman M, Saewan SA, Khashroum AO (2022) Miso: a traditional nutritious & health-endorsing fermented product. Food Sci Nutr 10(12):4103–4111. https://doi.org/10.1002/fsn3.3029
74. Kothe CI, Rasmussen JA, Mak SST, Gilbert MTP, Evans J (2024) Exploring the microbial diversity of novel misos with metagenomics. Food Microbiol 117:104372. https://doi.org/10.1016/j.fm.2023.104372
75. Allwood JG, Wakeling LT, Bean DC (2021) Fermentation and the microbial community of Japanese koji and miso: a review. J Food Sci 86(6):2194–2207. https://doi.org/10.1111/1750-3841.15773
76. Ibrahim SA, Ayivi RD, Zimmerman T, Siddiqui SA, Altemimi AB, Fidan H, Esatbeyoglu T, Bakhshayesh RV (2021) Lactic acid bacteria as antimicrobial agents: food safety and microbial food spoilage prevention. Foods 10(12):3131. https://doi.org/10.3390/foods10123131
77. Zhao S, Hao C, Niu C, Lu J, Wang L, Shi Y, Li Q (2024) Insight into the keystones of Chinese broad bean paste fermentation: brewing techniques, chemosensory characteristics, and microbial community. Trends Food Sci Technol 154:104775. https://doi.org/10.1016/j.tifs.2024.104775
78. Zhang L, Zeng Y, Yu J, Li Y, Zeng P (2024) Metagenomic and metatranscriptomic analyses of microbial genera and volatiles produced during the fermentation of doubanjiang meju. Food Biosci 62:105100. https://doi.org/10.1016/j.fbio.2024.105100
79. Xu M, Guo Y, Song X, Li L, Xu Z, Zhao J, Zhao J, Lin H, Dong S, Lu J, Ding W, Liu P, Tang J (2024) Analysis on microbial communities and characteristic flavor metabolic of PXDB-meju by partially substituting wheat flour with soybean flour and gluten flour. Curr Res Food Sci 9:100904. https://doi.org/10.1016/j.crfs.2024.100904
80. Li Z, Zhao C, Dong L, Zhang F, Wang X, Zhao S, Li L (2024) Impact of NaCl on physicochemical properties, microbial community, and pathogen surveillance in the Chinese traditional fermented broad bean (Vicia faba L.) paste. LWT Food Sci Technol 199:116071. https://doi.org/10.1016/j.lwt.2024.116071
81. Gao Y, Hu M, Meng W, Wen W, Zhang P, Fan B, Wang F, Li S (2024) Study on the quality of soybean proteins fermented by *Bacillus subtilis* BSNK-5: insights into nutritional, functional, safety, and flavor properties. Food Chem 443:138523. https://doi.org/10.1016/j.foodchem.2024.138523
82. Que Z, Ao L, Zhou R, Yang J, Zeng S, Shen X, Shen C, Zhang S, Wu C (2025) Taste and microorganism characteristics of Pixian Doubanjiang and the correlation between microbial communities and taste formation. Food Biosci 68:106773. https://doi.org/10.1016/j.fbio.2025.106773
83. Niu C, Yang L, Zheng F, Liu C, Wang J, Xu X, Li Q (2022) Systematic analysis of the aroma

profiles produced by *Zygosaccharomyces rouxii* Y-8 in different environmental conditions and its contribution to doubanjiang (broad bean paste) fermentation with different salinity. LWT Food Sci Technol 158:113118. https://doi.org/10.1016/j.lwt.2022.113118

84. Li Z, Dong L, Huang Q, Wang X (2016) Bacterial communities and volatile compounds in Doubanjiang, a Chinese traditional red pepper paste. J Appl Microbiol 120(6): 1585–1594. https://doi.org/10.1111/jam.13130
85. Yang M, Huang J, Zhou R, Jin Y, Wu C, Zhao N (2024) Regulatory mechanisms of the microbial starter fortification on the community and flavor characteristics in Doubanjiang. Food Biosci 61:104913. https://doi.org/10.1016/j.fbio.2024.104913
86. Zhang Y, Zhu X, Wang N, Liu X, Wang L, Ning K (2025) Synergy of traditional practices and modern technology: advancing the understanding and applications of microbial resources and processes in fermented foods. Trends Food Sci Technol 157:104891. https://doi.org/10.1016/j.tifs.2025.104891
87. Jin G, Zhu Y, Xu Y (2017) Mystery behind Chinese liquor fermentation. Trends Food Sci Technol 63:18–28. https://doi.org/10.1016/j.tifs.2017.02.016
88. Huang Q, Liu Y, He Z, Mao Y, Wu H, Tian L, Xiang S, Long L, Li Y, Guan T (2024) Environmental temperature variations drive the changes of microbial communities to affect Baijiu flavor quality: case study of Qingxiangxing Baijiu. Food Biosci 59:104064. https://doi.org/10.1016/j.fbio.2024.104064
89. Xu S, Zhang M, Xu B, Liu L, Sun W, Mu D, Wu X, Li X (2022) Microbial communities and flavor formation in the fermentation of Chinese strong-flavor Baijiu produced from old and new Zaopei. Food Res Int 156: 111162. https://doi.org/10.1016/j.foodres.2022.111162
90. Luo L-J, Song L, Han Y, Zhen P, Han D-Y, Zhao X, Zhou X, Wei Y-H, Yu H-X, Han P-J, Bai F-Y (2023) Microbial communities and their correlation with flavor compound formation during the mechanized production of light-flavor Baijiu. Food Res Int 172:113139. https://doi.org/10.1016/j.foodres.2023.113139
91. Zhang H, Zhang H, Du H, Zhang Y, Zhang M, Yu X, Xu Y (2025) Unraveling the multiple interactions between phages, microbes and flavor in the fermentation of strong-flavor Baijiu. Bioresour Bioprocess 12(1):14. https://doi.org/10.1186/s40643-025-00852-1
92. Zeng X, Zou Y, Zheng J, Qiu S, Liu L, Wei C (2023) Quorum sensing-mediated microbial interactions: mechanisms, applications, challenges and perspectives. Microbiol Res 273: 127414. https://doi.org/10.1016/j.micres.2023.127414
93. Wang L, Zhang Y, Xu J, Shi Q, Peng Y, Long C, Li L, Yin Y (2023) Listening to enteric bacteria from the perspective of antibiotic alternatives in animal husbandry. Innov Life 1(2):100022. https://doi.org/10.59717/j.xinn-life.2023.100022
94. Tu W, Cao X, Cheng J, Li L, Zhang T, Wu Q, Xiang P, Shen C, Li Q (2022) Chinese Baijiu: the perfect works of microorganisms. Front Microbiol 13:919044. https://doi.org/10.3389/fmicb.2022.919044
95. Tong W, Wang H, Li J, Zhang L, Huang Z, Yang Y, Qiao Z, Luo H, Huang D (2024) Strengthening the microbial community and flavor structure of jiupei by simulating strong-aroma baijiu fermentation with DQA21. J Sci Food Agric 104(9):5338–5349. https://doi.org/10.1002/jsfa.13373
96. Jiang S, Ma Y, Dai J (2024) Synthetic yeast genome project and beyond. Innov Life 2(1): 100059. https://doi.org/10.59717/j.xinn-life.2024.100059
97. Liang L, Ma Y, Jiang Z, Sam FE, Peng S, Li M, Wang J (2023) Dynamic analysis of microbial communities and flavor properties in Merlot wines produced from inoculation and spontaneous fermentation. Food Res Int 164:112379. https://doi.org/10.1016/j.foodres.2022.112379
98. Liu D, Zhang P, Chen D, Howell K (2019) From the vineyard to the winery: how microbial ecology drives regional distinctiveness of wine. Front Microbiol 10:2679. https://doi.org/10.3389/fmicb.2019.02679
99. Gao J, Geng H, Chai R, Wu T, Huang W, You Y, Zhan J (2023) Fungal community composition and its relationship with volatile compounds during spontaneous fermentation of Cabernet Sauvignon from two Chinese wine-growing regions. Foods 13(1):106. https://doi.org/10.3390/foods13010106
100. Tan Y, Du H, Zhang H, Fang C, Jin G, Chen S, Wu Q, Zhang Y, Zhang M, Xu Y (2022) Geographically associated fungus-bacterium interactions contribute to the formation of geography-dependent flavor during high-complexity spontaneous fermentation. Microbiol Spectrum 10(5):e01844–e01822.

https://doi.org/10.1128/spectrum.01844-22

101. 蒲领平, 邓杰, 卫春会, 任志强, 黄治国 (2023) 贝莱斯芽孢杆菌对小曲酒酿造过程中酵母菌生长及风味物质代谢的影响. 中国酿造 42(2):53–57. https://doi.org/10.11882/j.issn.0254-5071.2023.02.009
102. Tantikachornkiat M, Morgan SC, Lepitre M, Cliff MA, Durall DM (2020) Yeast and bacterial inoculation practices influence the microbial communities of barrel-fermented Chardonnay wines. Aust J Grape Wine Res 26(3):279–289. https://doi.org/10.1111/ajgw.12438
103. Jin R, Song J, Liu C, Lin R, Liang D, Aweya JJ, Weng W, Zhu L, Shang J, Yang S (2024) Synthetic microbial communities: novel strategies to enhance the quality of traditional fermented foods. Compr Rev Food Sci Food Saf 23(4):e13388. https://doi.org/10.1111/1541-4337.13388
104. Zhang B, Liu D, Liu H, Shen J, Zhang J, He L, Li J, Zhou P, Guan X, Liu S, Shi K (2024) Impact of indigenous Oenococcus oeni and *Lactiplantibacillus plantarum* species co-culture on Cabernet Sauvignon wine malolactic fermentation: kinetic parameters, color and aroma. Food Chem: X 22:101369. https://doi.org/10.1016/j.fochx.2024.101369
105. Englezos V, Giacosa S, Rantsiou K, Rolle L, Cocolin L (2017) Starmerella bacillaris in winemaking: opportunities and risks. Curr Opin Food Sci 17:30–35. https://doi.org/10.1016/j.cofs.2017.08.007
106. Daferner M, Pütz R, Steiner C, Rainer G (2025) Nature in the glass': principles and practices in low-intervention winemaking. Appl Geogr 178:103610. https://doi.org/10.1016/j.apgeog.2025.103610
107. James A, Yao T, Ke H, Wang Y (2023) Microbiota for production of wine with enhanced functional components. Food Sci Human Wellness 12(5):1481–1492. https://doi.org/10.1016/j.fshw.2023.02.008
108. Akaike M, Miyagawa H, Kimura Y, Terasaki M, Kusaba Y, Kitagaki H, Nishida H (2020) Chemical and bacterial components in sake and sake production process. Curr Microbiol 77(4):632–637. https://doi.org/10.1007/s00284-019-01718-4
109. Nishida H (2024) Kuratsuki bacteria interactions with sake yeast and effect on taste. Appl Microbiol 4(3):1309–1319. https://doi.org/10.3390/applmicrobiol4030090
110. Takahashi T, Ohara Y, Sueno K (2017) Breeding of a sake yeast mutant with enhanced ethyl caproate productivity in sake brewing using rice milled at a high polishing ratio. J Biosci Bioeng 123(6):707–713. https://doi.org/10.1016/j.jbiosc.2017.01.014
111. Kobayashi T, Abe K, Asai K, Gomi K, Juvvadi PR, Kato M, Kitamoto K, Takeuchi M, Machida M (2007) Genomics of *Aspergillus oryzae*. Biosci Biotechnol Biochem 71(3):646–670. https://doi.org/10.1271/bbb.60550
112. Naganuma K, Youji N, Susumu K, Takuya H, Kayo H, Naoko A, Masayuki H, Yamamura H (2023) Traditional microbial control methods used in sake brewing effectively suppress predominant bacteria emerging during production of rice koji. Biotechnol Biotechnol Equip 37(1):2271566. https://doi.org/10.1080/13102818.2023.2271566
113. Tian S, Zeng W, Fang F, Zhou J, Du G (2022) The microbiome of Chinese rice wine (Huangjiu). Curr Opin Food Sci 5:325–335. https://doi.org/10.1016/j.crfs.2022.01.007
114. Peng Q, Li L, Xie G (2025) Impact of glutinous rice varieties from different regions on microbial community structure, metabolic profiles, and flavor characteristics of Chinese rice wine (Huangjiu). Foods 14(7):1261. https://doi.org/10.3390/foods14071261
115. Zhao Y, Li M, Zhan P, Wang P, He W, Tian H (2024) A quality comparison for Xiecun Huangjiu with different aging stages based on chemical profile, aroma composition and microbial succession. Food Chem: X 21:101132. https://doi.org/10.1016/j.fochx.2024.101132
116. Peng Q, Zheng H, Li J, Li S, Huang J, Xu Y, Xie G (2024) Impact of Bacillus subtilis on Chinese yellow rice wine (Huangjiu) fermentation: method variations and flavor analysis. Food Chem 460:140658. https://doi.org/10.1016/j.foodchem.2024.140658
117. Yu H, Wu S, Li Q, Chen C, Chen Q, Tian H (2025) Novel insights into delayed bitterness control in traditional Huangjiu: regulating bacterial community by inoculated with *Saccharomyces cerevisiae* SC-6. Food Chem: X 25. https://doi.org/10.1016/j.fochx.2024.102065
118. Peng Q, Zheng H, Quan L, Li S, Huang J, Li J, Xie G (2025) Development of a flavor-oriented synthetic microbial community for pour-over rice wine: a comprehensive microbial community analysis. Food Microbiol 126:104677. https://doi.org/10.1016/j.fm.2024.104677

Chapter 13

Microbial Omics Analysis for Multispecies Symbioses in Specialty Fermentations and Nonfood Fermentations

Fan Wang, Deshui Liu, Haibo Zhang, and Qian Wang

Abstract

This chapter describes the microbial diversity and interactions underlying the production of fermented dairy products, meats, tea, coffee, chocolate, and tobacco. It summarizes how raw materials, processing parameters, and environmental factors shape microbial communities and their metabolic activities, ultimately determining flavor, texture, and functional qualities. Microbial cooperation and succession drive key biochemical transformations that influence product characteristics. With advances in microbial omics, strain identification and functional analysis now enable more precise control of fermentation, improved quality and safety, and the development of novel functional products. The chapter also highlights future directions in high-dimensional omics and targeted fermentation.

Key words Microbial omics, Microbial diversity, Microbial interactions, Specialty fermented products, Fermented tobacco

1 Introduction

Fermentation has long been a cornerstone of human food culture, biotechnology, and agricultural practices. Across diverse regions and historical periods, communities have relied on complex microbial ecosystems to transform raw materials into products with enhanced flavor, stability, nutrition, and functional benefits [1]. Although many traditional fermentation processes evolved through empirical knowledge, modern scientific advances have revealed that these transformations are driven by intricate multispecies microbial symbioses [2]. These communities, comprising bacteria, yeasts, and fungi, carry out cooperative metabolic activities that shape the chemical, sensory, and functional characteristics of fermented foods and plant-derived products. Despite this recognition, fundamental questions remain regarding how microbial members assemble, interact, and contribute to product quality across

Tao Huang (ed.), *Microbial 'Omics: Methods and Protocols*, Methods in Molecular Biology, vol. 3033,
https://doi.org/10.1007/978-1-0716-5264-0_13,

different fermentation systems. Addressing these knowledge gaps requires a comprehensive, comparative perspective.

In recent decades, microbial omics—particularly metagenomics, metatranscriptomics, metabolomics, and metabolic modeling—has revolutionized our ability to explore microbial community structure and function in complex environments [3–6]. Such tools have uncovered unexpectedly high microbial diversity and revealed that even small variations in raw materials, processing parameters, or environmental conditions can significantly shift microbial communities and their metabolic outputs. For example, milk source, geographic origin, and processing significantly affect cheese microbiomes [7], while fermentation temperature and milk composition drive community dynamics in yogurt [8]. Likewise, kefir grains host spatially structured consortia whose succession correlates with flavor and functionality [9]. Similar patterns are observed in fermented meats [10], tea [11], coffee [12], chocolate [13], and tobacco leaves [14]. Each system represents a unique ecological niche shaped by species interactions, metabolic cooperation, and environmental constraints. Despite methodological progress, understanding of interspecies dynamics, metabolic cooperation, and strain-level diversity remains fragmentary [15, 16].

This chapter aims to synthesize current knowledge on multispecies microbial symbioses across diverse fermentation systems; to highlight the ecological and functional principles that govern microbial community assembly, metabolism, and interaction; to outline industrial applications and emerging research directions; and to emphasize the essential role of microbial omics in advancing both traditional and modern fermentation science. By framing fermented foods and plant-based products as dynamic microbial ecosystems, this chapter connects traditional fermentation knowledge with modern microbiology, ecology, and biotechnology, providing a foundation for future research and industrial innovation.

2 Fermented Dairy Products

2.1 Cheese

Cheese, one of the oldest fermented foods, showcases a highly diverse microbial community that significantly influences its overall characteristics, including flavor, texture, and aroma. Lactic acid bacteria (LAB) such as *Lactococcus*, *Streptococcus*, *Pediococcus*, *Lactobacillus*, and *Leuconostoc* dominate cheese microbiomes; yeasts such as *Candida*, *Pichia*, *Hansenula*, *Saccharomyces*, *Kluyveromyces*, and mold genera such as *Penicillium*, *Aspergillus*, *Rhizopus*, and *Mucor* are also detected in cheese [7]. Factors affecting this microbial community composition include milk source, processing conditions, starter cultures, and environmental factors such as

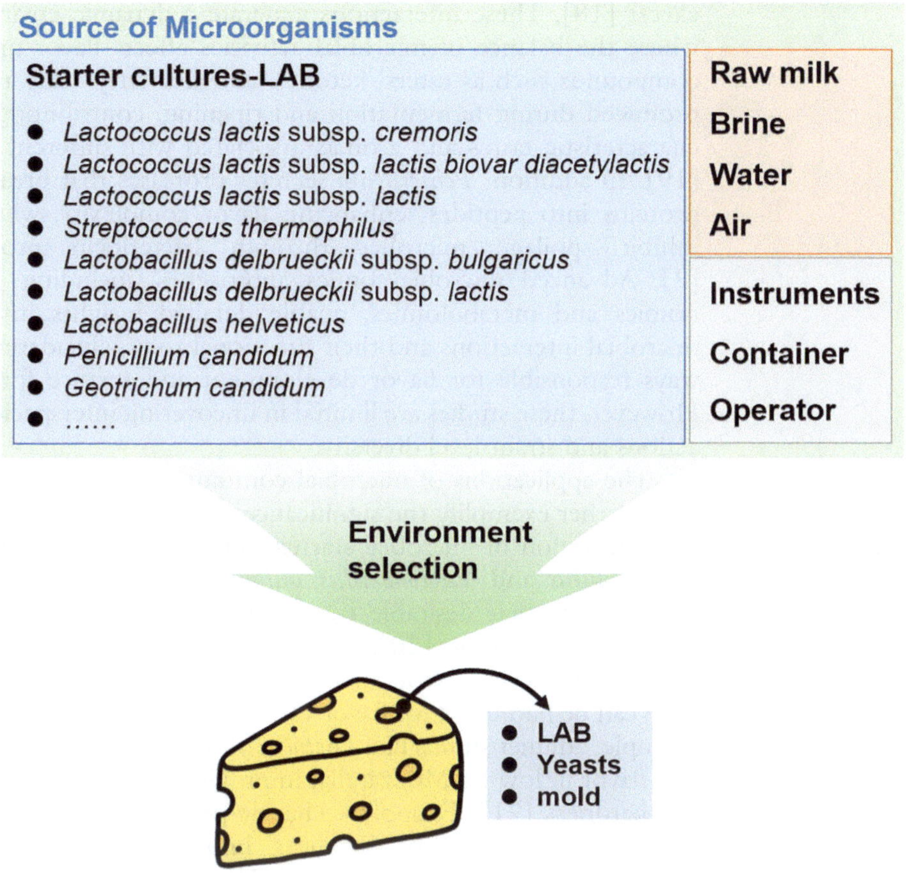

Fig. 1 Formation of microbial community on cheese

temperature and humidity (Fig. 1) [17]. Raw milk, for instance, contains a diverse range of microorganisms that can introduce spontaneous fermentation dynamics, while pasteurized milk leads to more controlled and specific microbial growth. In addition, there is a positive relationship between microbiome richness and pH, while milk and cheese type, geographic location, and pasteurization collectively explain 65% of the variation in microbial community composition [7].

Interactions within these microbial communities are critical in defining the sensory attributes of cheese. LAB can produce metabolites that stimulate mold growth, while molds can enhance flavor complexity through the breakdown of fats and proteins, releasing volatile compounds [18]. *Lactococcus lactis* produces lactic acid, creating an acidic environment that favors acid-tolerant species like *Lactobacillus*. *Streptococcus thermophilus* secretes proteases to relieve nitrogen limitation for *Lactococcus* and boosts de novo nucleotide biosynthesis, while *Lactococcus cremoris* limits diacetyl and acetoin formation, which can result in off-flavors when in excess [18]. These interactions generate a dynamic environment where the balance of microbial activities affects flavor profiles—compounds such as esters, ketones, and free fatty acids are often produced during fermentation and ripening, contributing to the characteristic tastes and aromas associated with different cheeses [19]. In addition, *Penicillium* secretes proteases that break down proteins into peptides, enhancing flavor complexity, while LAB inhibit spoilage microbes through bacteriocin production [2]. Advanced microbial omics approaches, including metagenomics and metabolomics, enable detailed insights into these microbial interactions and their functional roles, elucidating pathways responsible for flavor development and texture formation. However, these studies are limited in uncovering interspecies interactions and strain-level diversity.

The applications of microbial communities in cheese production further exemplify the significance of microbial omics analysis. The utilization of microbial starter cultures (such as *Streptococcus thermophilus* and *Lactobacillus bulgaricus*), which are carefully selected for their desirable traits, ensures consistent production and quality of cheese [20]. These starter cultures can be tailored to enhance specific flavors or textural properties, while nonstarter LAB can be harnessed for flavor variety during ripening stages. For example, adjunct strains like *Lactobacillus casei* enhance proteolysis and flavor in low-fat Mozzarella, improving meltability and reducing hardness [21]. Exopolysaccharide-producing strains, such as *Lactococcus lactis* subsp. *Cremoris*, improve moisture retention and texture in reduced-fat cheeses [22]. Moreover, the exploration of novel strains or wild microbial consortia from traditional cheesemaking practices offer an avenue for developing innovative cheese varieties with unique profiles [23]. Techniques such as bioinformatics aid in the identification and selection of probiotics that can provide health benefits, offering an avenue for functional cheeses that not only satisfy culinary preferences but also contribute positively to consumer health [24]. The continuous integration of microbial omics data into cheese production processes will undoubtedly lead to enhanced quality control, flavor optimization, and ultimately the discovery of new cheeses that reflect the rich tapestry of microbial life involved in their creation.

2.2 Yogurt

Yogurt is a fermented dairy product using a combination of the two LAB species *Streptococcus thermophilus* and *Lactobacillus delbrueckii* subsp. *Bulgaricus*. The predominant microbial species in yogurt are these two LAB [25]. A small number of *Enterobacter*, *Lactococcus*, and *Aeromonas* are also present in yogurt, while *Kluyveromyces*, *Trichosporon*, *Clavispora*, *Candida*, *Iodophanus*, *Apiotrichum*, and *Issatchenkia* are the most abundant fungal genera [26]. The microbial community composition can be influenced by several factors, including the type of milk used, fermentation temperature, pH, and the duration of fermentation. Studies suggest that variations in milk composition—such as fat content and protein structure—can affect the metabolic activities of LAB, thus altering community dynamics and fermentation efficiency. Additionally, the presence of prebiotics and flavoring agents can further shape the microbial landscape, promoting a diverse consortium of microorganisms that includes not only LAB but also yeasts and molds that can be introduced through environmental contamination or directly from starter cultures [8]. The inoculation rate of starter cultures also plays a significant role. A higher inoculation rate accelerates fermentation but may reduce the diversity of the microbial community.

The interactions between microbial communities in yogurt are intricate and pivotal for enhancing its quality and functional properties. Cooperative interactions, such as cross-feeding, result in mutual benefits, where one species' metabolic by-products serve as substrates for another, thereby optimizing growth and activity. *Streptococcus thermophilus* generates formic acid, promoting *L. bulgaricus* growth, while the latter's proteolytic activity releases nutrients that fuel cross-feeding [27]. The acidic environment also helps eliminate undesired sensor strains [28]. Additional probiotic strains, such as *Lactobacillus plantarum* and *Bifidobacterium* spp., are often incorporated to enhance functionality [29]. Volatile compounds produced during fermentation, such as diacetyl (buttery aroma) and acetaldehyde (contributing "fresh" notes), arise from metabolic pathways involving multiple microbial contributors, resulting in a complex bouquet of flavors that define the sensory experience of yogurt. This multispecies symbiosis underlines the importance of conducting comprehensive microbial omics analyses to elucidate functional dynamics and the interplay of different microorganisms, thereby optimizing flavor and texture attributes.

The applications of microbial communities in yogurt extend beyond traditional fermentation processes. Advances in microbial omics analysis enable the identification of key microbial players responsible for desirable attributes such as probiotic properties, textural qualities, and flavor complexities [8]. Furthermore, genetic insights into microbial strains can guide the development of tailored starter cultures through selective breeding or genetic engineering, fostering the production of specialty yogurts enriched with health-promoting compounds or unique flavors [30]. *Lactobacilli* and *Bifidobacteria* are the primary probiotics used in yogurt [31]. Hybrid membrane vesicles (hMVs), inspired by yogurt components, have been engineered to combat *Helicobacter pylori* by mimicking host-pathogen interactions, highlighting the translational potential of yogurt-derived microbes [32]. Integrating omics tools will further enable precise manipulation of microbial communities to optimize health benefits, flavor profiles, and industrial scalability, positioning yogurt as both a cultural staple and a frontier in microbial biotechnology.

2.3 Kefir

Kefir, a traditional fermented dairy product originating from the Caucasus region, is produced through the symbiotic fermentation of milk by LAB, acetic acid bacteria (AAB), and yeasts in kefir grains. Different types of kefir have different production processes (Fig. 2). The predominant genera include *Lactobacillus*, *Lactococcus*, *Acetobacter*, *Saccharomyces*, *Kluyveromyces*, *Kazachstania turicensis*, and *Torulaspora delbrueckii* [33]. Kefir grains are a complex biofilm composed of a polysaccharide-protein matrix, housing a diverse microbial community. The microbial composition of kefir grains and kefir may differ. For instance, *Lactobacillus kefiranofaciens* dominates in kefir grains, whereas *Lactococcus lactis* is more prevalent in kefir [34]. Studies utilizing advanced microbial omics techniques have revealed that the specific composition and abundance of these microbes can be influenced by several factors, including the source of milk, and fermentation technology. For example, extended fermentation (5–7 days) promotes microbial succession and metabolite diversification, favoring acid-tolerant species like *Lactobacillus*. In addition, the microbial composition of goat milk kefir differs from that of donkey milk kefir [9]. In addition, a higher grain-to-milk ratio may increase the number of LAB in kefir, and the use of different starter cultures may also lead to variability in the microbial landscape [35]. Furthermore, environmental conditions such as oxygen levels, humidity, acidity, and nutrient availability have been shown to impact microbial interactions and population dynamics within the kefir ecosystem. Packaging materials (e.g., metallized polyester) and lyophilization also impact microbial viability and activity, with optimal preservation maintaining metabolic functions like lactose conversion and lactic acid production [36].

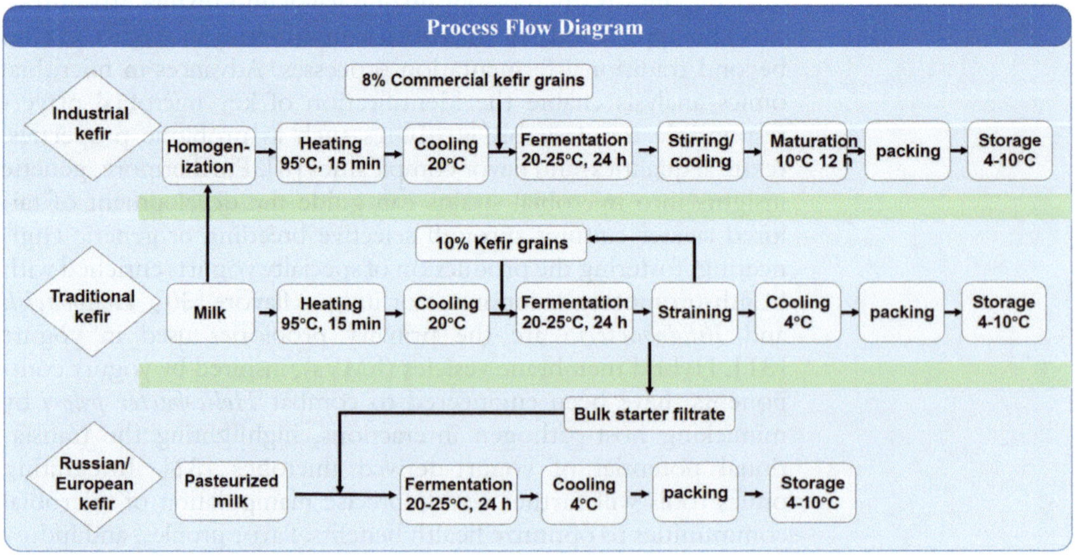

Fig. 2 Methods for production of kefir

Understanding the interactions among kefir's microbial constituents is fundamental to deciphering the product's sensory and functional characteristics. The cross-feeding of LAB and yeasts in kefir not only facilitates fermentation but also enhances the production of flavor compounds that contribute to the beverage's complex taste profile. For instance, LAB ferment lactose in milk into lactic acid and other metabolites, which are then utilized by yeasts. Yeasts, in turn, produce ethanol and carbon dioxide, which affect the growth and metabolism of LAB. AAB can metabolize lactic acid produced by LAB into acetic acid, contributing to the unique flavor of kefir [37]. Additionally, these metabolic interactions can enhance nutrient bioavailability and probiotic functionality, promoting health benefits such as improved gut health and immune responses. LAB dominate acidification and antimicrobial activity, while yeasts contribute to flavor via volatile compounds (e.g., esters, alcohols) [9]. The flavor profile of kefir, which is a key aspect of its sensory appeal, can thus be viewed as a product of intricate microbial interactions governed by a dense and diverse community.

Kefir itself is a functional food with significant health benefits. Its microbial communities possess probiotic properties, capable of regulating the gut microbiota, enhancing immunity, and improving lactose digestion, which provides opportunities for biotechnological exploitation, including the development of functional foods and nutraceuticals [38]. The probiotics derived from kefir, including *Lactobacillus* and other LAB, are increasingly investigated for their therapeutic potential, including the modulation of gut microbiota and enhancement of gastrointestinal health [39]. Various studies have explored the potential of kefir's microbial metabolites as antimicrobial agents, highlighting their ability to inhibit the growth of pathogenic bacteria and fungi. Furthermore, the applications of microbial communities in kefir also include the exploration of strains to standardize fermentation processes and improve fermentation efficiency and product quality. Research on the microbial communities of kefir holds important theoretical and practical significance for understanding microbial symbiosis mechanisms and developing functional foods.

3 Fermented Meats

Fermented meat products, renowned for their distinctive flavors and preservation qualities, rely heavily on the complex interplay of microbial communities during the solid-state fermentation (SSF) process. The primary microbial community composition in fermented meats generally includes LAB, such as *Lactobacillus* and *Pediococcus*, alongside *Staphylococcus*, yeasts and molds, including species of *Penicillium*, *Debaryomyces*, and *Candida* (Fig. 3). Various factors influence the microbial community composition, including the

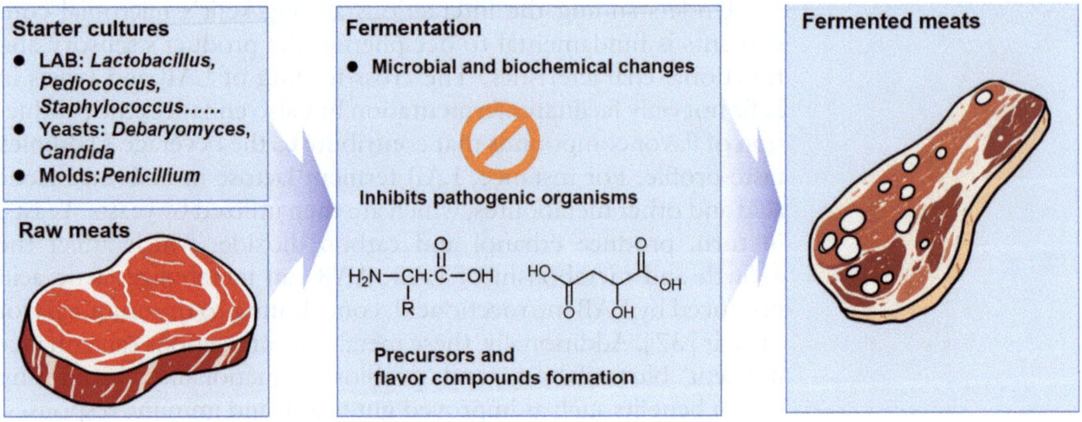

Fig. 3 Diagram of meat fermentation

type of meat used, the preparation methods, the environmental conditions during fermentation (temperature, humidity, and oxygen availability), and the initial microbial load introduced through natural contamination or starter cultures [10].

The interactions among these microbial communities are critical in determining the fermentation outcome, influencing both the functional properties of the microorganisms and the overall flavor profile of the meat [40]. LAB are primarily responsible for the production of lactic acid, which lowers the pH and creates an environment that inhibits pathogenic organisms. Concurrently, LAB and *staphylococci* synergistically break down proteins and lipids in the meat into smaller molecules, such as amino acids, peptides, and free fatty acids, which serve as precursors for flavor compounds. Through metabolic pathways like the Maillard reaction and lipid oxidation, these precursors are transformed into volatile flavor substances, such as aldehydes, ketones, and esters, which impart the distinctive flavors to fermented meats. Yeasts and molds such as *Debaryomyces hansenii* and *Penicillium* spp. contribute to the rich spectrum of flavors through their metabolic by-products, such as esters, phenols, and organic acids (Fig. 3). This synergistic relationship results in the development of complex flavor notes characteristic of fermented meats, such as tanginess, umami, and aroma. The interplay of these microbial populations also dictates the formation of bioactive compounds and antioxidants, which not only enhance the sensory qualities but also potentially impart health benefits to consumers.

The application of microbial communities in fermented meats is not limited to flavor development but also extends to improving product safety and nutritional value. The use of starter cultures allows for greater control over the fermentation process, leading to consistent product quality and minimizing the risk of spoilage and foodborne illnesses [10]. LAB can degrade nitrites and nitrates in meat products, reducing the risks associated with these compounds. Moreover, researchers are increasingly turning to microbial omics analyses to study the genetic and metabolic pathways employed by these communities, allowing for enhanced selection and optimization of starter cultures tailored to specific flavor and safety profiles. For example, the addition of specific probiotic strains to fermented meats is gaining traction, promoting gut health while delivering the traditional flavors consumers expect [41]. Overall, the intricate relationships among microbial communities in fermented meats not only shape the organoleptic properties but also pave the way for future innovations in the food industry, where traditional practices meet modern scientific insights for superior product development.

4 Fermented Tea, Coffee, and Chocolate

4.1 Fermented Tea

Fermented tea, particularly varieties such as black tea, Fu brick tea, kombucha and pu-erh, is a complex product shaped by diverse microbial communities that significantly influence its flavor, aroma, and health benefits. The composition of microbial communities varies across different types of fermented teas and generally comprises a variety of bacteria and fungi. For example, in Fu brick tea, *Eurotium cristatum* and *Aspergillus* spp. are dominant fungi; in Liu Bao tea, *Aspergillus* spp. predominates; and in Pu-erh tea, *Penicillium chrysogenum* is the dominant fungus; *Klebsiella*, *Paenibacillus*, *Cohnella*, *Pantoea*, *Aspergillus*, and *Cyberlindnera* are main microbial genera in Qingzhuan tea fermentation [11]. The composition of these microbial communities can be affected by several factors, including the horticulture of tea leaves, the fermentation conditions (temperature, pH, and oxygen availability), fermentation time, and the initial microbial inoculum [42]. Research indicates that environmental parameters such as the geographic location where the tea is grown, as well as the specific processing methods employed, can lead to significant variability in microbial profiles [43].

Microbial metabolism in fermented tea plays a critical role in determining its sensory properties and overall functionality. For example, *Aspergillus* contributes to the aroma quality of Qingzhuan tea through pile fermentation and shows a significant positive correlation with tea aroma quality; *Wickerhamomyces* produces alcohols, esters, and terpenes, contributing to a floral aroma; *Agathobacter* enhances acid production [11]. *Acidovorax* in baked raw tea (BT) is positively correlated with alcohols and hydrocarbons, leading to a 47% increase in volatile alcohol levels and imparting a fresh and refreshing quality. In contrast, *Geobacter* in pan-fried raw tea (PFT) negatively correlates with aldehydes. The presence of *Streptococcus* and *Ligilactobacillus* in PFT significantly increases volatile acid content, boosting it from 5% to 25% [43]. *Eurotium cristatum* secretes extracellular enzymes (e.g., oxidases, hydrolases) that degrade polyphenols and flavonoids, reducing bitterness while increasing free amino acids and tea pigments [44].

On an industrial scale, the understanding of microbial dynamics in fermented tea can inform the optimization of fermentation processes, leading to improved product consistency and quality. For instance, studies have shown that the addition of *Eurotium cristatum* can increase the content of volatile flavor compounds in fermented tea, such as ketones, aldehydes, esters, and alcohols, thereby elevating its quality [44]. Researchers have developed microbial inoculants containing strains like *Aspergillus*, which are used in the production of Pu-erh tea to accelerate fermentation and improve its quality [45]. Additionally, fermented tea contains various bioactive substances, such as tea polyphenols, polysaccharides, and theaflavins, whose transformations are closely linked to functional core microorganisms with antioxidant, antibacterial, and antiviral properties [46, 47]. The ongoing study of microbial communities in fermented tea is paving the way for novel applications, including the development of functional beverages that serve tailored health benefits, thus broadening their appeal in the burgeoning market for health-oriented food products.

4.2 Coffee

In coffee processing, coffee fermentation significantly influences the flavor and quality of coffee, necessitating a comprehensive understanding of the microbial communities involved and their interactions. The microbial community composition during coffee fermentation is primarily diverse, consisting of yeasts (such as *Saccharomyces cerevisiae* and *Pichia*), bacteria (such as *Leuconostoc*, *Lactobacillus*, *Erwinia*, *Bacillus*, *Klebsiella*, and *Pseudomonas*), and filamentous fungi (*Aspergillus*). These microorganisms originate from coffee cherries, soil, water, air, and other environmental factors. Factors such as the geographical origin of the coffee beans, the type of processing method employed (wet, dry, or semi-washed), and environmental conditions (temperature, humidity) play critical roles in shaping this microbial landscape. Microbial community composition during coffee fermentation is also influenced by abiotic stresses such as anoxia and osmotic stress during fermentation or dehydration during drying [12].

During coffee fermentation, bacteria, yeasts, and filamentous fungi work synergistically to degrade the mucilage layer of coffee cherries. Bacteria, such as *Erwinia* and *Klebsiella*, produce pectinase and other enzymes to break down pectin in the mucilage, while yeasts like *Saccharomyces cerevisiae*, *Wickerhamomyces anomalus*, *Pichia kudriavzevii*, and *Papiliotrema flavescens* also secrete pectinase enzymes, collectively promoting the degradation of the mucilage layer. In addition, the synergistic relationship between yeasts and LAB can enhance the production of organic acids, alcohols, and esters, which contribute significantly to the sensory attributes of the final coffee product. Flavor compounds such as volatile phenols and fatty acids, which arise from the microbial degradation of polysaccharides and lipids in the coffee pulp, are particularly influential in establishing the distinct sensory characteristics associated with different coffee varieties. Coffee fermented with *Lactobacillus plantarum* produces compounds like decanol, 2-undecanone, and phenol, 2-methyl-, resulting in a fruity, acidic flavor and a caramel-honey-like aroma, with high cupping scores [48].

Harnessing specific microbial strains (starter cultures) can lead to controlled coffee fermentation processes that yield consistent flavor profiles, enhancing product differentiation. For instance, the selective use of particular yeasts or LAB can be employed to favor the production of desirable aroma compounds while suppressing those that may lead to off-flavors [49]. Additionally, advancements in microbial omics analysis—such as metagenomics and metabolomics—are providing deeper insights into the functional roles of microbial consortia during coffee fermentation, enabling the identification of key microbial players linked to flavor generation. Mining functional strains from coffee fermentation, such as *Pichia pastoris*, would further provide a versatile platform for the production of value-added chemicals [50]. As consumers increasingly seek unique and high-quality coffee experiences, the understanding and application of these microbial communities not only promote sustainable and efficient fermentation practices but also allow for innovative approaches to coffee production [12].

4.3 Chocolate

The production of chocolate hinges on an SSF process of cocoa beans involving multispecies microbial consortia, which drive biochemical transformations critical to flavor development and product quality (Fig. 4). The main microbial community responsible for cocoa beans fermentation comprises a diverse array of organisms, including yeasts such as *Saccharomyces cerevisiae*, LAB like *Lactobacillus* spp., and AAB such as *Acetobacter* spp. [13]. Factors influencing microbial community composition include environmental conditions such as temperature, humidity, and the intrinsic properties of the beans, such as their initial sugar content. Other elements such as the fermentation duration, the presence of specific microbial strains, and even the geographical origin of the cocoa beans contribute to the microbial diversity observed in the fermentation process.

During cocoa fermentation, microorganisms interact dynamically. Yeasts typically initiate the fermentation process by converting the sugars present in the cocoa pulp into alcohol, which subsequently provides a suitable environment for LAB to thrive. These LAB contribute to lactic acid production, further acidifying the environment and preparing the substrate for AAB growth. AAB's

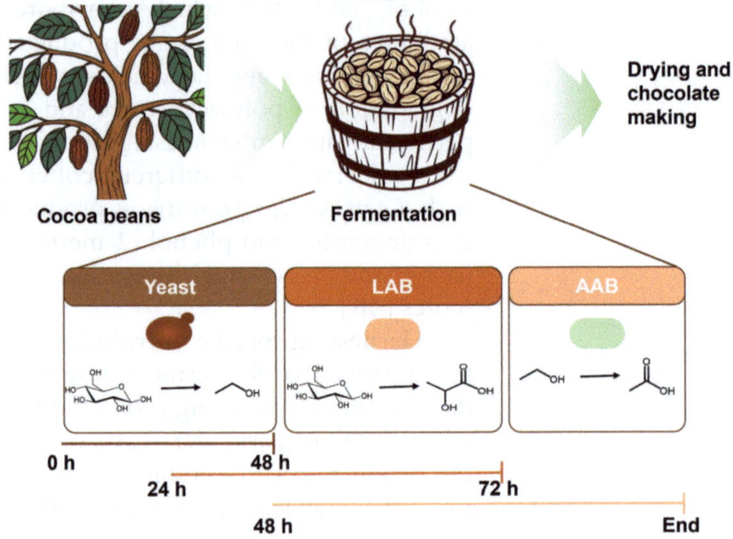

Fig. 4 The chocolate making process and the microorganisms involved in cocoa fermentation

acetic acid production not only sterilizes the bean but also triggers enzymatic reactions within the cotyledon, liberating amino acids and reducing sugars for Maillard reactions during roasting—key to chocolate's aroma (Fig. 4) [51]. This intricate interplay not only impacts the fermentation kinetics but also influences the development of volatile compounds essential for flavor generation. The metabolic activities of microorganisms also activate endogenous enzymes, such as proteases, invertases, and glycosidases, which degrade proteins, sugars, and phenolic compounds, thereby shaping the flavor precursors of cocoa. Furthermore, yeast-generated esters (e.g., ethyl acetate) impart fruity notes; LAB produce lactic acid, which contributes a smooth acidity to cocoa nibs; however, excessive lactic acid can overwhelm the palate. Acetic acid production by AAB adds volatile acidity, enhancing flavor intensity, but in excess, it can impart overly acidic, vinegar-like notes. Diacetyl, produced by both LAB and yeasts, offers mild buttery notes at low concentrations but may seem artificial at high levels [13].

In recent years, there has been a surge in the application of microbial communities and individual microorganisms in chocolate production, thereby enhancing fermentation efficiency and chocolate quality. Studies have shown that inoculating with a cocktail of yeasts and LAB can increase fermentation rates by up to 24% compared to spontaneous fermentation while improving the sensory properties of cocoa beans. The manipulation of microbial communities through such strategies has the potential to optimize the fermentation conditions, minimize undesirable microbial contaminations, and improve the flavor complexity of the final chocolate product. Additionally, studies have revealed that *Acetobacter* species inhibits other microorganisms, such as filamentous fungi, which impair grain flavor and quality by influencing the synthesis of dipeptides [13].

5 Fermented Tobacco

Tobacco leaves fermentation refers to the fermentation treatment of cigar and flue-cured tobacco leaves. The microbial community composition plays a critical role in determining the overall quality, flavor, and aromatic profiles of the finished products. However, the microbial communities in cigars and flue-cured tobacco leaves are diverse and dynamically shaped by environmental conditions, raw material properties, and processing stages [52–55]. For example, among the bacterial community, genera such as *Streptococcus*, *Pseudomonas*, *Corynebacterium*, and *Enterococcus* are frequently noted during cigar fermentation, while *Pseudomonas*, *Sphingomonas*, *Bacillus*, and *Methylobacterium* appear during flue-cured tobacco fermentation. Fungal species, particularly those from the genera *Alternaria*, *Aspergillus*, *Penicillium*, and *Symmetrospora*, are the main species during the tobacco fermentation [54]. Additionally, the chemical composition of the tobacco leaves themselves (e.g., sugar and nitrogen content) also affect the microbial community composition, and microbial inoculation is often tailored to boost desired microbial pathways, directly influencing final product characteristics [14, 56].

The interactions between these diverse microbial communities are multifaceted, constituting a network of synergistic and antagonistic relationships that shape the fermentative environment. For example, during the fermentation of cigar tobacco leaves, the bacterial co-occurrence networks of middle and top tobacco leaves during fermentation for 3, 6, and 9 months show significantly fewer connections compared to those at fermentation for 0 and 12 months [14]. The predicted metabolic network related to the formation of the main characteristic flavors of tobacco leaves is showed in Fig. 5. This interplay not only enhances the safety and stability of the fermented product but also directly influences the olfactory and gustatory qualities through biosynthesis of flavor compounds, such as esters, alcohols, and phenolic compounds. Fungal genera such as *Aspergillus* enhance flavor by producing hydrolytic enzymes (e.g., pectinases and amylases), which elevate reducing sugar content by 289% and generate volatile compounds like hydroxylated derivatives, detected via FTIR and GC-MS analyses [57]. *Bacillus* can break down and utilize macromolecular substances like starch, protein, cellulose, and carotene in the tobacco leaves to produce aromatic compounds, thereby enhancing the tobacco's aroma [58]. Microbial metabolites can exhibit complex interactions with tobacco's inherent chemical compounds,

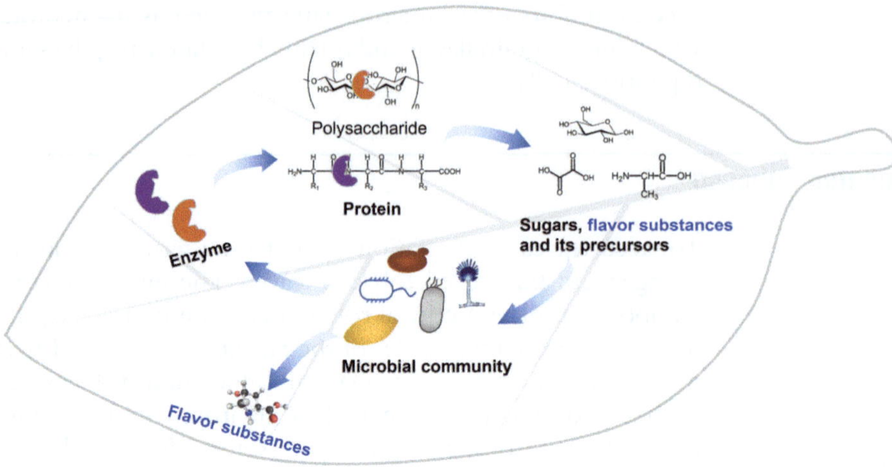

Fig. 5 The predicted metabolic network related to the formation of the main characteristic flavors of tobacco leaves

amplifying specific flavors or masking undesirable ones, emphasizing the intricate balance of microbial contributions.

The applications of microbial communities in fermented tobacco extend beyond traditional fermentation practices into innovative biotechnology approaches. Utilizing microbial consortia, researchers and tobacco manufacturers are exploring targeted fermentation designed to enhance flavor profiles and ensure consistency across production batches [14]. Biotechnological advancements in microbial omics allow for the analysis of the community structure and functional potential of specific microorganisms through metagenomics and metabolomics techniques [59], facilitating a more precise manipulation of fermentation processes. Beyond traditional bioinformatics, the integration of artificial intelligence and machine learning has become essential for mining high-dimensional multi-omics data and predicting the behavior of complex microbial consortia in fermentation processes [60]. By identifying key microbial players responsible for desirable flavor compounds, tobacco producers can incorporate these specific strains into their fermentation regimes, optimizing the production of premium cigars or flue-cured varieties. Furthermore, the integration of microbial communities into bioprocessing strategies serves as a sustainable alternative to chemical additives, fostering a more ecological approach to flavor enhancement while maintaining quality and preserving traditional fermentation methods.

6 Summary and Perspectives

Across the specialty and nonfood fermentations discussed in this chapter, microbial diversity and interaction form the dual foundation of ecosystem function and product identity. Multispecies consortia of bacteria, yeasts, and filamentous fungi assemble under substrate composition, processing parameters, and environmental conditions, occupying structured microhabitats (cheese rinds, kefir grains, tea and tobacco piles, meat surfaces) that create niches and successional dynamics. Within these networks, cross-feeding, metabolic handoffs, and environmental modification drive the formation of complex flavor profiles, textures, bioactive compounds, and safety attributes. Further, high-dimensional omics—metagenomics, transcriptomics, metabolomics, single-cell and spatial tools—combined with isotope-guided metabolic flux analysis will be essential to link gene content with real carbon and energy flows in such consortia. Quantitative models and artificial intelligence will help mine these data, identify keystone taxa and critical interactions, and predict fermentation trajectories and host outcomes. Ultimately, shifting from description to directed fermentation—through interaction-aware consortia design and omics-informed process control—will transform specialty fermentations into curated microbial resource libraries, serving food and beverage innovation, agriculture, environmental bioremediation, animal production, and even precision medicine.

Conflict of Interest

The authors have no conflicts of interest to declare that are relevant to the content of this chapter.

References

1. Marco ML, Sanders ME, Gänzle M, Arrieta MC, Cotter PD, De Vuyst L, Hill C, Holzapfel W, Lebeer S, Merenstein D, Reid G, Wolfe BE, Hutkins R (2021) The International Scientific Association for Probiotics and Prebiotics (ISAPP) consensus statement on fermented foods. Nat Rev Gastroenterol Hepatol 18(3):196–208. https://doi.org/10.1038/s41575-020-00390-5
2. Wolfe BE, Button JE, Santarelli M, Dutton RJ (2014) Cheese rind communities provide tractable systems for in situ and in vitro studies of microbial diversity. Cell 158(2):422–433. https://doi.org/10.1016/j.cell.2014.05.041
3. Liu Y, Kang Z, Zhang W, Ma C, Xu P, Gao C (2024) Metabolic functions of misunderstood D-2-hydroxyglutarate. Innov Life 2(4): 100101. https://doi.org/10.59717/j.xinn-life.2024.100101
4. Bahadur A, Sajjad W, Banerjee A, Ilahi N, Din SZU, Chen T, Zhang W, Chen S (2024) Could microbes inhabiting extreme desert environments be a gateway to life on the Martian surface? Innov Life 2(4):100091. https://doi.org/10.59717/j.xinn-life.2024.100091
5. Rong B, Zhang Q, Zhang X, Zhang N, Shen Z, Pang Y, Lin X, Liu D, Yang X (2024) Hyocholic acid: a novel therapeutic strategy for metabolic syndrome. Innov Life 2(4):

100093. https://doi.org/10.59717/j.xinn-life.2024.100093

6. Chen S, Xu J, Cao J, Bahadur A, Wu M, Wang Z, Chen J, Wang J, Shi Y (2024) Depth-related microbial communities and functional genes in alpine permafrost. Innov Life 2(3):100081. https://doi.org/10.59717/j.xinn-life.2024.100081
7. Reuben RC, Langer D, Eisenhauer N, Jurburg SD (2023) Universal drivers of cheese microbiomes. iScience 26(1):105744. https://doi.org/10.1016/j.isci.2022.105744
8. de Souza M, Drunkler DA, Colla E (2024) Probiotic functional yogurt: challenges and opportunities. Fermentation 10(1):6. https://doi.org/10.3390/fermentation10010006
9. Zhang T, Chang M, Zhou Y, Wang M, Yan M, Hou X, Liu R, Yuan Y, Yue T (2024) Dynamic alterations of flavor, functional nutrients, and microbial community during fermentation of different animal milk kefirs. Food Res Int 186:114305. https://doi.org/10.1016/j.foodres.2024.114305
10. Ashaolu TJ, Khalifa I, Mesak MA, Lorenzo JM, Farag MA (2023) A comprehensive review of the role of microorganisms on texture change, flavor and biogenic amines formation in fermented meat with their action mechanisms and safety. Crit Rev Food Sci Nutr 63(19):3538–3555. https://doi.org/10.1080/10408398.2021.1929059
11. Wu S, Wang W, Zhu W, Chen W, Xu W, Sui M, Jiang G, Xiao J, Ning Y, Ma C, Fang X, Wang Y, Huang Y, Lei G (2022) Microbial community succession in the fermentation of Qingzhuan tea at various temperatures and their correlations with the quality formation. Int J Food Microbiol 382:109937. https://doi.org/10.1016/j.ijfoodmicro.2022.109937
12. Shen X, Wang Q, Wang H, Fang G, Li Y, Zhang J, Liu K (2025) Microbial characteristics and functions in coffee fermentation: a review. Fermentation 11(1):5. https://doi.org/10.3390/fermentation11010005
13. Campos SM, Martínez-Burgos WJ, dos Reis GA, Ocán-Torres DY, dos Santos Costa G, Rosas Vega F, Alvarez Badel B, Sotelo Coronado L, Lima Serra J, Soccol CR (2025) The role of microbial dynamics, sensorial compounds, and producing regions in cocoa fermentation. Microbiol Res 16(4):75. https://doi.org/10.3390/microbiolres16040075
14. Tao J, Chen S, Jiang Z, Wang C, Zhang E, Liang H, Xu Y, Cao P, Ding N, Zhang M, He W, Chen Q (2024) Fermentation process of tobacco leaves drives the specific changes of microbial community. BMC Microbiol 24(1):534. https://doi.org/10.1186/s12866-024-03702-w
15. Wang X, Li Z, Chen N, Luo X (2024) Co-evolve strategy for the discovery of genetic "dark matter". Innov Life 2(2):100071. https://doi.org/10.59717/j.xinn-life.2024.100071
16. Zhou L-W (2023) Microbial taxonomy with DNA sequence data as nomenclatural type: how far should we go? Innov Life 1(2):100017. https://doi.org/10.59717/j.xinn-life.2023.100017
17. Neviani E, Gatti M, Gardini F, Levante A (2025) Microbiota of cheese ecosystems: a perspective on cheesemaking. Foods 14(5):830. https://doi.org/10.3390/foods14050830
18. Melkonian C, Zorrilla F, Kjærbølling I, Blasche S, Machado D, Junge M, Sørensen KI, Andersen LT, Patil KR, Zeidan AA (2023) Microbial interactions shape cheese flavour formation. Nat Commun 14(1):8348. https://doi.org/10.1038/s41467-023-41059-2
19. Lecomte M, Cao W, Aubert J, Sherman DJ, Falentin H, Frioux C, Labarthe S (2024) Revealing the dynamics and mechanisms of bacterial interactions in cheese production with metabolic modelling. Metab Eng 83:24–38. https://doi.org/10.1016/j.ymben.2024.02.014
20. Wu T, Guo S, Kwok L-Y, Zhang H, Wang J (2025) Temperature-dependent metabolic interactions between *Streptococcus thermophilus* and *Lactobacillus delbrueckii* ssp. bulgaricus in milk fermentation: insights from gas chromatography–ion mobility spectrometry metabolomics. J Dairy Sci 108(1):242–256. https://doi.org/10.3168/jds.2024-25153
21. Merrill RK, Oberg CJ, WR MM, Kalab M, DJ MM (1996) Microstructure and physical properties of a reduced fat mozzarella cheese made using lactobacillus casei ssp. casei adjunct culture. LWT Food Sci Technol 29(8):721–728. https://doi.org/10.1006/fstl.1996.0112
22. Costa NE, Wang L, Auty ME, Hannon JA, McSweeney PLH, Beresford TP (2012) Rheological, microscopic and primary chemical characterisation of the exopolysaccharide produced by Lactococcus lactis subsp. cremoris DPC6532. Dairy Sci Technol 92(3):219–235. https://doi.org/10.1007/s13594-012-0059-4
23. de Vos WM (2011) Systems solutions by lactic acid bacteria: from paradigms to practice. Microb Cell Factories 10(1):S2. https://doi.org/10.1186/1475-2859-10-S1-S2
24. Rolim FRL, Freitas Neto OC, Oliveira MEG, Oliveira CJB, Queiroga RCRE (2020) Cheeses

as food matrixes for probiotics: in vitro and in vivo tests. Trends Food Sci Technol 100: 138–154. https://doi.org/10.1016/j.tifs.2020.04.008

25. Alsaleem KA, Hamouda MEA (2024) Optimizing probiotic low-fat yogurt: the benefits of incorporating defatted rice bran for enhanced quality and functionality. Food Sci Nutr 12(12):10242–10254. https://doi.org/10.1002/fsn3.4558
26. Islam SMR, Tanzina AY, Foysal MJ, Hoque MN, Rumi MH, Siddiki AMAMZ, Tay AC-Y, Hossain MJ, Bakar MA, Mostafa M, Mannan A (2021) Insights into the nutritional properties and microbiome diversity in sweet and sour yogurt manufactured in Bangladesh. Sci Rep 11(1):22667. https://doi.org/10.1038/s41598-021-01852-9
27. Sieuwerts S (2016) Microbial interactions in the yoghurt consortium: current status and product implications. SOJ Microbiol Infect Dis 4(2):1–5
28. Gao Y-Z, Wang Y, Ji M, Zhou N-Y, Huang WE (2023) A whole-cell hydrogen peroxide biosensor and its application in visual food analysis. Innov Life 1(1):100011. https://doi.org/10.59717/j.xinn-life.2023.100011
29. Tang C, Zhou W, Shan M, Lu Z, Lu Y (2022) Yogurt-derived *Lactobacillus plantarum* Q16 alleviated high-fat diet-induced non-alcoholic fatty liver disease in mice. Food Sci Human Wellness 11(5):1428–1439. https://doi.org/10.1016/j.fshw.2022.04.034
30. Robitaille G, Tremblay A, Moineau S, St-Gelais D, Vadeboncoeur C, Britten M (2009) Fat-free yogurt made using a galactose-positive exopolysaccharide-producing recombinant strain of *Streptococcus thermophilus*. J Dairy Sci 92(2):477–482. https://doi.org/10.3168/jds.2008-1312
31. Cogan TM, Beresford TP, Steele J, Broadbent J, Shah NP, Ustunol Z (2007) Advances in starter cultures and cultured foods. J Dairy Sci 90(9):4005–4021. https://doi.org/10.3168/jds.2006-765
32. Liu L, Guo Y, Wang B, Pan Y, Zhao C, Xiao Z, Liu Y, Yu G, Chen X, Rao L (2025) Yogurt-inspired hybrid membrane vesicles for the prevention and treatment of *Helicobacter pylori* infection. Cell Biomater 4(1):100072. https://doi.org/10.1016/j.celbio.2025.100072
33. Alraddadi FAJ, Ross T, Powell SM (2023) Evaluation of the microbial communities in kefir grains and kefir over time. Int Dairy J 136:105490. https://doi.org/10.1016/j.idairyj.2022.105490
34. Tingirikari JMR, Sharma A, Lee H-J (2024) Kefir: a fermented plethora of symbiotic microbiome and health. J Ethnic Foods 11(1):35. https://doi.org/10.1186/s42779-024-00252-4
35. Baars T, van Esch B, Diks M, van Ooijen L, Zhang Z, Dekker P, Boeren S, Garssen J, Hettinga K, Kort R (2025) Bacterial diversity, bioactive peptides, and enhanced immunomodulatory effects in raw milk kefir made with defined starter cultures versus backslopping. Int Dairy J 164:106202. https://doi.org/10.1016/j.idairyj.2025.106202
36. Witthuhn RC, Schoeman T, Cilliers A, Britz TJ (2005) Impact of preservation and different packaging conditions on the microbial community and activity of Kefir grains. Food Microbiol 22(4):337–344. https://doi.org/10.1016/j.fm.2004.09.001
37. Blasche S, Kim Y, Mars RAT, Machado D, Maansson M, Kafkia E, Milanese A, Zeller G, Teusink B, Nielsen J, Benes V, Neves R, Sauer U, Patil KR (2021) Metabolic cooperation and spatiotemporal niche partitioning in a kefir microbial community. Nat Microbiol 6(2):196–208. https://doi.org/10.1038/s41564-020-00816-5
38. Salvatore MM, Maione A, Imparato M, Buonanno A, de Alteriis E, Naviglio D, Gallo M, Andolfi A, Guida M, Salvatore F, Galdiero E (2025) Development and metabolomic characterization of novel functional fermented juices fortified with a probiotic consortium. Appl Food Res 5(1):100905. https://doi.org/10.1016/j.afres.2025.100905
39. Sun Y, Guo S, Wu T, Zhang J, Kwok L-Y, Sun Z, Zhang H, Wang J (2023) Untargeted mass spectrometry-based metabolomics approach unveils biochemical changes in compound probiotic fermented milk during fermentation. NPJ Sci Food 7(1):21. https://doi.org/10.1038/s41538-023-00197-z
40. Wang Y, Han J, Wang D, Gao F, Zhang K, Tian J, Jin Y (2022) Research update on the impact of lactic acid bacteria on the substance metabolism, flavor, and quality characteristics of fermented meat products. Foods 11(14): 2090. https://doi.org/10.3390/foods11142090
41. Munekata PES, Pateiro M, Tomasevic I, Dominguez R, da Silva Barretto AC, Santos EM, Lorenzo JM (2022) Functional fermented meat products with probiotics-a review. J Appl Microbiol 133(1):91–103. https://doi.org/10.1111/jam.15337
42. Chen Y, Li D, Lang Z, Xu Y, Luo Z (2024) An emerging role beyond genetics: DNA

methylation in horticultural quality shaping. Innov Life 2(1):100050. https://doi.org/10.59717/j.xinn-life.2024.100050

43. Guo Y, Pan Y, Feng X, Guo H, Liu L, Zhang K, Xie H, Zhu B, Gong S, Chu Q, Fang H, Chen P (2025) Reshaped local microbiology metabolism by raw tea according to pile fermentation in the dark tea. J Adv Res 78:211-224. https://doi.org/10.1016/j.jare.2025.02.039
44. Liangzhen J, Luo W, Tao Y (2022) Research progress on *Eurotium cristatum* and its fermentation application. Sci Technol Food Ind 43(4):454–462. https://doi.org/10.13386/j.issn1002-0306.2021030124
45. Zheng J, Yu L, Aaqil M, Wang Q, Peng W, Zhuang L, Gong W, Zheng T, Zhao M, Wang C, Jiang X, Yan L, Yang R (2024) Enhanced fermentation of Pu-erh tea with *Aspergillus niger*: quality and microbial community analysis. Molecules 29(23):5647. https://doi.org/10.3390/molecules29235647
46. Aloo OS, Kim D-G, Vijayalakshmi S, Aloo DO, Ochola CO, Oh D-H (2024) Polyphenol constituents and impacts of fermented teas (*Camellia sinensis*) in human wellness. Food Biosci 60:104389. https://doi.org/10.1016/j.fbio.2024.104389
47. Jakubczyk K, Łopusiewicz Ł, Kika J, Janda-Milczarek K, Skonieczna-Żydecka K (2024) Fermented tea as a food with functional value—its microbiological profile, antioxidant potential and phytochemical composition. Foods 13(1):50. https://doi.org/10.3390/foods13010050
48. Todhanakasem T, Van Tai N, Pornpukdeewattana S, Charoenrat T, Young BM, Wattanachaisaereekul S (2024) The relationship between microbial communities in coffee fermentation and aroma with metabolite attributes of finished products. Foods 13(15):2332. https://doi.org/10.3390/foods13152332
49. Pereira GVM, de Mello SV, Wiele N, da Silva VA, de Carvalho Neto DP, Souza AFD, Nogueira dos Santos DV, Ruiz IR, Rogez H, Soccol CR (2024) How yeast has transformed the coffee market by creating new flavors and aromas through modern post-harvest fermentation systems. Trends Food Sci Technol 151:104641. https://doi.org/10.1016/j.tifs.2024.104641
50. Zhou Q, Yang J, Zhang W, Jiang W, Chen J, Chen L, Jiang Y, Xin F (2025) *Komagataella phaffii*: a versatile platform for the production of value-added chemicals. Green Carbon 3(4):407–418. https://doi.org/10.1016/j.greenca.2025.01.004
51. Hirko B, Mitiku H, Getu A (2023) Role of fermentation and microbes in cacao fermentation and their impact on cacao quality. Syst Microbiol Biomanuf 3(4):509–520. https://doi.org/10.1007/s43393-023-00160-9
52. Wang F, Zhao H, Xiang H, Wu L, Men X, Qi C, Chen G, Zhang H, Wang Y, Xian M (2018) Species diversity and functional prediction of surface bacterial communities on aging flue-cured tobaccos. Curr Microbiol 75(10):1306–1315. https://doi.org/10.1007/s00284-018-1525-x
53. Zhang M, Guo D, Wang H, Wu G, Shi Y, Zhou J, Zhao E, Zheng T, Li X (2024) Analyzing microbial community and volatile compound profiles in the fermentation of cigar tobacco leaves. Appl Microbiol Biotechnol 108(1). https://doi.org/10.1007/s00253-024-13043-3
54. Wang F, Jin Y, Chen X, Zhang Y, Jiang X, Zhang G, Chen G, Yang M, Leng F, Li H, Wu L, Zhang H (2022) The diversity, structure and function of microbial communities changes across aging process of tobacco leaves. Environ Res Commun 4(9):095012. https://doi.org/10.1088/2515-7620/ac9352
55. Wang F, Men X, Zhang G, Liang K, Xin Y, Wang J, Li A, Zhang H, Liu H, Wu L (2018) Assessment of 16S rRNA gene primers for studying bacterial community structure and function of aging flue-cured tobaccos. AMB Express 8(1):182. https://doi.org/10.1186/s13568-018-0713-1
56. Wang F, Jiang X, Liu Y, Zhang G, Zhang Y, Jin Y, Shi S, Men X, Liu L, Wang L, Liao W, Chen X, Chen G, Liu H, Ahmad M, Fu C, Wang Q, Zhang H, Lee SY (2024) Tobacco as a promising crop for low-carbon biorefinery. Innovation 5(5):100687. https://doi.org/10.1016/j.xinn.2024.100687
57. Zhang-de L, Hao-rui C, Hong L, Yun X, Li-zheng N, Zhen-xing N, Long-yan Y, Zhi-zhong H, Pei-hong S (2018) Fermentation effects of three Aspergillus spp. fungi derived from Liupao tea on the application in tobacco. J South Agric 49(10):2055–2061. https://doi.org/10.3969/j.issn.2095-1191.2018.10.23
58. Zhang L, Li W, Peng Z, Zhang J (2025) Effect of microbial community on the formation of flavor components in cigar tobacco leaves during air-curing. BMC Microbiol 25(1):56. https://doi.org/10.1186/s12866-025-03774-2
59. Li C, Gillings MR, Zhang C, Chen Q, Zhu D, Wang J, Zhao K, Xu Q, Leung PH, Li X, Liu J, Jin L (2024) Ecology and risks of the global plastisphere as a newly expanding microbial

habitat. Innovation 5(1):100534. https://doi.org/10.1016/j.xinn.2023.100543

60. Luo M, Yang W, Bai L, Zhang L, Huang J-W, Cao Y, Xie Y, Tong L, Zhang H, Yu L, Zhou L-W, Shi Y, Yu P, Wang Z, Yuan Z, Zhang P, Zhang Y, Ju F, Zhang H, Wang F, Cui Y, Zhang J, Jia G, Wan D, Ruan C, Zeng Y, Wu P, Gao Z, Zhao W, Xu Y, Yu G, Tian C, Jin LN, Dai J, Xia B, Sun B, Chen F, Gao Y-Z, Wang H, Wang B, Zhang D, Cao X, Wang H, Huang T (2024) Artificial intelligence for life sciences: a comprehensive guide and future trends. Innov Life 2(4):100105. https://doi.org/10.59717/j.xinn-life.2024.100105

Chapter 14

Omics-Based Insights into Polysaccharide Biosynthesis in Macrofungi

Xing Luo, Ji-Hang Jiang, and Li-Wei Zhou

Abstract

To address the limited understanding of polysaccharide biosynthesis in medicinal macrofungi, we present an integrated multi-omics protocol. Our method details the entire workflow, from experimental design and de novo genome assembly to transcriptomic and metabolomic profiling. Key genes are pinpointed through weighted gene co-expression network analysis and KEGG pathway mapping, followed by experimental validation using RT-qPCR. This established pipeline provides a robust framework for elucidating biosynthetic pathways and guiding strain improvement.

Key words Strategic bioresource, Medicinal macrofungi, Bioactive compounds, Metabolic network

1 Introduction

Macrofungi, well-known also as mushroom-forming fungi, are a group of fungi bearing visible fruiting bodies to naked eye [1]. Many species of macrofungi possess various bioactive compounds and have been utilized as medicinal resources for thousand years [2–5]. In terms of their high medicinal values and differentiated geographic distribution, macrofungi can be considered as important strategic bioresources [6].

Among the various metabolites produced by macrofungi, polysaccharides are the most abundant and widely used as medicinal supplements. Nevertheless, the biosynthetic pathways of polysaccharides in macrofungi have not well explored, and thus the key enzymes and their encoding genes in charge of the pathways are not clear. This knowledge gap restricts the scientific utilization of macrofungal polysaccharides. Indeed, the functions of many genes in macrofungi are unknown and rarely explored, partially due to the lack of high-quality omics data [7–12].

Therefore, it is necessary to generate such knowledge from as many as macrofungal species and then summarize a general

Tao Huang (ed.), *Microbial 'Omics: Methods and Protocols*, Methods in Molecular Biology, vol. 3033,
https://doi.org/10.1007/978-1-0716-5264-0_14,

pathway of polysaccharide biosynthesis, especially in the case that macrofungi are highly diverse from biodiversity and metabolic perspectives. To achieve this aim, here we provide a basic protocol for clarifying the biosynthetic pathway of polysaccharides in macrofungi, which can be expanded following the specific characters of each targeted macrofungal species.

2 Materials

2.1 Biological Material and Cultivation

1. Biological Material: This protocol is designed for a wide range of macrofungi.
2. Culture Media.

   Activation medium: Potato dextrose agar (PDA).

   Liquid fermentation medium: Glucose (35 g/L), yeast extract (10 g/L), potassium dihydrogen phosphate (1.5 g/L), magnesium sulfate (1.0 g/L), and VB_1 (0.05 g/L).
3. Cultivation conditions: Inoculate liquid fermentation medium with active mycelial plugs or a homogenized seed culture. Incubate in a rotary shaker at 28 °C and 140 r/min.

2.2 Reagents and Kits

1. DNA extraction: Genomic DNA extraction kit.
2. Total RNA extraction: Trizol reagent.
3. Library preparation kits (genome and transcriptome).
4. Metabolite extraction solvents: HPLC/MS grade methanol, acetonitrile, and water. A common and effective ratio is 2:2:1 (v/v/v).
5. RT-qPCR reagents: Reverse transcription reagent kit with gDNA Eraser, SYBR Green Premix, and gene-specific primers (a stable reference gene must be identified and validated for the specific fungal species and experimental conditions).

2.3 Equipment

1. Quantifluor-ST fluorometer (Promega, Madison, Wisconsin, USA).
2. Agilent 2100 Bioanalyzer system (Agilent Technologies Inc., California, USA).
3. High-throughput sequencers (e.g., Illumina, PacBio).
4. Vanquish UHPLC system (Thermo Fisher Scientific, USA).
5. Orbitrap Exploris 120 mass spectrometer (Thermo Fisher Scientific, USA).
6. CFX96 Touch Real-Time PCR Detection System (Bio-Rad, USA).

3 Methods

3.1 Experimental Design and Sample Preparation

1. Defining Growth Stages and Critical Time Points for Sampling.

Based on the data obtained from periodic measurements of mycelial biomass (dry weight) and polysaccharide content (e.g., every 2–3 days), key time points for sampling were selected.

2. Sample Preparation.

Following filtration, the mycelium was rinsed three times with sterile distilled water and subsequently homogenized into a fine powder under liquid nitrogen. The homogenized powder was aliquoted into nuclease-free centrifuge tubes for dedicated DNA, RNA, and metabolomics analyses, with all aliquots stored at −80 °C (*see* **Note 1**).

3.2 De Novo Genome Sequencing and Assembly

1. High-Quality Genomic DNA Extraction and QC.

High-molecular-weight genomic DNA was extracted from mycelia using the genomic DNA extraction kit. The quality and integrity of the DNA were verified by 1% agarose gel electrophoresis to ensure it suitable for long-read sequencing [13]. Concentration and purity were assessed using both a NanoDrop spectrophotometer and a Qubit fluorometer. Only DNA samples with A260/A280 ratios between 1.8 and 2.0 and showing minimal degradation were used for constructing both Illumina and PacBio sequencing libraries [14].

2. Library Construction.

Both second- and third-generation sequencing approaches were employed by constructing both a short-insert genomic library (300–350 bp) using the Illumina Library Preparation Kit and a long-insert SMRTbell library (20 kb) using the PacBio SMRTbell® Express Template Prep kit [14].

3. Sequencing Platform Selection and Coverage Depth Recommendations.

Sequencing was performed on both the Illumina NovaSeq PE150 and PacBio Sequel platforms to leverage a hybrid assembly strategy. For the ~50 Mb genome, we targeted a minimum coverage depth of 50X for each platform [7, 13].

4. Genome Assembly Using Long-Read Assemblers.

The genome was assembled using Canu (v1.8) with standard parameter settings [15].

5. Assembly Polishing with Short Reads.

The initial assembly was further refined and error-corrected using Pilon (v.1.24) with short reads [16]. The genome quality was evaluated by BUSCO (v6.0.0) (*see* **Note 2**).

6. Genome Annotation: Gene Prediction and Functional Annotation.

The ab initio prediction of protein-coding gene was carried out by the BRAKER2 (v2.1.5) pipeline which integrates evidence from protein databases to generate reliable and accurate gene models [17].

Gene functions were annotated with the references of nine databases, viz., Gene Ontology (http://geneontology.org/), Kyoto Encyclopedia of Genes and Genomes (KEGG) (https://www.kegg.jp/), Eukaryotic Orthologous Groups (https://www.creative proteomics.com/services/kogannotation-analysis-service.htm), Non-Redundant Protein Database (NR) (https://www.ncbi.nlm.nih.gov/protein/),

Transporter Classification Database (http://www.tcdb.org), Fungal Cytochrome P450 Database (http://p450.riceblast.snu.ac.kr/cyp.php), Swiss-Prot (https://www.uniprot.org/), Pfam (http://pfam.xfam.org/), and Carbohydrate-Active enZYmes (CAZymes) Database (http://www.cazy.org/). All predicted coding genes were aligned with these nine databases using DIAMOND (v2.0.2) with the cutoff values of E-value no more than 1×10^{-5}, identity not less than 40%, and coverage not less than 40%. When a single gene retrieved more than one result meeting the cut-off values from databases, this gene was annotated following the result with the best score [7].

3.3 Transcriptome Sequencing and Analysis

1. Total RNA Extraction, Library Preparation, and Sequencing.

Total RNA was extracted from mycelial powder using the Trizol Reagent. After purity check, 3 μg of total RNA were prepared for cDNA synthesis using an RNA library preparation kit. The quantity and purity of the libraries were evaluated with the Quantifluor-ST fluorometer and Agilent 2100 Bioanalyzer system. Finally, RNA sequencing was performed on an Illumina NovaSeq 6000 platform [18].

2. Quality Control and Read Mapping.

Use FastQC (v0.11.9) for raw read quality control and fastp (v0.22.0) for adapter trimming and quality filtering.

Align the clean reads to the assembled genome using HISAT2 (v2.1.0) [19].

3. Transcript Quantification and Differentially Expressed Gene Analysis.

Assemble transcripts and quantify gene expression levels in Fragments Per Kilobase of transcript per Million mapped reads (FPKM) using StringTie (v2.2.1) [20].

Perform differential expression analysis with DESeq2 (v1.38.3) in R, using a threshold of $|\log_2\text{FoldChange}| > 1$ and p-value <0.05 [21].

3.4 Metabolomic Profiling

1. Metabolite Extraction from Mycelia.

Weigh 25 mg of mycelial powder, and add 500 μL of extraction solution (methanol:acetonitrile:water = 2:2:1). Vortex, homogenize, sonicate in an ice water bath, and incubate at −40 °C for 1 h. Centrifuge at 13,800× g for 15 min at 4 °C. Collect the supernatant for LC-MS analysis [22].

2. LC-MS Analysis.

Analyze each sample in both ESI^+ and ESI^- modes. Use a gradient elution program. Acquire data in data-dependent acquisition (DDA) or information-dependent acquisition (IDA) mode for MS/MS spectra.

3. Data Processing and Differentially Accumulated Metabolites Identification.

Process raw data using the XCMS (v4.6.0) package in R for peak picking, alignment, and integration. Identify metabolites by matching accurate mass and MS/MS spectra against public databases (HMDB, MassBank, Metlin, mzCloud and a proprietary database). Identify DAMs using multivariate statistics (OPLS-DA) with criteria of variable importance projection (VIP) >1 and p-value <0.05 [23].

3.5 Integrative Bioinformatics Analysis

1. Pathway Mapping and Visualization.

Generate integrated pathway visualizations by mapping the expression levels (FPKM) of DEGs and the relative abundance of DAMs onto polysaccharide synthesis pathway.

2. Weighted Gene Co-expression Network Analysis (WGCNA).

Construct a signed co-expression network from the filtered gene expression matrix using the WGCNA R package (v1.73). Select a soft-thresholding power that results in an approximate scale-free topology (*see* **Note 3**).

Calculate module-trait associations by correlating module eigengenes with the measured traits: the abundance of carbohydrate metabolites associated with polysaccharide synthesis (e.g., glucose-6-phosphate, UDP-glucose).

Identify the most relevant modules (correlation coefficients >0.8 and a p-value <0.05), and extract the genes within them [18].

3. KEGG Enrichment Analysis of Key Module Genes.

The ClusterProfiler software (v4.6.0) was used for KEGG pathway enrichment analysis of key module genes, focusing on pathways with significant enrichment (p-value <0.05).

3.6 Screening and Validation of Key Genes

1. Key Genes Screening.

Based on the KEGG enrichment analysis results, pathways significantly associated with polysaccharide synthesis were identified. Subsequently, genes enriched in these pathways were compared with those encoding enzymes involved in polysaccharide synthesis, leading to the identification of key genes regulating polysaccharide synthesis.

2. RT-qPCR Validation.

Design primers for the key genes and the pre-validated reference gene. Perform RT-qPCR on the same RNA samples used for RNA-seq. The relative expression patterns obtained by RT-qPCR (calculated via the $2^{-\Delta\Delta CT}$ method) should show a strong and significant correlation (e.g., Pearson r >0.8 or 0.9) with the FPKM values from the RNA-seq data. This confirms the reliability of the transcriptomic data for the genes of interest [24].

4 Notes

1. The cornerstone of successful multi-omics integration is the simultaneous collection of all sample types. To ensure this, all aliquots for DNA, RNA, metabolomic, and polysaccharide analyses must be obtained from the same homogenized mycelial powder.
2. Proceed with functional annotation only after rigorously assessing the de novo genome assembly, with a focus on key benchmarks including completeness (evaluated via BUSCO) and contiguity (reflected by a high contig N50).
3. During WGCNA, determine the soft-thresholding power (β) using the pickSoftThreshold function. The appropriate power is defined as the lowest value for which the scale-free topology fit index (R^2) exceeds 0.85.

Acknowledgments

The research was financed by the National Key Research and Development Program of China (No. 2022YFC2601200).

Conflict of Interest The authors have no conflicts of interest to declare that are relevant to the content of this chapter.

References

1. Wang XW, Liu SL, Zhou LW (2021) Citizen science initiative in taxonomy of macrofungi in China 1.0. Mycosystema 40(4):844–850. https://doi.org/10.13346/j.mycosystema.200283
2. Wu F, Zhou LW, Yang ZL, Bau T, Li TH, Dai YC (2019) Resource diversity of Chinese macrofungi: edible, medicinal and poisonous species. Fungal Divers 98(1):1–76. https://doi.org/10.1007/s13225-019-00432-7
3. Zhou LW, Ghobad-Nejhad M, Tian XM, Wang YF, Wu F (2022) Current status of 'Sanghuang' as a group of medicinal mushrooms and their perspective in industry development. Food Rev Intl 38(4):589–607. https://doi.org/10.1080/87559129.2020.1740245
4. Cheng Y, Zhou LJ, Jiang JH, Tian XM, Zhou LW (2023) *Phylloporia* (Hymenochaetales, Basidiomycota), a medicinal wood-inhabiting fungal genus with much potential for commercial development. Food Rev Intl 39(5):2776–2789. https://doi.org/10.1080/87559129.2021.1967382
5. Zhou LW (2020) Systematics is crucial for the traditional Chinese medicinal studies and industry of macrofungi. Fungal Biol Rev 34(1):10–12. https://doi.org/10.1016/j.fbr.2019.10.002
6. Bai M, Zhou L, Tong Y, Yin F (2023) Risk assessment and warning system for strategic biological resources in China. Innov Life 1(1):100004–100002. https://doi.org/10.59717/j.xinn-life.2023.100004
7. Jiang JH, Wu SH, Zhou LW (2021) The first whole genome sequencing of *Sanghuangporus sanghuang* provides insights into its medicinal application and evolution. J Fungi 7(10):787. https://doi.org/10.3390/jof7100787
8. Shen ZQ, Jiang JH, Li CT, Li Y, Zhou LW, Shen ZQ, Jiang JH, Li CT, Li Y, Zhou LW (2023) Genome re-annotation and transcriptome analyses of *Sanghuangporus sanghuang*. J Fungi 9(5). https://doi.org/10.3390/jof9050505
9. Zhou LW (2023) Microbial taxonomy with DNA sequence data as nomenclatural type: how far should we go? Innov Life 1(2):100017–100012. https://doi.org/10.59717/j.xinn-life.2023.100017
10. Liang J, Huang JE, Wu L, Cai L (2023) The Chinese microbial type-strain genome project. Innov Life 1(3):100045–100042. https://doi.org/10.59717/j.xinn-life.2023.100045
11. Nagy LG (2023) Mushroom functional genomics springs up. Innov Life 1(1):100005–100002. https://doi.org/10.59717/j.xinn-life.2023.100005
12. Wang XW, Zhou LW (2024) Mycologists' newest unsuccessful effort towards permitting DNA sequences as the nomenclatural types. Innov Life 2(4):100098. https://doi.org/10.59717/j.xinn-life.2024.100098
13. Chen G, Mostafa S, Lu Z, Du R, Cui J, Wang Y, Liao Q, Lu J, Mao X, Chang B, Gan Q, Wang L, Jia Z, Yang X, Zhu Y, Yan J, Jin B (2023) The Jasmine (*Jasminum sambac*) genome provides insight into the biosynthesis of flower fragrances and jasmonates. Genom Proteom Bioinf 21(1):127–149. https://doi.org/10.1016/j.gpb.2022.12.005
14. Jiang D, Li X, Li Y, Zhou S, Zhou Q, Liu X, Shen X (2023) Chromosome-level assembly of flowering cherry (*Prunus campanulata*) provides insight into anthocyanin accumulation. Genes 14(2):389. https://doi.org/10.3390/genes14020389
15. Koren S, Walenz BP, Berlin K, Miller JR, Bergman NH, Phillippy AM (2017) Canu: scalable and accurate long-read assembly via adaptive k-mer weighting and repeat separation. Genome Res 27(5):722–736. https://doi.org/10.1101/gr.215087.116
16. Walker BJ, Abeel T, Shea T, Priest M, Abouelliel A, Sakthikumar S, Cuomo CA, Zeng Q, Wortman J, Young SK, Earl AM (2014) Pilon: an integrated tool for comprehensive microbial variant detection and genome assembly improvement. PLoS One 9(11):e112963. https://doi.org/10.1371/journal.pone.0112963
17. Brůna T, Hoff KJ, Lomsadze A, Stanke M, Borodovsky M (2021) BRAKER2: automatic eukaryotic genome annotation with GeneMark-EP+ and AUGUSTUS supported by a protein database. NAR Genom Bioinform 3(1):lqaa108. https://doi.org/10.1093/nargab/lqaa108
18. Luo X, Jiang JH, Li QZ, Zou SN, Zhou LW (2025) Multiomics and enzyme activity

analyses provide insights into polysaccharide biosynthesis in *Laetiporus cremeiporus*. Int J Biol Macromol 318:145166. https://doi.org/10.1016/j.ijbiomac.2025.145166

19. Kim D, Langmead B, Salzberg SL (2015) HISAT: a fast spliced aligner with low memory requirements. Nat Methods 12(4):357–360. https://doi.org/10.1038/nmeth.3317
20. Pertea M, Pertea GM, Antonescu CM, Chang TC, Mendell JT, Salzberg SL (2015) StringTie enables improved reconstruction of a transcriptome from RNA-seq reads. Nat Biotechnol 33(3):290–295. https://doi.org/10.1038/nbt.3122
21. Love MI, Huber W, Anders S (2014) Moderated estimation of fold change and dispersion for RNA-seq data with DESeq2. Genome Biol 15(12):550. https://doi.org/10.1186/s13059-014-0550-8
22. Luo X, Jiang JH, Liu SL, Gao JY, Zhou LW (2024) Metabolomics analysis of rice fermented by medicinal fungi providing insights into the preparation of functional food. Food Chem 459:140372. https://doi.org/10.1016/j.foodchem.2024.140372
23. Liao G, Liu Q, Xu X, He Y, Li Y, Wang H, Ye B, Huang C, Zhong M, Jia D (2021) Metabolome and transcriptome reveal novel formation mechanism of early mature trait in kiwifruit (*Actinidia eriantha*). Front Plant Sci 12:760496. https://doi.org/10.3389/fpls.2021.760496
24. Livak KJ, Schmittgen TD (2001) Analysis of relative gene expression data using real-time quantitative PCR and the $2^{-\Delta\Delta CT}$ method. Methods 25(4):402–408. https://doi.org/10.1006/meth.2001.1262

Chapter 15

Preparation and Characterization of Metal-Tannic Acid Coated Microorganisms

Yuemei Zhang, Yuan Qin, Yuping Zhou, Yongcheng Zhu, and Huilong Luo

Abstract

Microorganisms have broad applications in disease treatment, bioremediation, and gut microbiota modulation. However, their survival rate and targeted delivery efficiency in complex environments, such as the gastrointestinal tract, remain major practical challenges. Metal-tannic acid complexes exhibit excellent biocompatibility, tunable degradability, and surface functionalization potential, making them suitable for constructing protective coatings on microbial surfaces. This chapter describes a robust and straightforward protocol for fabricating a conformal metal-phenolic network (MPN) coating on microorganisms via the sequential self-assembly of iron ions (Fe^{3+}) and tannic acid. The method includes detailed procedures for material preparation, the rapid coating process, comprehensive characterization, and functional assessment. This simple, versatile, and efficient strategy offers a promising platform for advancing probiotic encapsulation and engineered microbial therapeutics.

Key words Microbial therapy, Coating, Bacteria, Microorganisms, Metal-tannic acid complexes

1 Introduction

Microorganisms play vital roles in diverse fields, including human health, environmental remediation, and biomanufacturing [1, 2]. Nevertheless, their viability and functionality are often severely compromised when they are exposed to harsh conditions, such as gastric acid, bile salts, enzymes, or antibiotics, during oral delivery or in challenging environments [3–5]. To enhance microbial stability and targeting capabilities, various encapsulation strategies have been developed [5–7]. Among these, coatings formed by metal–phenolic networks (MPNs) have gained significant attention due to their mild preparation conditions, simplicity, and versatility [4, 8–11].

Yuemei Zhang, Yuan Qin and Yuping Zhou contributed equally with all other contributors.

Tao Huang (ed.), *Microbial 'Omics: Methods and Protocols*, Methods in Molecular Biology, vol. 3033,
https://doi.org/10.1007/978-1-0716-5264-0_15,

Tannic acid (TA), a natural plant polyphenol, is particularly advantageous for MPN formation due to its abundance, low cost, and GRAS (Generally Recognized As Safe) status. Its numerous pyrogallol and catechol groups serve as potent ligands for metal coordination [12, 13]. Metal ions, such as iron ions (Fe^{3+}), can rapidly self-assemble with TA under physiological conditions to form stable coordination complexes, which create a protective, matrix-like coating around microbial cells [12, 14, 15]. This nanoscale coating acts as a physical barrier that shields the encapsulated microbe from environmental insults. A key characteristic of the Fe-TA network is its pH-responsiveness: It remains stable in the acidic gastric environment but disassembles at the neutral pH of the intestine, thereby enabling the controlled release of viable microbes at the desired site [14, 16, 17]. Consequently, this coating not only improves microbial resistance to acidic and enzymatic degradation but also facilitates subsequent surface functionalization with molecules such as targeting antibodies or mucoadhesive polymers [12, 14].

Here, we present a detailed, step-by-step protocol for fabricating metal-TA coated microorganisms, with a focus on using iron ions (Fe^{3+}) as a model system. We expand upon basic methods by including optimized parameters for microbial strains, along with comprehensive methods for characterizing the coating and evaluating microbial viability, storage stability, and release profiles. This protocol is designed to be accessible for researchers aiming to develop advanced microbial delivery systems for both fundamental research and translational applications.

2 Materials

Prepare all solutions using ultrapure water (prepared by purifying deionized water to attain a sensitivity of 18 MΩ·cm at 25 °C) and analytical grade reagents. All solutions intended for contact with live microorganisms should be sterilized by filtration through a 0.22 μm membrane, unless otherwise specified. All chemical waste disposal must be performed in strict accordance with relevant regulations.

2.1 Reagents

1. TA (Sigma-Aldrich, Cat. No. 403040). Store desiccated at 4 °C. Prepare a 2 mg/mL stock solution in deionized water fresh on the day of use, as it can oxidize over time.
2. Ferric chloride hexahydrate ($FeCl_3·6H_2O$, Sigma-Aldrich, Cat. No. 236489). Prepare a 1 mg/mL stock solution in deionized water. The solution may appear slightly yellow and cloudy; filter sterilization is recommended.
3. Phosphate-buffered saline (PBS), pH 7.4, sterile.

4. Appropriate microbial culture medium (e.g., Lysogeny Broth (LB) for *E. coli*, de Man, Rogosa and Sharpe (MRS) broth for *Lactobacilli*).
5. Agar plates corresponding to the culture medium for viability plating.
6. Deionized water.
7. Ethanol (70% and 96–100% v/v) for sterilization and SEM sample preparation.

2.2 Equipment

1. Centrifuge (e.g., Eppendorf 5430 R) with a rotor capable of holding 1.5 mL or 2 mL microcentrifuge tubes.
2. Shaking incubator capable of maintaining 37 °C and various agitation speeds.
3. Scanning electron microscope (SEM, e.g., Hitachi SU8000).
4. Transmission electron microscope (TEM, e.g., JEOL JEM-1400Plus) for higher resolution imaging of the coating thickness and integrity.
5. Confocal laser scanning microscope (CLSM, e.g., Leica SP8).
6. Flow cytometry (e.g., Beckman Cytomics FC 500).
7. UV-Vis spectrophotometer for monitoring microbial growth (e.g., Thermo Scientific™ One/One Plus).
8. Zetasizer Nano ZS (Malvern Panalytical) for dynamic light scattering (DLS) size measurements and zeta potential analysis.
9. Fourier transform infrared spectrometer (FTIR, e.g., Thermo Scientific Nicolet iS5) for confirming the formation of Fe^{3+}-phenolate coordination bonds.
10. Vortex mixer and microcentrifuge.
11. Vertical Flow Clean Bench.
12. pH meter.

3 Methods

Carry out all procedures at room temperature (20–25 °C) unless otherwise specified. Perform all steps involving live microorganisms under sterile conditions in a biosafety cabinet using standard aseptic technique.

3.1 Microbial Culture and Harvesting

1. From a −80 °C glycerol stock, streak the target microorganism (e.g., *Lactobacillus rhamnosus* GG or *Escherichia coli* Nissle 1917) onto an appropriate agar plate. Incubate the plate under optimal conditions (e.g., 37 °C, anaerobic for *Lactobacilli*; 37 °C, aerobic for *E. coli*) for 24–48 h to obtain isolated colonies.

2. Inoculate a single, well-isolated colony into 5–10 mL of sterile liquid medium in a culture tube. Incubate overnight (12–16 h) under optimal growth conditions with shaking at 200 rpm, where appropriate, to establish a pre-culture.
3. Subculture the overnight pre-culture into a fresh flask containing 50–100 mL of sterile medium at a dilution of 1:100. Incubate with shaking at 200 rpm at 37 °C until the optical density at 600 nm (OD_{600}) reaches 0.6–0.8, indicating mid-logarithmic phase (*see* **Note 1**). Transfer the culture to sterile centrifuge tubes, and harvest cells by centrifugation at 4000× g for 10 min at 4 °C. Carefully discard the supernatant, and wash the cell pellet three times with sterile, ice-cold phosphate-buffered saline (PBS, pH 7.4) to remove residual media components that might interfere with subsequent coating formation.
4. Resuspend the final cell pellet in sterile PBS to a standardized concentration of 10^9 CFU/mL. This concentration can be estimated using OD_{600} measurements using a preestablished calibration curve of OD_{600} versus colony-forming units (CFU) determined by plate counting.

3.2 Metal–TA Coating

The following protocol is optimized for a 1 mL microbial suspension. Scale volumes accordingly while maintaining reagent ratios.

1. Prepare a 3 mg/mL TA solution in deionized water and filter-sterilize (0.22 μm pore size). Prepare a 1 mg/mL ferric chloride ($FeCl_3$) solution in deionized water and filter-sterilize. Protect the TA solution from light, especially if fluorescent labeling is intended.
2. In a sterile 1.5 mL microcentrifuge tube, combine 0.8 mL of the standardized microbial suspension (10^9 CFU/mL) with 100 μL of the $FeCl_3$ solution. Incubate the mixture at room temperature with gentle shaking (approximately 100 rpm) for 5 s to facilitate the initial adsorption of TA molecules onto the microbial surface (*see* **Note 2**).
3. Add 100 μL of the TA solution dropwise to the mixture over a period of 10–15 s while maintaining gentle vortexing. The immediate formation of a bluish-black or dark purple coloration indicates the successful coordination between Fe^{3+} ions and the pyrogallol/catechol groups of TA, signifying the initiation of the MPN formation (*see* **Note 3**).
4. Continue the incubation for an additional 20 s at room temperature with gentle agitation to allow for the maturation and stabilization of the Fe-TA coating around the microbial cells.
5. Pellet the coated microorganisms by centrifugation at 8000× *g* for 5 min. Carefully discard the supernatant. Wash the pellet

twice with 1 mL of sterile PBS to remove any unbound or loosely associated Fe-TA complexes, ensuring a clean and stable final coating (*see* **Note 4**).

6. Resuspend the final pellet in 1 mL of sterile PBS or an appropriate storage buffer for immediate use in downstream applications or characterization assays (*see* **Note 5**).

3.3 Characterization

1. Scanning electron microscopy (SEM): Fix coated and uncoated (control) samples with 2.5% glutaraldehyde in PBS for 2 h at 4 °C. Wash the fixed cells three times with PBS, and dehydrate them through a graded ethanol series (30%, 50%, 70%, 80%, 90%, 95%, and 100% v/v), allowing 10–15 min per step. Critical point dries the samples using liquid CO_2 to preserve native morphology. Mount samples on SEM stubs, sputter-coat them with a thin layer of gold or platinum, and image using an SEM at an accelerating voltage of 5–10 kV. A successful coating is indicated by a distinct alteration in surface topography, such as increased roughness or a particulate layer, compared to the smooth surface of uncoated cells.
2. Transmission electron microscopy (TEM): For ultrastructural analysis and coating thickness measurement, embed fixed and dehydrated cells in Spurr's resin according to standard protocols. Collect them on copper grids, and stain with uranyl acetate and lead citrate. Image with a TEM. The presence of a continuous, electron-dense layer surrounding the cell wall confirms coating formation and allows for nanoscale measurement of its thickness.
3. Confocal laser scanning microscopy (CLSM): To visually confirm the presence and uniformity of the coating, the samples were stained with 25 μg/mL BSA-FITC, which specifically binds to the polyphenolic components, and subsequently examined them using CLSM (*see* **Note 6**).
4. Hydrodynamic size and zeta potential: Dilute coated and uncoated microbial samples approximately 1:100 in PBS or 1 mM KCl. Use a Zetasizer Nano ZS to determine the hydrodynamic diameter via dynamic light scattering (DLS) and the surface zeta potential via laser Doppler velocimetry. A slight increase in hydrodynamic diameter and a significant negative shift in zeta potential (attributed to the deprotonated phenolate groups of the TA component) provide strong evidence of successful surface modification.
5. Fourier transform infrared (FTIR) spectroscopy: Lyophilize coated and uncoated microbial pellets. Mix a small amount of the lyophilized powder with potassium bromide (KBr), and press into a transparent pellet. Acquire FTIR spectra in the range of 4000–400 cm^{-1}. The formation of the Fe^{3+}-phenolate

complex is corroborated by the appearance of a characteristic peak around 1650–1670 cm^{-1} (attributed to C=O stretching) and a broadening of the absorption band in the 1300–1450 cm^{-1} region (associated with C–O bonding in the complex).

6. Microbial viability assessment: Perform serial dilutions (e.g., 10^{-1} to 10^{-8}) of both the original uncoated suspension and the final coated suspension in PBS. Plate 100 μL of each dilution onto appropriate agar plates in triplicate. Incubate plates under optimal conditions for 24–48 h. Enumerate the CFU, and calculate the percentage viability: % Viability = (CFU(coated)/CFU(uncoated)) × 100%. A well-optimized coating protocol should yield viability exceeding 80%.
7. In vitro release kinetics study: To quantitatively monitor the protective behavior and pH-responsive release of microbes from the Fe-TA coating, employ a fluorescence-based or enzymatic activity tracking method. For fluorescent tracking, pre-label the microbial cells with a lipophilic membrane dye (e.g., DiO or PKH67) prior to the coating process. Incubate the labeled and coated cells in simulated gastric fluid (SGF, pH 2.0) and simulated intestinal fluid (SIF, pH 7.4) under gentle agitation at 37 °C. At predetermined time points, centrifuge samples, and measure the fluorescence intensity using a fluorescence microplate reader. A low fluorescence signal in SGF indicates effective retention (protection), while a significant increase in fluorescence upon transition to SIF demonstrates the pH-triggered release of viable, metabolically active cells. Alternatively, the release of intracellular enzymes (e.g., lactate dehydrogenase for bacteria) can be quantified in the supernatant as an indicator of cell membrane integrity and coating protection.
8. Cell metabolic activity assessment: Evaluate the physiological state and metabolic activity of the coated microbes using a redox indicator assay, such as the resazurin reduction assay (also known as the AlamarBlue assay). Resazurin, a blue, non-fluorescent compound, is reduced to resorufin, a pink and highly fluorescent molecule, by metabolically active cells. Wash coated and uncoated cells, and resuspend them in fresh, prewarmed medium containing 10% (v/v) resazurin sodium salt solution. Incubate the mixture at 37 °C for 1–4 h, protecting it from light. Measure the fluorescence (excitation, 560 nm; emission, 590 nm) or the absorbance (570 nm and 600 nm) using a microplate reader. Calculate the percentage reduction compared to a negative control (medium with resazurin only). A high percentage of resazurin reduction by coated cells, comparable to that of uncoated cells, confirms that the coating

process preserves the metabolic vitality and enzymatic activity of the encapsulated microorganisms, which is crucial for their intended probiotic or therapeutic function.

4 Notes

1. The physiological state is critical. Mid-log phase cells have healthier, more uniform cell walls that facilitate consistent coating. Stationary phase cells may lead to heterogeneous or incomplete coatings.
2. Avoid vigorous shaking to prevent cell damage, and ensure uniform coating.
3. The optimal ratio of Fe^{3+}-TA: Cells may vary depending on the microbial strain and its surface properties. We recommend a preliminary experiment titrating the TA concentration (e.g., 0.5, 1.0, 2.0 mg/mL) while keeping the Fe^{3+} concentration constant, and vice versa, to maximize viability and coating efficiency.
4. Coated microorganisms should be used within 48 h when stored at 4 °C in PBS. For longer-term storage, consider resuspending in a cryoprotectant like 20% glycerol and freezing at −80 °C. However, note that freeze-thaw cycles may compromise coating integrity.
5. If coating efficiency is low (evidenced by minimal color change in the pellet or a large amount of complex in the supernatant), ensure that the washing steps post-coating are sufficient and that the cells are well-dispersed during the coating reaction.
6. For CLSM imaging, label Fe-TA complexes with fluorescent dyes after coating. The fluorescence signal localized on the cell periphery provides clear, visual confirmation of the coating.

5 Discussion

The Fe-TA coating protocol presented in this study offers a robust, straightforward, and efficient approach for enhancing the stability of microorganisms under challenging conditions. The primary strength of this method lies in its simplicity and the utilization of biocompatible, natural components. The entire process is conducted under mild aqueous conditions, which is rapid and applicable to a diverse array of prokaryotic and eukaryotic microorganisms. The Fe-TA coating forms a conformal layer around microbial cells. This nanoscale shield plays a crucial role in significantly improving the survival of probiotic strains, during transit through gastric environments [14, 17, 18]. Moreover, the pH-responsive disassembly of the MPN in the neutral intestine ensures precise temporal and spatial control of the protective function [19–21]. This controlled release mechanism allows viable microbes to be released exactly where they are intended to colonize and perform their functions [22].

Beyond its protective function, the resulting MPN coating serves as a versatile platform for further functionalization. The phenolic groups of TA can be exploited to conjugate targeting ligands, such as antibodies specific to intestinal epithelia, or other therapeutic agents, including prebiotics or enzymes [10, 23, 24]. This functionalization creates multifunctional "microbial robots" with enhanced capabilities. However, it is essential for researchers to note certain limitations. In some sensitive microbial strains, the coating may lead to a slight reduction in final viability, which necessitates optimization of the coating protocol. Additionally, the potential interactions of the coating with complex food matrices or other gut microbiota members in vivo require further in-depth investigation.

This method shows particular promise in applications such as probiotic encapsulation, engineered microbial therapeutics, and microbial-based biosensors. The potential adaptations could involve the incorporation of additional functional polymers to create multilayered coatings. Alternatively, the use of other metal-phenolic combinations, such as Zn^{2+} or Al^{3+}, or Epigallocatechin gallate (EGCG), could be explored to fine-tune release kinetics and environmental responsiveness. These adaptations could expand the application scope beyond the gastrointestinal (GI) tract, including potential uses in soil amendment or aquaculture.

In summary, the current metal-TA coating protocol provides a novel, simple, and effective strategy for microbial protection. While it has demonstrated significant advantages, addressing the identified limitations and exploring further adaptations will be crucial for maximizing its potential in a wide range of applications.

Acknowledgments

This work was funded by the Youth S&T Talent Support Programme of Guangdong Provincial Association for Science and Technology (SKXRC2025383), the National Natural Science Foundation of China (32501495), the Guangdong Provincial Characteristic Innovation Project (2025KTSCX103), the Guangzhou Municipal Science and Technology Project (2024A03J0289), the Plan on Enhancing Scientific Research in GMU (Yongcheng Zhu), the Guangdong University Innovation Team Project (2024KCXTD029), and the Zhejiang Provincial Natural Science Foundation of China (No. LTGD23C040008, LBY23H200006).

Author Contribution Statements Y.Z., Y.Q. and Y.P.Z. contributed equally to this work. Y.Z., Y.Q. and Y.P.Z. conducted literature collection, critical analysis, and drafted the manuscript, and H.L. and Y.C.Z. supervised the project, provided conceptual guidance, and revised the manuscript. All authors discussed the content, approved the final version, and agreed to its submission.

Competing Interests The authors declare that they have no conflict of interest.

References

1. García G, Carlin M, Cano RJ (2025) Holobiome harmony: linking environmental sustainability, agriculture, and human health for a thriving planet and one health. Microorganisms 13(3):514
2. Marco ML (2021) Defining how microorganisms benefit human health. Microb Biotechnol 14(1):35–40
3. Peng P, Feng T, Yang X, Nie C, Yu L, Ding R, Zhou Q, Jiang X, Li P (2023) Gastrointestinal microenvironment responsive nanoencapsulation of probiotics and drugs for synergistic therapy of intestinal diseases. ACS Nano 17(15):14718–14730. https://doi.org/10.1021/acsnano.3c02646
4. Pan J, Gong G, Wang Q, Shang J, He Y, Catania C, Birnbaum D, Li Y, Jia Z, Zhang Y, Joshi NS, Guo J (2022) A single-cell nanocoating of probiotics for enhanced amelioration of antibiotic-associated diarrhea. Nat Commun 13(1):2117. https://doi.org/10.1038/s41467-022-29672-z
5. Fan G, Wasuwanich P, Rodriguez-Otero MR, Furst AL (2022) Protection of anaerobic microbes from processing stressors using metal-phenolic networks. J Am Chem Soc 144(6):2438–2443. https://doi.org/10.1021/jacs.1c09018
6. Yang X, Yang J, Ye Z, Zhang G, Nie W, Cheng H, Peng M, Zhang K, Liu J, Zhang Z (2022) Physiologically inspired mucin coated Escherichia coli Nissle 1917 enhances biotherapy by regulating the pathological microenvironment to improve intestinal colonization. ACS Nano 16(3):4041–4058
7. Zhu L, Yu T, Wang W, Xu T, Geng W, Li N, Zan X (2024) Responsively degradable nanoarmor-assisted super resistance and stable colonization of probiotics for enhanced inflammation-targeted delivery. Adv Mater 36(18):2308728
8. Wang M, Zhang S, Liu S, Nian L, Wang S, Cao C (2025) Inulin and metal-phenolic network single-cell encapsulation: boosting the tolerance and intestinal colonization capability of Bifidobacterium lactis. Food Hydrocolloids 169:111619
9. Fan G, Cottet J, Rodriguez-Otero MR, Wasuwanich P, Furst AL (2022) Metal–phenolic networks as versatile coating materials for biomedical applications. ACS Appl Bio Mater 5(10):4687–4695
10. Zhao Z, Pan DC, Qi QM, Kim J, Kapate N, Sun T, Shields CW, Wang LL, Wu D, Kwon CJ, He W, Guo J, Mitragotri S (2020) Engineering of living cells with polyphenol-functionalized biologically active nanocomplexes. Adv Mater 32(49):e2003492. https://doi.org/10.1002/adma.202003492
11. Ejima H, Richardson JJ, Caruso F (2017) Metal-phenolic networks as a versatile platform to engineer nanomaterials and biointerfaces. Nano Today 12:136–148
12. Ejima H, Richardson JJ, Liang K, Best JP, van Koeverden MP, Such GK, Cui J, Caruso F (2013) One-step assembly of coordination complexes for versatile film and particle engineering. Science 341(6142):154–157. https://doi.org/10.1126/science.1237265
13. Shin M, Kim K, Shim W, Yang JW, Lee H (2016) Tannic acid as a degradable mucoadhesive compound. ACS Biomater Sci Eng 2(4):687–696
14. Luo H, Wu F, Wang X, Lin S, Zhang M, Cao Z, Liu J (2023) Encoding bacterial colonization and therapeutic modality by wrapping with an adhesive drug-loadable nanocoating. Mater Today 62:98–110

15. Markova Z, Novak P, Kaslik J, Plachtova P, Brazdova M, Jancula D, Siskova KM, Machala L, Marsalek B, Zboril R (2014) Iron (II, III)–polyphenol complex nanoparticles derived from green tea with remarkable ecotoxicological impact. ACS Sustain Chem Eng 2(7):1674–1680
16. Liu J, Li W, Wang Y, Ding Y, Lee A, Hu Q (2021) Biomaterials coating for on-demand bacteria delivery: selective release, adhesion, and detachment. Nano Today 41:101291
17. Luo H, Xie J, Su X, Wang P, Chen H, Kuang X, Liu J (2024) Tannic acid-based metal-phenolic networks as a versatile platform to mediate cell therapy. Sci China-Mater 67(12):3833–3848
18. Wang C, Fulati A, Kimura K, Li X, Richardson JJ, Naito M, Miyata K, Ichiki T, Ejima H (2025) Encapsulation of small extracellular vesicles into selectively disassemblable shells of PEGylated metal-phenolic networks. Adv Healthc Mater:2405188
19. Shen L, Zhang Y, Feng J, Xu W, Chen Y, Li K, Yang X, Zhao Y, Ge S, Li J (2022) Microencapsulation of ionic liquid by interfacial self-assembly of metal-phenolic network for efficient gastric absorption of oral drug delivery. ACS Appl Mater Interfaces 14(40): 45229–45239
20. Park J, Wu Y, Le Q-V, Kim JS, Xu E, Lee J, Oh Y-K (2025) Self-disassembling nanoparticles as oral nanotherapeutics targeting intestinal microenvironment. Nat Commun 16(1):3365
21. Qin Y, Wang J, Qiu C, Hu Y, Xu X, Jin Z (2019) Self-assembly of metal–phenolic networks as functional coatings for preparation of antioxidant, antimicrobial, and pH-sensitive-modified starch nanoparticles. ACS Sustain Chem Eng 7(20):17379–17389
22. Kim C-J, Ercole F, Chen J, Pan S, Ju Y, Quinn JF, Caruso F (2021) Macromolecular Engineering of Thermoresponsive Metal–Phenolic Networks. J Am Chem Soc 144(1):503–514
23. Hasan A, Rani S, Hussain A (2025) Harnessing engineered probiotics for gastrointestinal diseases therapy. In: Cell membrane engineering for advancing cell-based and nanotechnology-based therapies. Springer, pp 275–302
24. Fang T, Liu S (2024) Metal-phenolic network directed coating of single probiotic cell followed by photoinitiated thiol-ene click fortification to enhance Oral therapy. Small 20(19): 2308146

Chapter 16

Omics Technologies-Driven Research and Applications of Synthetic Biology

Dai-Xu Wei, Zheng-Dong Qi, Weixu Cai, Zhimeng Yuan, and Jiahao Zhang

Abstract

As a core discipline for the engineering modification of biological systems, synthetic biology is currently confronted with key challenges such as poor functional stability of synthetic communities, insufficient metabolic pathway adaptability of chassis strains, and difficulties in simulating interactions in natural microecosystems. Traditional single technologies fail to achieve precise design and regulation of synthetic biological systems. However, microbiomics, which integrates multi-omics, spatial omics, and culturomics, provides a full-chain solution for synthetic biology covering "functional prediction-dynamic regulation-functional verification-community construction" by decoding multidimensional molecular information, spatial distribution patterns, and cultivation interaction mechanisms of biological systems. This chapter systematically elaborates on how integrated multi-omics guides the optimization of metabolic pathways in synthetic biological systems, how spatial omics deciphers the spatial interaction mechanisms of synthetic communities, and how innovations in culturomics break through the cultivation bottlenecks of synthetic strains. Ultimately, it realizes the in-depth integration of omics technologies and synthetic biology, thereby promoting the industrial application of synthetic biological systems in fields such as biomanufacturing, environmental remediation, and medical health.

Key words Integrated multi-omics, Synthetic biology, Spatial omics, Long-read sequencing, Microbial coculture, Genome-guided cultivation

1 Introduction

Synthetic biology aims to achieve specific functions through the engineering design and construction of biological systems, and it holds broad application prospects in biomanufacturing, medical health, environmental remediation, and other fields. Nevertheless, this field is confronted with systematic challenges, including low functional stability of synthetic microbial communities, insufficient adaptability between metabolic pathways and chassis organisms, and difficulties in simulating the interaction mechanisms of natural microorganisms. Traditional single-level research methods are

Tao Huang (ed.), *Microbial 'Omics: Methods and Protocols*, Methods in Molecular Biology, vol. 3033,
https://doi.org/10.1007/978-1-0716-5264-0_16,

incapable of comprehensively deciphering the complexity of biological systems. Thus, an urgent need exists for the introduction of a systematic and multidimensional technical system.

The rapid advancement of omics technologies has provided a novel research paradigm for synthetic biology, spanning from mechanism elucidation to system optimization. Integrated multi-omics reveals cross-hierarchical regulatory networks through the synergistic analysis of genomics, transcriptomics, proteomics, and metabolomics data; spatial omics deciphers microbial distribution and interactions while preserving the original spatial structure; and culturomics breaks through the resource bottleneck of uncultured microorganisms via technologies such as microfluidics and in situ cultivation. These technologies collectively form a full-chain solution encompassing "functional prediction-dynamic regulation-verification-construction," propelling synthetic biology from component assembly toward system-level decipherment and engineering reconstruction.

This chapter systematically expounds on the fundamental principles and methodological progress of four major technical systems, namely, long-read sequencing, integrated multi-omics, spatial omics, and culturomics. It focuses on exploring their key applications in metabolic pathway optimization, analysis of spatial interactions in microbial communities, and resource exploration of chassis cells, with the aim of providing theoretical support and technical approaches for the rational design of synthetic biological systems and the integration of industry and education.

2 Long-Read Sequencing Technology

Long-read sequencing, also referred to as third-generation sequencing (TGS) technology [1], is characterized by single-molecule sequencing and real-time sequencing capabilities, enabling direct sequencing of individual DNA or RNA molecules ranging from tens of thousands to hundreds of thousands of base pairs in length [2]. Although second-generation sequencing technology has achieved significant improvements over Sanger sequencing, its limitation of short-read lengths renders it incapable of addressing certain specific biological questions [3]. In contrast, long-read sequencing technology facilitates the assembly and identification of complex genomic regions, detection of gene isoforms, and analysis of methylation. It substantially shortens the average time required for library construction, simplifies the de novo genome assembly process, and thus can be more conveniently applied to resolve issues related to complex regions and detect structural variations. This technology promotes advancements in fields such as transcriptomics analysis, epitranscriptomics, bacterial genome assembly, and viral metagenomics and has been employed in human genome sequencing.

The representative technologies of long-read sequencing include Oxford Nanopore Technology (ONT) and the single-molecule real-time (SMRT) sequencing technology developed by Pacific BioSciences.

2.1 Single-Molecule Real-Time (SMRT) Sequencing Technology

SMRT sequencing represents the first mature single-molecule real-time sequencing technology, wherein sequencing reactions occur on the core component known as the single-molecule real-time sequencing chip (SMRT Cell). The chip surface contains 150,000 to 1000,000 nanostructures termed zero-mode waveguides (ZMWs), each with a diameter of approximately 70 nm and a depth of around 100 nm (or smaller), which is shorter than the wavelength of visible light. This structural feature enables the system to detect chemical reactions of individual molecules.

Double-stranded DNA samples are fragmented and ligated with hairpin-shaped adapters to form circular templates. The circular DNA templates then bind with DNA polymerase to form a sequencing complex, which is immobilized at the bottom of the ZMWs within the detection area [4]. When four types of dNTPs (A, T, C, G) labeled with fluorescent groups are introduced into the SMRT Cell, the DNA polymerase captures the dNTPs and incorporates them into the catalytic site. During this process, the dNTPs temporarily remain in the detection area, where the fluorescent groups are excited by laser. A high-sensitivity camera at the bottom of the ZMWs integrated with a confocal system for detecting fluorescent signals of incorporated nucleotides, real-time monitors and records the color and intensity of the fluorescence. The incorporated base sequence is determined based on the fluorescence color [5].

The replication processes occurring in all ZMWs within the SMRT Cell are recorded by light pulses, and the pulses corresponding to each ZMW can be interpreted as a base sequence. Factors affecting polymerase speed, such as DNA secondary structure or base modifications, are reflected in kinetic signals, enabling PacBio technology to directly detect epigenetic modifications, such as DNA methylation (e.g., 5mC).

SMRT technology encompasses two modes: the traditional continuous long read (CLR) mode and the emerging high fidelity (HiFi) mode (also known as circular consensus sequencing, CCS). CLR utilizes linear templates, with the polymerase continuously synthesizing along the template. It features low raw accuracy and single-pass reading, resulting in a relatively high random error rate. This mode is primarily employed for de novo genome assembly, leveraging ultra-long reads to span complex repetitive regions, and for linking and calibrating short-read data. In contrast, CCS uses circular templates with adapters. Once the polymerase binds to the circular template, it continuously synthesizes multiple rounds around the template, generating multiple coverage reads for the same DNA segment. A highly consistent consensus sequence is then generated through algorithms. After approximately 25 cycles, the accuracy reaches QV40, comparable to that of Illumina sequencing; after about 50 cycles, the accuracy attains QV50, making it suitable for a broader range of applications.

Taking advantage of the characteristics of DNA synthesis, SMRT technology can identify DNA molecules up to 50 kb or longer in length. By confining the reaction within ZMWs, using phosphate-labeled dNTPs, and employing DNA polymerase with high processivity, PacBio SMRT technology achieves real-time, long-read, and high-fidelity sequencing of individual DNA molecules.

2.1.1 SMRT Platforms and Throughput

In early 2011, Pacific Biosciences (PacBio) launched its PacBio RS sequencer, marking the advent of the first mature single-molecule real-time (SMRT) sequencing technology. The initial sequencer produced relatively short average read lengths (approximately 1.5 kb) with a high error rate (around 13%). Following years of technological advancements, PacBio ultimately introduced a new sequencer, the Sequel system. This system increased the average read length by more than tenfold and improved the single-run throughput by approximately 100 times. Over time, PacBio's Sequel system has undergone continuous upgrades, leading to the development of two new Sequel platforms: the Sequel II and Sequel IIe systems. Among these, the Sequel IIe system stands as the most powerful platform in the Sequel series. Owing to improvements in circular templates and polymerase processivity, it is now possible to sequence molecules of approximately 1–2 kb in length multiple times.

2.2 Oxford Nanopore Technology

In 2014, Oxford Nanopore Technologies (ONT) launched nanopore sequencing technology, which is based on the principle of sequencing single-stranded DNA or RNA molecules through nanopores embedded in a membrane. This concept was first proposed in the late 1980s. With the rapid development of ONT chemistry, sequencing throughput has significantly increased, enabling the application of nanopore sequencing technology in numerous research fields.

The core of Oxford Nanopore technology lies in the direct measurement of current changes induced by the passage of single-stranded DNA or RNA molecules through nanoscale pores. These nanopores are typically engineered transmembrane proteins (e.g., α-hemolysin) embedded in a robust lipid bilayer or synthetic membrane, allowing only single-stranded nucleic acid molecules to pass through. Motor proteins bind to DNA or RNA to control the translocation speed of nucleic acid molecules through the nanopore, enabling the detection system to capture the signals generated by each base or base combination. Under a constant ionic current, as nucleic acid molecules are guided through the nanopore by motor proteins, differences in the chemical structures (size, shape, charge distribution) of different bases result in varying degrees of obstruction to the ionic current. Typically, five to six consecutive bases collectively determine the current signal at a given moment. Algorithms then real-time translate these current signals into nucleotide sequences, a process referred to as "Basecalling" [6].

Technological evolution has enabled the transition from 1D-to-1D^2 sequencing modes. 1D sequencing involves sequencing only one strand (the template strand) as it passes through the nanopore, which was the standard mode in the early stages. In contrast, 1D^2 sequencing represents a significant innovation by ONT. A specialized adapter is ligated to the end of the template strand; after the template strand has completely traversed the pore, its complementary strand is captured by motor proteins and subsequently sequenced through the same pore. By performing consensus alignment of the sequencing results from both strands, random errors are significantly reduced, and the accuracy of the final sequence is improved.

The accuracy of raw reads in ONT was a major challenge in its early development. However, through the iteration of chemical reagents and continuous optimization of Basecaller algorithms based on deep learning, the raw single-molecule accuracy has improved from approximately 85% in the early stages to 98–99%. By combining 1D^2 sequencing or generating consensus sequences through increased sequencing depth, the consensus accuracy can exceed 99.9%.

2.2.1 ONT Platforms and Throughput

With the advancement of ONT technology, various platforms ranging from portable to high-throughput systems have been developed. Theoretically, read length is only limited by the integrity of the input DNA molecules, while in practice, it depends on DNA quality. Key platforms include the following:

- *MinION Handheld Sequencer:* A portable platform capable of generating up to 50 Gb of data over a 72-hour run.
- *Flongle:* A low-throughput, cost-effective flow cell compatible with the MinION, suitable for rapid, small-sample analyses such as rapid pathogen screening or quality control.
- *GridION:* A benchtop, multifunctional platform that can operate multiple MinION flow cells simultaneously and integrates computing units.
- *PromethION:* An ultrahigh-throughput platform that can accommodate up to 48 flow cells, each containing 3000

channels. A single run provides up to 144,000 channels, and theoretically, a 48-hour run can generate up to 15 Tb of data.

2.3 Advantages of TGS

Single-molecule and real-time sequencing: Data can be generated within minutes of the start of sequencing and transmitted continuously. This allows users to monitor experiment quality in real time during the run or terminate the run early once sufficient data volume is achieved, saving time and costs.

Rich information content: Continuous long reads can directly span entire repetitive regions, inversions, or gene fusion sites, enabling haplotype-resolved detection of large structural variations. Ultra-long reads greatly simplify the genome assembly process, particularly for highly repetitive and heterozygous animal and plant genomes, yielding highly contiguous assembly results. Additionally, TGS contains more phylogenetic information, enabling more accurate species classification and differentiation.

Portability: TGS enables on-site, real-time gene sequencing during field environmental gene surveys.

Advantages in RNA sequencing: ONT allows direct sequencing of native RNA molecules without reverse transcription.

2.4 TGS Applications and Synthetic Biology

2.4.1 De Novo Genome Assembly

De novo assembly strategies based on whole-genome sequencing have revolutionized genomics and synthetic biology research, significantly expanding the scope of DNA research. Long-read sequencing technology can span repetitive sequences and complex genomic regions to directly obtain complete contiguous sequences, thereby substantially improving the completeness and accuracy of assembly results [7]. In metagenomic studies, this technology enables direct sequencing and assembly of long DNA fragments from environmental samples, facilitating more accurate reconstruction of complete genomes in the system and correction of errors in existing reference genomes. For ultra-long synthetic genes, entire metabolic pathways, or genes containing repetitive sequences, Sanger sequencing becomes inefficient and costly. In such cases, PacBio HiFi or Oxford Nanopore technologies demonstrate significant advantages, as they can sequence entire insert fragments in a single pass and directly obtain end-to-end complete sequences.

2.4.2 Whole-Exome Sequencing and Targeted Region Sequencing

In terms of targeted sequencing, the CRISPR-Cas9-based targeting strategy developed by Pacific Biosciences provides an efficient tool for the validation of large-scale synthetic biology components. This technology requires only 5000 ng of non-amplified genomic DNA and enables specific enrichment of target regions (e.g., integration sites of artificially designed metabolic pathways) from complex genomic backgrounds through double-guide RNA-mediated

Cas9 digestion, followed by ligation of SMRTbell sequencing adapters. This strategy supports parallel sequencing of dozens to hundreds of targets using multiple pairs of guide RNAs, greatly improving validation throughput. It is suitable for high-throughput quality control of standardized biological part libraries or diverse strain libraries.

Oxford Nanopore Technologies offers unique advantages for the testing phase of synthetic biology due to its flexibility and real-time capabilities. Its PCR-based and PCR-free targeted sequencing strategies can adapt to different throughput and timeliness requirements. The non-amplification strategy involves direct ligation of sequencing adapters after fragmentation and end repair, requiring 1000 ng of double-stranded DNA, and is suitable for parallel validation of multiple targets and samples, with a single chip yielding up to 50 Gb of data. In contrast, the PCR-based strategy requires only 100 ng of starting DNA, offers multiplex detection capabilities, and can rapidly respond to the need for rapid validation of specific components during the construction phase of synthetic biology, accelerating the design-test cycle. Collectively, these targeted sequencing technologies constitute a precise and efficient genotyping validation platform in synthetic biology, providing critical data support for rational design and engineering optimization.

2.4.3 Protein Sequencing-Related Applications

Given the growing recognition of the size of the protein interactome and the increasing demand for protein function information, methods for detecting protein-protein interactions remain an important area of development. For instance, high-throughput protein interactome mapping methods using two-hybrid systems have been applied to full-length proteins in numerous species. In 2025, Stephanie Schaefer and colleagues applied full-length SARS-CoV open reading frames to the study of interactions with the human ACE2 protein, demonstrating that bacterial systems can be used to investigate full-length protein interactions. This study represents the first direct application of third-generation sequencing technologies, such as Oxford Nanopore, to generate full-length protein-protein fusion proteins [8], realizing the direct application of TGS in protein interaction sequencing.

2.4.4 Genetic Variation Analysis and Evolution Research

Long-read sequencing technology can directly identify large-scale genetic variations that are difficult to detect using short-read platforms, including insertions, deletions, inversions, translocations, and copy number variations. It also enables "haplotype phasing" of genetic variations (e.g., single nucleotide polymorphisms), i.e., clarifying their parental chromosomal origin, thereby accurately resolving alternative splicing events, gene fusions, and novel gene discoveries. Examples include the systematic identification of rare single-nucleotide polymorphisms, single-nucleotide variations, haplotypes, and structural variations.

In the context of synthetic biology, TGS facilitates the acquisition of complete reference genomes, the reading of complete gene clusters or operons for the characterization of natural biological components, the validation of large-fragment DNA assembly, the detection of off-target effects, and the analysis of genome evolution. For example, Shi et al. (2021) improved the whole-genome sequence of *Phytophthora capsici* using ONT [9].

2.4.5 cDNA and RNA Sequencing

The long reads generated by third-generation sequencing platforms enable direct identification of full-length mRNAs and fusion transcripts without the need for assembly, facilitating the verification of whether exogenous genes are correctly transcribed into RNA. Among these technologies, PacBio achieves indirect sequencing of RNA molecules through its innovative Iso-Seq Express template preparation technology, which involves cDNA library construction and sequencing. This workflow uses 300 ng of total RNA as the starting template, synthesizes and amplifies the first-strand cDNA via reverse transcription, and then uses the resulting cDNA to construct SMRTbell libraries for subsequent sequencing, enabling the identification of full-length transcripts up to 10 kb in length.

Nanopore cDNA sequencing is an effective approach for transcriptome mapping and alternative splicing event analysis. Particularly for poly(A) + RNA samples with limited quantities, the PCR-cDNA sequencing protocol developed by ONT requires only 1 ng of starting RNA to obtain sequencing reads representing full-length cDNAs. Additionally, ONT offers a direct cDNA sequencing protocol without PCR amplification, which, however, requires a larger amount of starting material (100 ng of poly (A) + RNA). This workflow includes reverse transcription of poly (A) + RNA, ligation of cDNA end adapters, and direct sequencing. Although its data output is lower than that of the PCR-cDNA strategy, the avoidance of amplification bias makes the resulting data more suitable for gene expression difference analysis.

While RNA sequencing based on second-generation sequencing technology achieves high coverage and depth at the transcriptome level, it still relies on PCR amplification of templates. The advent of nanopore direct RNA sequencing technology has laid the foundation for in-depth research on transcriptomes and specific RNA types (e.g., mRNA). This innovative method directly sequences poly(A) + RNA molecules as they pass through the nanopore, eliminating the need for cDNA conversion steps required by all NGS platforms, thus serving as an ideal tool for mRNA research. Direct RNA sequencing is implemented by ligating adapters targeting the poly(A) tail and can be performed with or without a reverse transcription step, with no PCR amplification required throughout the process. Although omitting reverse transcription shortens the experimental time, it typically results in reduced data throughput. Notably, despite its technical ingenuity, this method requires a relatively large amount of starting RNA (500 ng of poly(A) + RNA), and its data output is generally lower than that of direct cDNA and PCR-cDNA strategies.

Direct RNA sequencing enables efficient identification of full-length mRNAs and novel transcripts, reveals alternative splicing patterns, and allows accurate quantification of isoforms, systematic annotation of polyadenylation sites, and splice sites. Furthermore, it enables direct detection of modified RNA bases (e.g., m^6A, m^5C) without relying on indirect methods such as antibody enrichment. This technical feature opens up revolutionary applications in epigenetics and synthetic biology. The nanopore sequencing platform also provides protocols for the detection of small RNAs (e.g., miRNAs and tRNAs), enabling accurate identification of their molecular subtypes and analysis of their expression profiles in various samples, thereby realizing miRNA type identification, genetic modification detection, and expression quantitative analysis [10].

2.4.6 Epigenetics and Metagenomics Research

Third-generation sequencing technology has demonstrated great potential in the fields of epigenetics and metagenomics, driving the advancement of synthetic biology. TGS enables systematic investigation of DNA chemical modifications and the modification mechanisms of DNA-binding proteins. Its whole-genome sequencing strategy not only allows direct detection of epigenetic marks but also provides more uniform coverage with lower GC bias. For example, PacBio enables high-throughput metagenomic sequencing through its HiFi reads; rRNA gene sequencing provides an effective tool for rapid and cost-efficient gene screening of microbial communities, simplifying the analysis workflow and delivering high-resolution phylogenetic information. Additionally, PacBio supports the assembly of complete genomes from microbial communities through multiplex sequencing strategies.

ONT offers a variety of metagenomic research protocols, including whole-genome sequencing, targeted sequencing, metatranscriptomics, and meta-epigenomics analysis. Recent studies have confirmed that ONT achieves superior taxonomic resolution at the genus and species levels compared to second-generation sequencing technology. Its rapid real-time sequencing capability not only supports accurate differentiation of bacterial clades and gene extraction but also enables functional profile analysis at the species level when combined with built-in bioinformatics tools. By resolving the complete genomes of individual members in synthetic microbial communities and monitoring their interaction dynamics, ONT lays the foundation for synthetic biological systems.

Furthermore, the portability of nanopore sequencing facilitates on-site real-time research of microbial populations, enabling direct acquisition of complete gene clusters or operons.

Third-generation sequencing technology provides unprecedented capabilities for deciphering transcriptomic and epigenetic landscapes in synthetic biology, strongly supporting the rational design and functional optimization of engineered living systems. Its long-read feature enables direct acquisition of full-length transcript information without assembly, providing precise molecular evidence for the expression regulation of artificial gene circuits. At the transcriptional level, PacBio's Iso-Seq technology can accurately resolve the complete transcription of exogenous genes in hosts, including transcription initiation, splicing patterns, and termination sites, thereby guiding the optimization of regulatory elements such as promoters and terminators. It can also detect the methylation status of genomic DNA in engineered strains.

3 Integrated Multi-omics Technology

As a core trend in the development of microbiome research, integrated multi-omics has evolved from the initial "multi-dimensional data superimposition" to a methodological system characterized by "cross-hierarchical mechanism integration" and "functional interpretation enhancement." A single omics approach can only reveal one aspect of a biological system: For example, metagenomics reflects genetic potential, metatranscriptomics reflects transient expression, proteomics characterizes functional execution, and metabolomics presents the final chemical outcomes. However, these hierarchical levels are tightly coupled, and no single dimension is sufficient to reveal the true dynamics of a microecosystem. With the rapid advancement of sequencing technology, mass spectrometry, and computational methodologies, researchers can now simultaneously observe the continuous trajectory of DNA-RNA-protein-metabolites within a single research system, significantly enhancing the ability to infer mechanisms.

The significance of multi-omics integration extends beyond comprehensive data acquisition; it endows researchers with a new paradigm for reconstructing biological processes using multidimensional evidence chains. It is this paradigm that has enabled multi-omics to demonstrate unprecedented potential in exploring disease mechanisms, resolving microecological stability, environmental bioremediation, and optimizing industrialized engineering strains. Building on the understanding of integration technologies, innovations at the application level are gradually becoming the core driver of progress in this field. Following an introduction to key technologies and analytical methods, this section will emphasize the critical application value of multi-omics in real research scenarios. Concurrently, the rise of multi-omics integration has been facilitated by the maturation of high-throughput technologies and the rapid reduction in costs, enabling researchers to obtain high-quality metagenomic, metatranscriptomic, proteomic, and metabolomic data simultaneously within a single project.

Therefore, this section aims to systematically introduce the technical basis, experimental design principles, key data processing workflows, application advantages, and current unresolved challenges of integrated multi-omics analysis.

In microbiome research, integrated multi-omics refers to the simultaneous acquisition of data from two or more "omics" levels from the same batch of samples or closely matched samples (e.g., metagenomics + metabolomics, or metagenomics + metatranscriptomics + metaproteomics), followed by joint analysis using statistical and computational methods to identify cross-hierarchical associations, potential causal chains, and functional modules [11]. Its core objective is to address key scientific questions such as "whether a gene is expressed despite its presence," "whether the expression product actually performs its function," "whether metabolic pathways are activated by the external environment," and "whether system outputs affect host physiology."

3.1 Key Technical Layers

3.1.1 Metagenomics

Metagenomics is an omics approach that utilizes high-throughput sequencing technology to perform culture-independent analysis of the total DNA from microbial communities in specific environments or host systems, enabling the characterization of their species composition, gene content, and potential functions [12]. Thus, metagenomics not only facilitates species identification but also resolves functional potential, metabolic pathway composition, and strain-level structural characteristics. Scientifically, metagenomics reflects the "potential functions" of microbial communities, providing an interpretive framework for subsequent expression, protein, and metabolic levels. Currently, widely used species and functional annotation workflows such as MetaPhlAn, Kraken, and HUMAnN enable researchers to establish dual hierarchies of species and functions from large-scale sequences. With the maturation of binning and assembly strategies, an increasing number of studies have achieved the recovery of genomes from uncultured microorganisms by constructing metagenome-assembled genomes (MAGs).

The recommended sequencing depth depends on sample complexity. For moderately complex gut samples, paired-end sequencing generating 6–12 Gbp of data per sample (approximately 30–50 million 150 bp paired-end reads) is commonly used, enabling species resolution down to the partial strain level. For studies targeting high-quality MAG recovery, the depth is recommended to be increased to 20–60 Gbp per sample, combined with long reads (e.g., 2 × 150 bp) or hybrid short-long reads (Illumina + Oxford Nanopore/PacBio) to improve assembly continuity.

After quality control of raw reads (using tools such as fastp), host DNA removal is recommended (via Bowtie2 mapping to the host genome). Assembly is typically performed using MEGAHIT or metaSPAdes, followed by binning with MetaBAT2, CONCOCT, or MaxBin2, and assessment of MAG completeness and contamination using CheckM. For functional annotation, HUMAnN3 combined with databases such as UniRef, KEGG, or MetaCyc is recommended. For relative abundance quantification, RPK/TPM normalization is employed, and compositional data transformation (e.g., centered log-ratio transformation, CLR) is recommended for subsequent statistical analysis.

Compared with the 16S rRNA sequencing method, metagenomics provides more than 100-fold expansion in functional information and improves the correct identification rate by approximately 20–40% in species resolution (especially for intra-genus differences). Under the MAG recovery strategy, near-complete genomes of unknown microorganisms can be obtained (with an average completeness often reaching 70–90%), enabling functional inference of uncultured bacteria. In clinical research, metagenomics can directly detect antibiotic resistance genes (ARGs) and virulence factors, serving as molecular evidence for rapid decision-making or epidemiological tracing.

Numerous researchers have utilized metagenomics to discover that patients with inflammatory bowel disease (IBD) exhibit a significant reduction in the short-chain fatty acid (SCFA) synthesis pathway and altered expression of bile acid transformation-related enzymes, thereby revealing the functional imbalance of microorganisms associated with chronic inflammation [13, 14].

3.1.2 Metatranscriptomics

Metatranscriptomics involves sequencing all RNA (especially mRNA) expressed by microbial communities in environmental or host-associated contexts, aiming to reveal the real-time functional activity and environmental response of microorganisms. Metatranscriptomic sequencing (RNA-seq) can characterize the genes that are actually active in microorganisms under specific environmental conditions [15].

The collection, preservation, and extraction of RNA are critical for the reliability of results. It is recommended that samples be immediately frozen (in liquid nitrogen or at −80 °C) or preserved in RNA stabilization reagents (e.g., RNAlater) after collection, with extraction completed within 24–48 hours. Since rRNA typically accounts for more than 80% of total RNA, rRNA depletion (using Ribo-Zero or custom probes) is required to enrich mRNA, or mRNA-selective library construction strategies can be employed. Sequencing is recommended to use 2 × 150 bp paired-end reads, with a recommended data volume of 50–100 million reads per sample (5–15 Gbp) to achieve sufficient coverage for community expression profile analysis. For studies focusing on complex environments or low-abundance transcripts, the depth should be appropriately increased. For mapping, pseudoalignment quantification using Salmon or kallisto can be performed, or reads can be mapped to MAGs/reference databases for taxon-resolved expression assignment (with attention to biases caused by multi-mapping).

The standard workflow includes quality control (fastp), rRNA read filtering (SortMeRNA), alignment to reference gene sets (Bowtie2/BBMap), and expression level normalization using TPM or counts + DESeq2-style methods for differential expression analysis. For time-series experiments, sampling intervals (e.g., 0, 0.5, 2, 6, 24 h) should be carefully designed to capture both rapid and medium-term responses.

Metatranscriptomics exhibits significant advantages in detecting short-term responses. For example, in antibiotic exposure experiments, the expression levels of relevant genes can increase by 10- to 50-fold (log2FC ≈ 3–5) within 30–60 min, a change that cannot be observed at the DNA level. In nutritional intervention studies, metatranscriptomics can directly link "species changes" with "pathway activation": For instance, fiber supplementation can upregulate the expression of key transcripts in certain SCFA synthesis pathways by three- to eightfold within 24–48 h, which is highly correlated with changes in fecal SCFA concentrations (correlation coefficient R ≈ 0.6–0.8).

Metatranscriptomics holds great value in antibiotic stress research. For example, a study published in *science* (2019) demonstrated that upon exposure to antibiotics, the gut microbiota immediately upregulates efflux pump genes associated with drug stress while downregulating genes related to energy metabolism. These rapid genetic changes cannot be reflected at the DNA level, highlighting the crucial role of metatranscriptomics in understanding the dynamic feedback mechanisms of microbial communities [16].

3.1.3 Metaproteomics

Metaproteomics is an omics technology that identifies and quantifies all proteins expressed by microbial communities under specific environmental conditions, revealing the actual execution status of functional proteins and metabolic pathways. Metaproteomic research obtains direct evidence of proteins in communities through mass spectrometry (liquid chromatography-mass spectrometry, LC–MS/MS), representing a method to characterize the actual execution of functions.

Protein extraction requires the selection of appropriate lysis and impurity removal methods (e.g., SDS/urea + bead-beating, SDS-free precipitation) based on the sample matrix (feces, soil, sediment, oral samples). Trypsin digestion is commonly used (16 h at 37 °C). Peptide purification and separation are performed using reverse-phase liquid chromatography (nanoLC), combined with high-resolution mass spectrometry (Orbitrap series or tim-sTOF) for MS/MS identification. A resolution (MS1) of ≥60,000 and MS2 resolution of ≥15,000 are recommended to improve the identification rate of low-abundance proteins.

In several studies, the integration of metaproteomics has improved the accuracy of disease classification models from 60–70% (using genomics alone) to 80–90%. This improvement is attributed to the ability of the protein layer to capture actual biochemical activities. For example, in dental caries research, the abundance of key glycolytic enzymes was found to be upregulated by two- to sixfold, which showed a high correlation with lactic acid production ($R \approx 0.9$), indicating that the protein layer provides higher biological credibility in functional validation.

3.1.4 Metabolomics

Metabolomics involves the qualitative and quantitative analysis of all detectable small-molecule metabolites in a biological system, reflecting the comprehensive output of the co-metabolic activities of microorganisms and hosts. Metabolomics captures the final outcome of microbe-host interactions through LC–MS, GC–MS, or nuclear magnetic resonance (NMR) [17]. Untargeted metabolomics enables the discovery of unknown metabolites.

Metabolite extraction should employ appropriate solvents based on the polarity of the target compounds (e.g., methanol/water for polar metabolites; tert-Butyl methyl ether (MTBE)/isopropanol for hydrophobic metabolites). Common mass spectrometry platforms include LC-MS/MS (positive/negative ion modes) and GC-MS (for volatile compounds); NMR is suitable for high-throughput analysis with accurate quantification but low sensitivity. For high-end LC-MS, a detection limit of the nM level can be achieved. For quantification, absolute quantification using stable isotope internal standards or semi-quantification based on external standard curves is recommended. Data processing is performed using XCMS (peak detection and alignment), MZmine, or MS-DIAL, and annotation can be carried out using databases such as GNPS, HMDB, and Metlin.

Parameters for peak detection and denoising need to be optimized according to the mass spectrometry platform (e.g., peak width, signal-to-noise ratio threshold, retention time window). Secondary confirmation (MS/MS fingerprint matching) is necessary during annotation to improve reliability. To determine the origin of metabolites (host vs. microorganism), isotope tracing or combined analysis with metabolism-related enzyme genes in metagenomes/MAGs can be used to enhance attribution credibility.

Metabolomics often directly links microbiome discoveries with host phenotypes. For example, in metabolic disease research, the correlation coefficients between specific microbial metabolites (e.g., SCFAs, secondary bile acids) and pathological indicators (insulin sensitivity, inflammatory markers) typically range from 0.6 to 0.8. In the field of drug metabolism, studies have shown that gut microorganisms can metabolize certain drugs, reducing their effective concentrations by 30–60%, which has direct implications for precision drug administration. Table 1 provides a comprehensive comparison of the performance parameters for the four aforementioned technologies.

3.2 Experimental Design and Sample Processing

The quality of integrated multi-omics research largely depends on preliminary experimental design. Due to significant differences in the stability and extraction methods of DNA, RNA, proteins, and metabolites, a unified and reproducible protocol must be established during the sampling phase. It is generally recommended to aliquot samples from the same parent sample collected at the same time and from the same source to maximize the consistency of information across different omics layers. This "matched sampling" approach has become a standard practice in multi-omics research. For example, in the HMP2 project, fecal samples from IBD patients were collected simultaneously to ensure consistency among metagenomics, metatranscriptomics, proteomics, and metabolomics data [13].

In terms of sample preservation, DNA is relatively stable, whereas RNA is highly prone to degradation and thus requires immediate addition of RNA stabilizers and rapid freezing after collection. Proteins and metabolites are also temperature-sensitive and should be processed as soon as possible under low-temperature conditions to avoid freeze-thaw cycles [14]. Consistent preservation conditions across omics layers help reduce pre-analytical bias, ensuring that information from different molecular layers truly reflects the same microbial ecological state.

During sample processing, internal standards (spike-ins) and quality control samples are critical tools for ensuring comparability across different omics, particularly in proteomics and metabolomics, where they can be used to monitor extraction efficiency and instrument drift [10]. To further reduce technical bias, it is necessary to accurately record batch information during the experimental workflow, randomize sample processing order, and apply statistical methods to correct for batch effects in subsequent analyses.

Overall, the core principles of multi-omics experimental design can be summarized as: consistent sample matching, unified preservation conditions, standardized operating procedures, and traceable batch information. Through a standardized pre-analytical workflow, the integrability and interpretability of cross-omics data can be ensured, thereby supporting subsequent mechanistic inference and systematic analysis.

Table 1
Comparison of technical performance parameters

Technical layer	Typical resolution	Recommended data volume/ sensitivity	Quantification method	Functional information level	Example improvement	Ref.
Metagenomics	Species–strain (MAG)	6–60 Gbp per sample	Relative quantification (RPK/TPM)	Potential function	20–40% higher species resolution than 16S rRNA sequencing, with a 100-fold increase in functional information	[13]
Metatranscriptomics	Gene/ transcript	5–15 Gbp (50–100 M reads)	TPM/DESeq2	Current active pathways	Detectable 10–50-fold expression changes, which cannot be captured by upstream DNA sequencing	[16]
Metaproteomics	Protein/ enzyme	MS sensitivity at fM level	MS1 intensity/ label-based quantification	Functional execution	10–20% improvement in prediction accuracy, with strong correlation to metabolism	[18]
Metabolomics	Small-molecule metabolites	Sensitivity at nm.–μM level	Absolute/relative quantification (internal standard)	Terminal phenotype	Microbial impact on drug metabolism can reach 30–60%	[19]

3.3 Data Preprocessing and Quality Control

Before integration analysis, multi-omics data must undergo systematic preprocessing and quality control. Due to the distinct data characteristics of different omics layers (metagenomics, transcriptomics, proteomics, metabolomics), preprocessing cannot follow a one-size-fits-all workflow but must adhere to mature standard procedures specific to each layer [20, 21]. Typically, sequencing-based data requires read quality control [22], host contamination removal, adapter trimming, and low-quality sequence filtering; mass spectrometry-based data requires peak extraction, denoising, peak alignment, and database matching for metabolites or proteins. Only after thorough cleaning and annotation of each omics layer can the reliability of subsequent integration be guaranteed.

Normalization is a crucial step in multi-omics preprocessing. High-throughput sequencing data is compositional, and total count constraints lead to spurious correlations, thus requiring log-ratio transformation. For metabolomics and proteomics, total normalization, internal standard correction, or batch correction methods are commonly used to reduce instrument drift and sample processing differences [23]. Particularly in multi-omics integration, differences in the scales of different omics can severely affect subsequent statistical models; therefore, the goal of normalization is to transform each omics layer into a comparable mathematical space.

Missing values are unavoidable in multi-omics data, including missing data at the omics layer level (e.g., RNA quality failure in a small number of samples) and feature-level missing values (e.g., undetected metabolites or proteins). Two main strategies are generally adopted to address this issue: one is to use integration methods capable of handling incomplete matrices, such as latent factor models like MOFA+ (Multi-Omics Factor Analysis); the other is to perform reasonable missing value imputation within a controllable range. The objective of missing value handling is not to "fill all data gaps" but to minimize the introduction of false trends or biases.

In terms of quality control, consistent cross-omics QC procedures are equally important, including sequencing quality assessment, mass spectrometer stability evaluation, internal standard deviation inspection, and outlier sample removal. By implementing strict QC for each omics layer, the stability and interpretability of downstream integration analysis can be significantly improved.

3.4 Core Application Fields

Integrated multi-omics has significantly transformed research paradigms in multiple fields, from disease mechanism elucidation and environmental microbial ecology to synthetic biology and engineering strain construction. Multi-omics integration enables research to move beyond "observing correlational phenomena" to understanding the cross-hierarchical biological processes underlying these phenomena. The following sections focus on three typical fields, using real cases to demonstrate how integrated multi-omics drives scientific understanding and application translation.

3.4.1 Human Disease Mechanisms and Biomarker Discovery

In complex diseases (e.g., inflammatory bowel disease, cancer, metabolic diseases), information from a single omics layer is often insufficient to explain etiologies or phenotypic differences. Multi-omics integration enables researchers to simultaneously validate hypotheses through the "genetic potential-expression activity-functional execution-metabolic output" chain, thereby constructing causal structures that more closely reflect biological reality.

Inflammatory bowel disease (IBD) serves as a typical example. Numerous studies have confirmed that IBD patients exhibit not only the absence or reduced copy number of genes involved in the SCFA synthesis pathway at the metagenomic level but also sustained downregulation of the expression of these pathways at the metatranscriptomic level. Further metabolomic evidence shows a significant reduction in SCFA production. The high consistency of multilayer data provides strong evidence for a causal chain, confirming that impaired SCFA metabolism is a core functional feature of IBD—an explanatory depth far beyond what single-layer data can provide.

A similar approach has been successfully applied to colorectal cancer research. Metagenomic sequencing revealed the enrichment of *Fusobacterium nucleatum* in cancer patients. However, what truly elevated this finding to a potential biomarker was the metabolomic discovery of significantly increased levels of polyamine metabolites (e.g., putrescine, cadaverine) in patients, which are highly associated with the metabolic network of *F. nucleatum*. This constructed a cross-layer evidence chain from "microorganism → metabolite → disease."

Multi-omics integration also plays a critical role in metabolic diseases. Taking trimethylamine-N-oxide (TMAO) as an example, metagenomics was used to identify TMAO-synthesizing microbial communities, metatranscriptomics to confirm the active expression of synthetic genes, and metabolomics to measure circulating TMAO levels. This systematic approach enabled researchers to comprehensively reveal how "microbial metabolites" contribute to the risk regulation of cardiovascular diseases.

3.4.2 Environmental Microbial Ecology and Bioremediation

In the field of environmental ecology, a single omics layer often fails to distinguish between "microbial communities with functional potential" and "active communities actually involved in environmental processes." The integration of metagenomics and metatranscriptomics can simultaneously reveal "who has the capability" and "who is actually performing the function." For example, in studies on the bioremediation of heavy metal-contaminated soil, metagenomics can identify bacteria harboring genes related to metal ion efflux, antioxidant repair, and sulfur metabolism.

Metatranscriptomics further verifies that these genes are significantly upregulated under contamination stress, thereby accurately identifying the core microbial communities truly involved in detoxification processes. The integration of multi-omics has shifted research from software-based functional prediction to function confirmation based on expression evidence, greatly improving the precision of bioremediation mechanism research.

3.4.3 Synthetic Biology and Engineered Strain Development

In industrial biotechnology, the construction of high-yield engineered strains frequently encounters challenges including product bottlenecks, precursor insufficiency, and metabolic flux deviation. Integrated multi-omics approaches enable the elevation of these issues from phenotypic observations to mechanistic interpretations. For instance, in the development of engineered yeast cell factories, transcriptomic analysis identifies insufficient expression of rate-limiting enzymes in key synthetic pathways, while metabolomic analysis quantifies precursor supply and downstream accumulant levels—collectively enabling clear identification of metabolic bottlenecks that restrict product yield. Researchers then perform rational reconstruction of these pathways based on such insights, achieving significant yield improvements. This cross-omics diagnostic model has become one of the most crucial strain optimization strategies in modern synthetic biology.

3.5 Extension and Discussion

While multi-omics integration has endowed life science research with structured, hierarchical, and causal analytical capabilities, it still faces challenges including data heterogeneity, batch effects, high sample matching difficulty, and substantial costs and computational resource consumption. Furthermore, key technical challenges in the coming years include extracting biologically meaningful signals from high-dimensional data, avoiding overfitting, and enhancing the interpretability of integrated models. With the integration of spatial omics and single-cell omics, the complexity of multi-omics integration will further increase—presenting both obstacles and opportunities for the next wave of breakthroughs. Future directions will shift from "single-layer alignment" to "cross-layer causal inference" and from "data stacking" to "mechanistic modeling," gradually exerting greater effects in clinical translation, industrial applications, and ecological assessment.

4 Spatial Omics Technology Deciphers the Functional Mechanisms of Synthetic Microbial Communities

In synthetic biology research, the functional realization of artificially constructed microbial communities (synthetic microbiota) highly depends on spatial interactions and chemical communication between strains. By resolving gene expression and metabolite distribution while preserving the original spatial structure, spatial omics technology provides a critical means to reveal functional heterogeneity in synthetic microbiota. Among these technologies, fluorescence in situ hybridization (FISH) enables precise localization of synthetic strains, while imaging mass spectrometry (IMS) deciphers the spatial distribution of metabolites. The synergistic application of these two technologies advances synthetic microbiota research from gene sequence design to in-depth exploration of spatial functional regulation.

4.1 Fluorescence In Situ Hybridization Technology: Spatial Localization and Functional Verification of Synthetic Microbiota

Through specific binding between fluorescently labeled probes and target nucleic acid sequences, FISH technology enables in situ identification and functional analysis of synthetic microorganisms. It addresses the limitation of traditional sequencing in linking "strain identity-spatial location-functional state" and has become a core tool for the construction and optimization of synthetic microbiota [24].

4.1.1 Technological Origin and Adaptive Evolution in Synthetic Biology

The development of FISH technology is highly aligned with the demand for microbiota decipherment in synthetic biology. In 1989, the rRNA-targeted FISH technology developed by DeLong et al. provided the first tool for spatial resolution of synthetic microbial communities. Although early DNA probes could meet the detection needs of simple synthetic microbiota, they exhibited insufficient specificity in complex systems.

The emergence of peptide nucleic acid (PNA) probes and locked nucleic acid (LNA) probes significantly improved detection performance: The neutral backbone structure of PNA enhances binding stability with high-GC regions in synthetic genes, while the rigid structure of LNA enables precise differentiation of closely related synthetic strains by increasing melting temperature [25]. Signal amplification technologies (e.g., catalyzed reporter deposition FISH) have reduced the detection limit by an order of magnitude [26], resolving the detection challenges of dormant engineered strains or low-abundance functional strains.

4.1.2 Probe Design: Adapting to the Precise Requirements of Synthetic Biology

The complexity of synthetic microbiota demands high specificity and stability of FISH probes, requiring systematic optimization of probe design based on the sequence characteristics of synthetic genes. Probes typically range from 15 to 25 nucleotides in length with a GC content of 40–60%, and secondary structure formation must be avoided to ensure binding efficiency. Target selection should be based on conserved regions of synthetic genes while avoiding homologous sequences in closely related strains to ensure interspecies specificity [27].

Thermodynamic parameters are central to probe optimization. According to the model proposed by Yilmaz et al. [28], the free energy change of probe-target binding should be below −13.5 kcal/mol. For low-affinity targets, optimization can be achieved through LNA modification or probe length extension. In experimental condition optimization, formamide concentration regulates hybridization stringency by adjusting melting temperature. For Gram-positive engineered strains, lysozyme treatment conditions need optimization, while archaeal synthetic strains may require hydrochloric acid or proteinase K to assist permeabilization. For targets with dense secondary structures in synthetic genes, helper probes can be designed to unfold spatial structures and improve main probe accessibility—a strategy particularly important for in situ detection of complex synthetic gene circuits [29].

4.1.3 Core Applications in Synthetic Biology

Spatial Verification of Synthetic Symbiotic Systems In metabolically complementary synthetic symbiotic systems, one engineered strain synthesizes essential amino acids, while the other provides growth factors. By labeling species-specific PNA probes with different fluorophores, confocal microscopy revealed a tightly intertwined spatial structure between the two strains, directly verifying the spatial basis of metabolic interactions and providing a foundation for the stable construction of synthetic microbiota.

In Situ Functional Analysis of Synthetic Gene Circuits For quorum-sensing regulated synthetic circuits, designing specific probes for reporter gene mRNA revealed that gene expression is specifically activated in high-density regions of biofilms. This clarifies the spatial regulatory characteristics of the circuit and guides the optimized design of inducer signaling molecules [30].

Environmental Colonization Tracking of Engineered Strains - After releasing synthetic engineered strains into soil or intestinal environments, the combined use of engineered strain-specific probes and environmental microorganism probes enables precise localization of engineered strains in rhizospheres or mucous membranes. This facilitates analysis of their spatial interactions with indigenous microbiota and evaluation of environmental adaptability and ecological safety of engineered strains [31, 32].

4.2 Imaging Mass Spectrometry: A Tool for Deciphering Chemical Communication in Synthetic Microbiota

By combining mass spectrometry analysis with spatial imaging, imaging mass spectrometry (IMS) enables label-free in situ detection of metabolites in synthetic microbiota. It reveals the molecular mechanisms of chemical communication between strains, providing direct evidence for the optimization of metabolic interactions in synthetic microbiota [33].

4.2.1 Technical Overview and Application Evolution in Synthetic Biology

IMS technology performs point-by-point scanning of samples using microprobes, detects ion signals at each position, and reconstructs molecular spatial distribution maps. Its "hypothesis-free" detection capability allows simultaneous analysis of hundreds of metabolites, perfectly matching the systematic decipherment needs of microbial chemical interactions in synthetic biology. Originating from materials science, this technology entered the biological field with the maturation of MALDI technology in the 1990s [34, 35] and has been innovatively applied in synthetic microbial community research since the twenty-first century.

The emergence of ambient ionization technologies such as desorption electrospray ionization (DESI) and nanoscale desorption electrospray ionization (nanoDESI) has enabled in vivo analysis of synthetic microbiota, providing possibilities for dynamic monitoring of metabolic interaction processes and promoting IMS to become a core technology in synthetic biology metabolic regulation research [36–38].

4.2.2 Core Technical Principles and Adaptive Characteristics in Synthetic Biology

The selection of IMS technologies in synthetic microbiota research depends on target molecule types, spatial resolution requirements, and sample characteristics. The core adaptive features of various technologies are shown in Table 2:

4.2.3 Typical Applications in Synthetic Biology

Deciphering Chemical Interactions in Synthetic Microbiota In a synthetic competitive system of *Bacillus* and *Streptomyces*, MALDI imaging revealed that surfactin produced by *Bacillus* accumulates at the interface between the two strains. This metabolite inhibits antibiotic synthesis in *Streptomyces* while inducing the expression of its own antimicrobial peptides [42]. This spatially resolved metabolic profile provides direct evidence for the design of "competition-defense" synthetic microbiota and guides the optimization of signaling molecule regulatory elements.

Spatial Localization of Metabolic Activity in Synthetic Strains Combining stable isotope-labeled substrates with NanoSIMS enables identification of metabolic "hotspot" strains in synthetic microbiota at the single-cell level. In degrading microbiota research, this method clarified the preferential degradation of target pollutants by specific engineered strains, providing data support for optimizing microbiota composition ratios [39, 40].

Precise Correlation of "Identity-Function" The combination of IMS and FISH technologies represents a major breakthrough in synthetic microbiota research—FISH identifies the phylogenetic identity of strains, while IMS detects their metabolites, enabling the "who is there-what is it doing" correlation analysis at the single-cell level. In synthetic nitrogen-fixing microbiota, this combined

Table 2
Main types of imaging mass spectrometry (based on ionization source)

Technology type	Ionization principle	Spatial resolution	Mass range (Da)	Advantages	Limitations	Ref.
MALDI	Ultraviolet laser + matrix-assisted desorption	50–500 μm	300–50,000	Broad mass range; suitable for proteins, peptides, and lipids	Requires matrix; background signal interferes with low-mass ions	[35]
Static SIMS	Low-energy ion beam surface bombardment	0.5–50 μm	100–1500	High resolution; suitable for elements and small molecules	Molecular fragmentation; limited mass range	[39]
Dynamic SIMS (NanoSIMS)	High-energy ion beam, high sensitivity	<50 nm	1–300	Nanoscale resolution; suitable for single-cell elemental and isotopic analysis	Limited to elements and small atomic clusters; unable to analyze macromolecules	[40]
DESI	Charged solvent spray desorption	100–1000 μm	100–5000	Atmospheric pressure operation; no sample pretreatment required	Low spatial resolution; sensitive to surface flatness	[36]
NanoDESI	Liquid microjunction extraction	10–1000 μm	100–10,000	Direct analysis of moist samples; suitable for living microorganisms	Relatively new technology; reproducibility needs improvement	[38]
LA-ICP	Laser ablation + inductively coupled plasma	1–10 μm	7–250	High-sensitivity metal analysis; quantitative imaging	Provides only elemental information; unable to analyze organic molecules	[41]

MALDI matrix-assisted laser desorption/ionization, *SIMS* secondary ion mass spectrometry, *NanoSIMS* nanoscale secondary ion mass spectrometry, *DESI* desorption electrospray ionization, *NanoDESI* nanoscale desorption electrospray ionization, *LA-ICP* laser ablation inductively coupled plasma technology clarified the distribution of nitrogen-fixing products among different engineered strains and verified the effectiveness of metabolic division-of-labor design [35, 43].

Current applications of IMS in synthetic biology still face challenges: Quantitative analysis requires matrix effect correction using synthetic standards, and identification of unknown metabolites requires coupling with tandem mass spectrometry technology. Addressing these issues will further enhance its application value in synthetic biology.

5 Microbial Fermentation Guided by Omics

5.1 Microfluidic Culture and Lab-on-a-Chip: Synthetic Biology Technology Platforms

Microfluidic culture lab-on-a-chip serves as a critical technical support platform for microbial resource exploration, engineered strain construction, and functional characterization in the synthetic biology technology system. Its core principle relies on the miniaturized and integrated design of microfluidic chips, which array hundreds to thousands of independent microculture chambers on a substrate of several square centimeters. This enables high-throughput simulation and recapitulation of in situ natural microbial habitats or artificially precisely controlled habitats required for synthetic biology research.

This technology effectively overcomes the inherent limitations of traditional pure culture methods in throughput scale, microenvironment controllability, and microecological simulation capabilities. It provides a core technical tool for the targeted exploration of "microbial dark matter" resources (accounting for over 99% of microorganisms) and high-efficiency screening of engineered strains in the field of synthetic biology [44].

The core advantages of this technology lie in its highly integrated microenvironment regulation capability and high-throughput parallel culture characteristics, which are highly aligned with the core technical needs of synthetic biology for precise regulation of microbial metabolic networks and high-efficiency screening of engineered strains. Through precise manipulation of microliter (μL) to picoliter (pL) scale fluids by microfluidic control units such as micropumps and microvalves, independent regulation of key parameters in each microculture unit—including nutrient substrate concentration gradients, signaling molecule ratios, dynamic pH ranges, and temperature stability—can be achieved. Additionally, it can simulate physicochemical gradients existing in natural environments or artificially designed targeted metabolic microenvironments.

This precise "habitat-targeted regulation capability" not only provides controllable growth conditions for difficult-to-culture microorganisms that rely on specific microecological niches or interspecies interactions but also offers a dedicated microenvironment for phenotypic precise optimization of genetically engineered strains constructed by synthetic biology methods [45].

Based on the aforementioned technical characteristics, microfluidic chip culture technology has formed three core technical strategies supporting synthetic biology research:

High-Throughput Single-Cell Isolation-Culture Coupling Strategy Environmental samples or synthetic biology gene-edited mutant libraries are processed by gradient dilution, and single cells are co-encapsulated with selective culture media in water-in-oil droplets or closed microchambers. This enables efficient isolation and independent culture of single cells, providing a technical pathway for high-throughput monoclonal screening of engineered strains.

Diffusion-Regulated Coculture Strategy Through precise fabrication of micro-nano channels between microchambers in chip microstructure design, controllable diffusion and exchange of small-molecule metabolites and signaling molecules between different culture units are achieved. This is highly suitable for the construction and decipherment needs of artificially constructed microbial cooperative metabolic networks in synthetic biology.

In Situ Screening-Sequencing Integration Strategy Targeted acquisition of microcolonies with desired phenotypes on the chip is achieved through optical tweezers capture or microvalve switching technology. Direct coupling with single-cell whole-genome amplification (WGA) and high-throughput sequencing platforms rapidly establishes genotype-phenotype correlation maps of engineered strains, providing critical data support for verification of gene editing efficiency and targeted optimization of functional modules in synthetic biology [46].

In summary, microfluidic culture lab-on-a-chip, with its core technical features of miniaturized integration, high-throughput parallelization, and precise regulation, has become an indispensable critical support technology in the synthetic biology technology chain. At the basic research level, it provides a unique innovative research paradigm for the decipherment of microbial metabolic networks, exploration of interaction mechanisms in artificial microbiota, and excavation of functional gene resources in the field of synthetic biology. At the application transformation level, it demonstrates significant application value in key synthetic biology application scenarios such as directed evolution screening of novel industrial enzymes, high-throughput breeding of artificially synthesized high-yield antibiotic strains, and precise verification of probiotic functional modules.

5.2 In Situ Culture and Diffusion Chambers: Microbial Resource Exploration in Synthetic Biology

In situ culture technology is a core means to break through the bottleneck of traditional microbial pure culture technology, and more importantly, it serves as an important technical support for targeted exploration of microbial resources and screening of functional strains in the field of synthetic biology. Its core technical logic is to provide adaptive growth conditions for "microbial dark matter" that is difficult to proliferate in artificial culture media by precisely simulating the natural growth microecological niche of microorganisms.

This section focuses on diffusion chambers and their derivative upgraded technology (iChip) in the in situ culture technology system, systematically deciphering their technical principles, standardized operation procedures, and application value in synthetic biology resource exploration. Traditional laboratory pure culture technology can only achieve effective isolation and recovery of less than 1% of environmental microorganisms—a phenomenon known as the "great plate count anomaly." The core reasons include the lack of signaling molecules produced by interspecies interactions, inability to adapt to natural nutrient gradients, and stress responses to artificial culture conditions, resulting in a large number of microorganisms remaining in a "viable but non-culturable (VBNC) state."

This predicament severely limits the development and utilization of novel functional genes, metabolic pathways, and engineered strain chassis resources in synthetic biology. In situ culture technology overcomes these limitations by reintroducing microbial samples into their natural habitats for cultivation, making full use of natural nutrient substrates, signaling molecules, and microenvironmental conditions in the environment. A landmark study published in *Science* by Kaeberlein et al. in 2002 confirmed that the cultivability of marine sediment microorganisms after in situ culture with diffusion chambers was increased by more than 300-fold compared with traditional plates, providing a critical technical pathway for synthetic biology to break through resource bottlenecks [47].

As a classic core device of in situ culture technology, the diffusion chamber features a "closed culture-open exchange" technical system in its design. It ensures cell isolation during microbial culture while enabling small-molecule substance exchange with the natural environment— a characteristic highly aligned with the needs of synthetic biology for microbial interaction signal capture and habitat adaptability research.

A typical diffusion chamber device adopts a "double-membrane sandwich" structure design, consisting of two polycarbonate filter membranes with a precisely controlled pore size of 0.03 μm, which sandwich an elastic gasket to form a closed culture chamber.

The filter membrane with this pore size specification exhibits precise selective permeability, allowing free diffusion of small-molecule nutrients, signaling molecules, and metabolic intermediates from the natural environment into the culture chamber while effectively blocking the migration of microbial cells in the environment to ensure the purity of the culture system.

In the standardized operation procedure, microorganisms isolated from environmental samples (such as intertidal marine sediments) are first mixed with semi-solid agar prepared with seawater, and 0.01% casein can be added to increase the number of colonies. The mixture is then injected into the diffusion chamber, sealed, and returned to a marine aquarium containing the original sediment for cultivation.

Research data from Kaeberlein et al. showed that this technology achieved a recovery rate of 22 $\pm$ 13% for marine sediment microorganisms, representing an order of magnitude improvement compared with 0.054 $\pm$ 0.051% for traditional plate culture. More notably for synthetic biology, the isolated MSC1 strain could only grow continuously in the diffusion chamber and exhibited cooperative interaction with MSC2—co-cultivation of the two strains enabled colony formation on plates, while single-strain culture resulted in growth arrest, confirming a signal-dependent growth mechanism. This discovery provides a direct natural reference model for synthetic biology to design inter-strain signal regulation modules and construct cooperative metabolic networks.

The iChip (isolation chip) technology represents a miniaturized and high-throughput upgraded version of the diffusion chamber technology. By introducing precise single-cell isolation design, it achieves integration of in situ culture and monoclonal isolation, greatly improving technical adaptability for chassis screening of synthetic biology engineered strains.

Its core structure has undergone multiple generations of optimization; the currently commonly used model employs a Thermo Fisher 96-well pipette tip box as the substrate, with 0.03 μm polycarbonate filter membranes bonded with silica gel on both sides to form micro diffusion chambers. Gradient dilution ensures an average of 1 cell per well, directly achieving monoclonal pure culture.

Compared with traditional diffusion chambers, iChip further increases the microbial recovery rate to 40–50%, and the phylogenetic novelty of isolated strains is significantly higher—in seawater and soil samples, the proportion of new species isolated by iChip at the genus level and above is much higher than that of plate culture.

The standardized iChip experimental procedure in synthetic biology research includes key steps:

Device preparation: Filter membranes are bonded to one side of a 96-well frame, followed by ultraviolet sterilization for standby use.

Sample processing: 1 g of soil is mixed with 10 mL of PBS and vortexed for 10 min to prepare a slurry; after standing, the supernatant is collected for gradient dilution.

Precise inoculation: Diluted bacterial solution is mixed with SMS agar, and 100 μL of the mixture is dispensed into each well using a multi-channel pipette; after solidification, the other side of the filter membrane is bonded.

In situ *culture:* The device is returned to the original environment for 1–4 weeks, with regular water replenishment to maintain moisture.

Recovery and identification: After sterile cleaning, the membrane is peeled off, and microcolonies are picked under a dissecting microscope, which can be directly subjected to 16S rRNA sequencing or subculture.

This technology has achieved industrial application: NovoBiotic Pharmaceuticals screened 10,000 strains using iChip and successfully discovered the novel antibiotic teixobactin, confirming its high efficiency in functional strain exploration [48].

From the perspective of core advantages supporting synthetic biology technology, the core value of in situ culture technology lies in its technical characteristics of "precise environmental simulation-functional phenotype activation," which can be explained by two key theoretical hypotheses:

Signal-dependent growth hypothesis: For example, the cooperative growth of MSC1 and MSC2 confirms that microorganisms require specific signaling molecules produced by neighboring cells to initiate growth, which fails in traditional pure culture due to the lack of such signals. This provides a natural template for synthetic biology to design cross-strain signal transduction pathways.

Progressive domestication effect: Repeated alternating culture between in situ and laboratory conditions can enable non-culturable microorganisms to gradually adapt to artificial conditions. Studies have shown that the domestication rate reaches 26% after a single round and increases to 91% after four rounds, and domestication can be accelerated by adding inducers such as short peptides. This effect provides an efficient technical pathway for synthetic biology to obtain engineered strain chassis with stable passage.

This technology system has limitations that require targeted optimization:

- Microorganisms' dependent on close interactions cannot be co-cultured due to membrane barrier, which can be improved by designing interconnected microchambers.

- Application in arid environments is limited, currently mitigated by adding moisturizing gels or sealed culture chambers.
- Oxygen diffusion control challenges in sediment culture can be solved by aquarium aeration systems.
- Contamination prevention can be achieved by setting sterile agar control wells and UV sterilization devices.

These optimization directions are highly aligned with the customized needs of synthetic biology for culture systems, promoting the development of in situ culture technology toward precision.

In general, the in situ culture technology system centered on diffusion chambers and iChip has significantly broken through the resource exploration bottleneck of traditional culture technology by precisely simulating and regulating the natural growth environment of microorganisms. It not only achieves an order of magnitude increase in the recovery rate of "microbial dark matter" but also successfully isolates a large number of strains with novel phylogenetic status and unique metabolic functions, providing abundant genetic resources and strain chassis for synthetic biology.

At the application level of synthetic biology, this technology has formed a complete chain from basic research to industrialization: It is used to decipher microbiota interaction mechanisms in basic research, supports the screening of novel antibiotics and industrial enzymes in application transformation, and can rapidly establish genotype-phenotype correlations when combined with single-cell sequencing.

In the future, integration with technologies such as microfluidic automated inoculation and real-time fluorescence monitoring can realize dynamic tracking of iChip culture processes; combination with synthetic biology gene editing technology is expected to achieve targeted modification of microbial functions in in situ environments. These developments will further unleash the potential of microbial resources and provide core technical support for fields such as metabolic engineering and environmental remediation microbiota construction.

5.3 Engineering Design and Application of Microbial Coculture Systems: Artificial Microbiota Construction in Synthetic Biology

Microbial coculture systems serve as the core technical carrier for synthetic biology to achieve "artificial microbiota functional programming." Through rational design of metabolic division of labor, signal transduction, and interaction networks between multiple species, they break through the inherent limitations of single-strain metabolic pathways, enabling complex product synthesis, efficient substrate conversion, and improved system stability.

Based on synthetic biology module design principles and microbial ecology interaction theories, this section systematically elaborates on the engineering construction methods, dynamic regulation strategies, and industrial application in biomanufacturing of coculture systems, focusing on the core supporting role of synthetic biology toolkits in microbiota function optimization.

The engineering construction of coculture systems centers on "modular splitting and cooperative reconstruction of metabolic networks," with synthetic biology tools such as gene editing, promoter engineering, and metabolic flux regulation providing precise technical pathways.

In studies using *Saccharomyces cerevisiae* as a model chassis, researchers constructed a library of 15 auxotrophic strains through CRISPR-Cas9-mediated gene knockout technology, covering deletion mutants of amino acid, nucleotide, and vitamin synthesis pathways. Simultaneously, constitutive promoters (TEF1p, ADH1p) and inducible promoter (GAL1p) of *S. cerevisiae* were used to construct target gene overexpression modules, ultimately forming a cross-feeding network of "auxotrophic strains-complementary metabolic modules."

This network can flexibly assemble a dual-member coculture system of leucine-auxotrophic and lysine-auxotrophic strains or a three-member coculture system by adding uracil-auxotrophic strains. Metabolite complementation enables stable coexistence of populations, providing a standardized paradigm for synthetic biology to design metabolic division of labor in artificial microbiota [49].

The optimization of coculture systems driven by synthetic biology relies on precise decipherment and regulation of key parameters of population dynamics. Global sensitivity analysis combined with 13C metabolic flux analysis confirmed that the metabolite exchange rate (φ_i) has an influence weight of 0.72 on population balance, significantly higher than traditional parameters such as initial inoculation ratio and culture temperature. This discovery provides quantitative basis for rational optimization of coculture systems.

Based on this, researchers precisely regulated metabolite secretion flux through promoter strength gradient engineering, constructing a gradient system using strong promoter (TEF1p), medium-strength promoter (ADH1p), and weak promoter (CYC1p). In the adenine-lysine dual-member coculture system, predictable regulation of the population ratio of the two strains from 1:1 to 3:1 was successfully achieved, with a simultaneous 2.3-fold increase in target product synthesis efficiency.

Furthermore, metabolite-responsive promoters (P_ADE2, P_LYS2) were used to construct dynamic sensor modules, which real-time sense population metabolic status and feedback-regulate the expression intensity of complementary genes, further enhancing the stability and reproducibility of the coculture system.

Naturally existing metabolic dependencies in microorganisms provide a natural "interaction scaffold" for synthetic biology to construct artificial coculture systems. Studies have confirmed that over 98% of sequenced microorganisms lack complete amino acid or vitamin synthesis pathways [45]. Synthetic biology constructs a cooperative network of "essential nutrient mutual supply-metabolic burden sharing" through rational design of metabolic defect combinations:

In the *Escherichia coli-S. cerevisiae* coculture system, the tryptophan synthesis gene (trpC) in *E. coli* and the phenylalanine synthesis gene (aroG) in *S. cerevisiae* were knocked out, enabling stable mutualism between the two strains through cross-supply of tryptophan and phenylalanine. Simultaneously, the complex product synthesis pathway was split between the two strains, reducing the metabolic burden of a single strain by more than 40%.

Further research confirmed that coculture systems involving exchange of high-energy-cost amino acids (such as tryptophan and tyrosine) exhibit significantly higher cooperative stability than those involving low-energy-cost amino acid exchange. This interaction selection mechanism based on "energy trade-off" provides theoretical guidance for designing coculture systems with long-term stable operation.

Coculture systems centered on acetate-producing bacteria demonstrate the technical potential of synthetic biology to achieve "cross-species metabolic coupling." In the coculture system constructed by *Clostridium ljungdahlii* and *Escherichia coli*, *C. ljungdahlii* fixes CO_2 to produce acetate through the Wood-Ljungdahl pathway (relying on the CODH/ACS enzyme complex for catalytic activity), while *E. coli* uses acetate as a carbon source to synthesize high-value products, forming an efficient carbon cycle network of "CO_2 fixation-acetate conversion" [50].

Through specific rRNA-FISH labeling and mCherry/GFP dual-fluorescent protein tracking technology, researchers observed high-frequency contact-dependent substance exchange between strains, with direct transfer frequencies of genetic materials such as proteins, tRNA, and plasmids reaching 18–50%. This "transcellular metabolic compensation" mechanism significantly improves carbon source utilization efficiency.

Additionally, the quorum-sensing receptor gene (clpR) of *C. ljungdahlii* was heterologously expressed in *E. coli*, endowing the system with acetate concentration sensing capability. When acetate accumulation exceeds the threshold, the expression of acid-producing genes in *C. ljungdahlii* is automatically downregulated, achieving dynamic balance of metabolic flux.

Coculture technology driven by synthetic biology has advanced from laboratory research to industrial scale application, with core breakthroughs manifested in the formation of a technical system featuring "modular design-standardized assembly-precise regulation."

- *Biofuel production field:* A three-member coculture system constructed by the cellulose degradation module of *Trichoderma reesei*, the pentose utilization module of *S. cerevisiae*, and the ethanol synthesis module of *E. coli* increased the utilization rate of corn stover substrate from 52% (traditional single strain) to 94%, with a 1.8-fold increase in ethanol yield.
- *Pharmaceutical intermediate synthesis field:* The *S. cerevisiae-Streptomyces* coculture system splits the paclitaxel precursor synthesis pathway into two modules—upstream mevalonate synthesis and downstream cyclization modification—undertaken by *S. cerevisiae* and *Streptomyces*, respectively. This solves the problem of metabolic flux imbalance caused by excessively long pathways in single strains, resulting in a 3.2-fold increase in precursor yield.

In the future, synthetic biology will further promote the evolution of coculture systems toward "intelligent responsive microbiota":

- Construction of environmental signal-sensing gene circuits (pH, temperature, substrate concentration) to realize autonomous switching of microbiota metabolic pathways.
- Utilization of the self-repair mechanism of synthetic microbial communities (SMCs) to improve anti-contamination capability in industrial fermentation processes.
- Combination with microfluidic chip technology to realize miniaturization and high-throughput screening of coculture systems, accelerating the development process of novel functional microbiota.

These technological innovations will enable coculture systems to play a core supporting role in carbon neutrality, biomanufacturing, and other fields, promoting the green upgrading of the biotechnology industry.

5.4 Cell Sorting-Assisted Cultivation

Cell sorting-assisted cultivation serves as a pivotal technical support for the precise exploration of microbial resources in synthetic biology. By coupling sophisticated separation technologies—such as flow cytometry sorting and microfluidic sorting—with targeted cultivation strategies, this approach enables the directional isolation and cultivation of microbial taxa with specific functions or phylogenetic characteristics from complex environmental samples. It is particularly instrumental in the efficient screening of rare, hard-to-culture microorganisms and synthetic biology engineering strains, offering unprecedented technical precision. This technology overcomes the limitations of blind screening in traditional cultivation methods, realizing the integration of "goal-oriented" microbial isolation and functional verification. Consequently, it lays a core technical foundation for chassis strain screening, functional gene mining, and metabolic pathway analysis in synthetic biology.

Modern cell sorting technologies have evolved into two core systems: flow cytometry and microfluidic sorting. Both achieve the unification of targeting and efficiency through integration with synthetic biology tools. Flow cytometry sorting, centered on fluorescent signal recognition, achieves precise sorting by combining diverse labeling strategies: Leveraging nucleic acid dyes to identify differences in cellular DNA content enables rapid isolation of microbial mutants with varying ploidy; utilizing catalytic reactions between specific fluorescent substrates and target enzymes allows targeted labeling of cells with specific metabolic activities; a more precise strategy involves fluorescent in situ hybridization (FISH), which uses oligonucleotide probes to target 16S rRNA of specific phylogenetic groups, enabling phylogeny-oriented isolation [51]. As a miniaturized advancement of flow technology, microfluidic fluorescence-activated cell sorting (μFACS) constructs T-shaped sorting junctions using polydimethylsiloxane (PDMS) chips fabricated via soft lithography, achieving cell manipulation through electroosmotic flow or pressure-driven mechanisms. Its ultrasmall detection volume of 250 femtoliters significantly reduces background fluorescence interference, and the disposable chip design fundamentally addresses the cross-contamination issue inherent in traditional flow sorting. For *Escherichia coli* labeled with green fluorescent protein (GFP), the enrichment factor via μFACS can reach 30-fold. Notably, its reverse sorting algorithm enables reverse flow recovery after high-speed screening of target cells, substantially improving the capture efficiency of rare functional strains.

Endowed with the advantage of miniaturized integration, microfluidic sorting technology achieves synergistic improvements in sorting throughput and microenvironment control, making it more suitable for high-throughput screening requirements in synthetic biology. Using microfluidic chips as core carriers, this technology constructs microscale channel and chamber arrays via photolithography or soft lithography. It achieves label-free cell sorting based on physical principles such as dielectrophoresis and acoustic manipulation or realizes precise capture and encapsulation of single cells through a microvalve-negative pressure linkage system. Compared with traditional flow sorting, microfluidic systems exhibit distinct advantages: The functional micropore laser sorting (FMLS) system constructs femtoliter droplet arrays through million-scale micropore arrays, achieving a capture efficiency exceeding 80% for *E. coli*, *Saccharomyces cerevisiae*, cyanobacteria, etc. Specifically, micropores with an 8 μm diameter achieve a 90.32% capture efficiency for *E. coli*, while 15 μm pores reach 81.98% for *Chlamydomonas*. By regulating the wettability of the chip surface, the synergistic effect of superhydrophilic metal layers and superhydrophilic microporous membranes eliminates liquid surface tension interference and reduces sorting energy consumption. Additionally, microfluidic systems maintain microenvironmental stability during sorting, effectively avoiding potential impacts of fluorescent labeling on cell viability, thus being more suitable for environment-sensitive, hard-to-culture microorganisms.

The technical core of cell sorting-assisted cultivation lies in maintaining post-sorting cell viability and directionally optimizing cultivation conditions, a process that must be closely aligned with the requirements of synthetic biology for strain viability and functional integrity. Laser irradiation, fluid shear stress, and oxidative stress during sorting may impair cell viability; therefore, orthogonal experiments are required to optimize key parameters. Studies on the FMLS system have demonstrated that a laser energy of 120 nJ represents the optimal sorting condition, achieving 100% sorting efficiency with minimal cell damage. Compared with the 300 nJ used in traditional laser ejection sorting and 11 μJ in liquid LIFT sorting, this energy parameter significantly reduces energy consumption. COMSOL simulations confirm that under this energy level, the temperature in the laser action area only increases transiently within a 2 μm range, and the temperature in the cell-residing region remains below 30 °C, ensuring viability. Experimental data show that under these parameters, the post-sorting survival rates of *E. coli* and *S. cerevisiae* reach 97.2% and 95.8%, respectively, representing increases of 75.2% and 32.8% compared with traditional laser sorting [52]. Furthermore, timely transfer of sorted cells into pre-warmed buffer can effectively mitigate the impact of temperature stress on cellular metabolic activity.

Directional optimization of post-sorting cultivation requires designing strategies based on the functional characteristics and niche properties of target microorganisms. The core concept involves simulating natural habitats while integrating functional activation requirements of synthetic biology. The application of low-nutrient concentration media can reduce the inhibitory effect of nutrient stress on rare microorganisms; adding quorum sensing signal molecules can activate growth-initiating genes in hard-to-culture microorganisms; and incorporating antioxidants such as glutathione can scavenge reactive oxygen species generated during sorting, improving cell recovery efficiency. For functional strains isolated from environmental samples, constructing auxiliary microbial communities using the aforementioned co-cultivation strategy can further enhance the cultivation success rate of target strains through interspecific metabolic complementarity, providing chassis materials with stable activity for subsequent synthetic biology gene editing.

As a typical example of technological integration, the functional micropore laser sorting (FMLS) system exhibits unique advantages in precise single-cell isolation for synthetic biology. The core component of this system is a wettability-regulated microporous jet chip (M-chip). Poly-p-xylylene C microporous membranes are fabricated via silicon master photolithography and chemical vapor deposition (CVD), followed by oxygen plasma treatment to achieve superhydrophilic modification of metal layers and microporous membranes. This constructs a hexagonal micropore array with 700,000 pores per 10 mm × 10 mm chip, enabling the dispersion of bacterial solutions into femtoliter droplets for single-cell capture. The system integrates bright-field, fluorescence, and Raman triple identification modules, enabling label-free targeted sorting via characteristic spectra. For instance, *Rhodotorula mucilaginosa* was successfully sorted from mixed samples using the characteristic peaks of carotenoids, and the sorted strains maintained stable target product yields. Experimental data confirm that the system achieves a capture efficiency exceeding 80% for *E. coli*, *S. cerevisiae*, cyanobacteria, and *Chlamydomonas*, with a 100% single-cell sorting efficiency that does not affect adjacent cells. The genomic coverage of sorted *E. coli* reaches 88.06%, significantly higher than the 38.46% achieved by traditional LIFT technology, meeting the genotyping requirements of synthetic biology [53]. Additionally, PCR verification of sorted cells confirms 100% specificity, and monoclonal colonies form after 36 h of cultivation, verifying the preservation of functional integrity.

Current cell sorting-assisted cultivation technologies still face challenges in adapting to the large-scale application needs of synthetic biology: Sorting throughput lags behind traditional screening methods and single-channel sorting speed cannot meet the requirements of large-scale mutant library screening; for microorganisms derived from extreme environments, unclear post-sorting recovery mechanisms result in low cultivation success rates; insufficient standardization of technical procedures leads to poor result reproducibility between laboratories. Future development should focus on three directions: developing multichannel parallel microfluidic sorting chips to improve throughput, integrating single-cell transcriptomics and metabolomics analysis technologies to realize rapid post-sorting functional identification, and constructing databases based on microbial niche data and cultivation parameters. Predicting optimal sorting-cultivation parameter combinations via artificial intelligence algorithms can further shorten the exploration cycle of hard-to-culture microorganisms, providing more abundant functional strain resources for synthetic biology.

Genome-guided medium design and optimization represent a core technical strategy in synthetic biology for breaking the bottleneck in the exploration of hard-to-culture microbial resources. By deciphering the metabolic coding information in microbial genome sequences, this approach reversely infers nutritional requirements, environmental adaptability, and interspecific interaction rules, enabling rational design and dynamic optimization of medium formulations. This technology completely transforms the "empirical trial-and-error" paradigm of traditional microbial cultivation, establishing a precise technical chain of "genome analysis—metabolic prediction—medium customization—cultivation verification." It provides key technical support for synthetic biology in screening functional chassis strains and analyzing metabolic pathways and exhibits irreplaceable value particularly in the cultivation of rare microorganisms and symbiosis-dependent microorganisms.

5.5 Genome-Guided Medium Design and Optimization: Precise Fermentation in Synthetic Biology

The core logic of this technology is based on the functional decoding of genomic information and rational verification in synthetic biology, achieving precise matching of cultivation conditions through multidimensional sequence analysis. Reconstructing metabolic pathways by annotating genomes using functional annotation databases such as COG and KEGG enables direct localization of missing nodes in key nutrient synthesis pathways. For example, if the gene cluster encoding thiamine synthase is absent from the genome, the microorganism can be identified as thiamine auxotrophic, requiring exogenous thiamine supplementation in the medium. Analyzing the sequence characteristics of stress response genes allows inference of optimal growth pH range and temperature, providing a basis for environmental parameter regulation. For symbiotic or parasitic microorganisms, the presence of specific transporter genes and interspecific signal transduction genes in the genome directly indicates nutritional dependence on hosts or signal requirements, guiding the selection of host strains and ratio regulation in cocultivation systems. Reverse genomics further enhances targeting: By predicting genes encoding surface-specific proteins of target microorganisms, specific antibodies are designed for targeted isolation, and cultivation conditions are optimized based on genomic predictions, forming an integrated "sorting-cultivation" strategy.

The practical value of genome-guided medium design has been verified in the cultivation of various hard-to-culture microorganisms, providing abundant chassis resources for synthetic biology. The TM7 bacteria in the human oral cavity serve as a typical case: Genomic analysis reveals deficiencies in multiple basic nutrient synthesis pathways, including amino acids, nucleotides, and vitamins, as well as encoding multiple host signal recognition-related genes, indicating symbiotic dependence on actinomycetes. Based on this, a medium was designed containing actinomycete hosts, brain-heart infusion, mucin, and other complex nutritional components. By maintaining the interaction balance between hosts and TM7 under anaerobic conditions, stable cultivation of TM7 was successfully achieved [54]. Similarly, SR1 bacteria exhibit defective independent metabolic capabilities due to genetic code recoding. After targeted isolation via surface protein-specific antibodies screened through reverse genomics, vitamin B complex and short-chain fatty acids were supplemented in the medium according to host nutritional requirements predicted by the genome. Ultimately, a host-target bacterium cocultivation system was established to achieve proliferation. These cases confirm that genomic data can precisely identify cultivation bottlenecks, providing a core design basis for the cultivation of symbiosis-dependent microorganisms.

Genome-guided medium design driven by synthetic biology has formed a standardized workflow, encompassing four key links: "analysis—design—screening—optimization." Taking the cultivation of a thiamine-auxotrophic environmental microorganism as an example, the workflow proceeds as follows: First, genome sequencing and pathway annotation identify the absence of the thiE gene in the thiamine synthesis pathway; based on this, 10 μg/L thiamine is added to the basal medium, and initial cultivation conditions (pH 7.0, temperature 30 °C) are set based on stress gene analysis; considering potential symbiotic characteristics, *Bacillus*, commonly found in the environment, is introduced to construct a cocultivation system with different host ratio gradients; a high-throughput cultivation array is established using microfluidic chips to simultaneously monitor cell density changes in each cultivation unit; metabolomic analysis of metabolites in the culture supernatant reveals that accumulated acetic acid inhibits growth; consequently, acetic acid deaminase-producing bacteria are added to the medium to adjust the metabolic environment, forming a closed-loop optimization. This workflow increases the cultivation success rate of the target strain from less than 5% (traditional methods) to 68%, laying a foundation for subsequent synthetic biology gene editing and modification.

Current technological development still faces challenges in adapting to large-scale synthetic biology applications: Incomplete genome annotation leads to deviations in metabolic pathway prediction, and the presence of partially uncharacterized genes may result in missed key nutritional requirements, affecting chassis strain screening efficiency; the complex host dependence mechanisms of symbiotic microorganisms make it difficult to fully decipher interaction networks based solely on genomic data, leading to insufficient stability of cocultivation systems; the high cost of high-throughput cultivation and screening platforms limits the verification efficiency of large-scale genomic data; the complex physicochemical gradients in natural habitats are difficult to simulate with simple media, affecting the recovery of extremophilic microorganisms. These challenges directly impact the efficiency and quality of chassis resource exploration in synthetic biology, driving technological development toward interdisciplinary integration.

In the future, genome-guided medium design will evolve toward "multi-omics integration + intelligent prediction +

customization," forming in-depth synergy with synthetic biology. The application of artificial intelligence technology constitutes a core breakthrough: Machine learning algorithms will analyze massive genomic data and successful cultivation cases to establish prediction models of nutrient composition-growth status, improving the precision of medium formulation design; integrating transcriptomic, metabolomic, and proteomic data will construct more comprehensive metabolic network models, compensating for prediction deficiencies of single genomic data; customized media will be developed for specific synthetic biology application scenarios to improve screening efficiency of target functional strains; public databases integrating genomic data, medium formulations, and cultivation results will be established to promote cross-laboratory collaboration. These developments will advance genome-guided medium design from "data-driven" to "intelligence-driven" [55].

In summary, genome-guided medium design and optimization breaks the blindness of traditional cultivation technologies by precisely deciphering microbial metabolic requirements, providing a key technical approach for synthetic biology to explore hard-to-culture microbial chassis strains. Its core value lies in realizing seamless translation from genome sequence to successful cultivation, converting a large amount of "microbial dark matter" into usable synthetic biology resources. With the integrated application of multi-omics technologies and artificial intelligence, this technology will play a more central supporting role in fields such as metabolic engineering and synthetic microbial community construction, driving the biotech industry toward precision and high efficiency.

Conflicts of Interest

The authors have no conflicts of interest to declare that are relevant to the content of this chapter.

References

1. Pushkarev D, Neff NF, Quake SR (2009) Single-molecule sequencing of an individual human genome. Nat Biotechnol 27(9): 847–850
2. Lee H, Gurtowski J, Yoo S, Marcus S, McCombie WR, Schatz M (2014) Error correction and assembly complexity of single molecule sequencing reads. bioRxiv:006395
3. Salzberg SL, Yorke JA (2005) Beware of mis-assembled genomes. Bioinformatics 21(24):4320–4321
4. Travers KJ, Chin C-S, Rank DR, Eid JS, Turner SW (2010) A flexible and efficient template format for circular consensus sequencing and SNP detection. Nucleic Acids Res 38(15): e159–e159
5. Harris TD, Buzby PR, Babcock H, Beer E, Bowers J, Braslavsky I, Causey M, Colonell J, DiMeo J, Efcavitch JW (2008) Single-molecule DNA sequencing of a viral genome. Science 320(5872):106–109
6. Jain M, Olsen HE, Paten B, Akeson M (2016) The Oxford Nanopore MinION: delivery of nanopore sequencing to the genomics community. Genome Biol 17(1):239
7. Goodwin S, Gurtowski J, Ethe-Sayers S, Deshpande P, Schatz MC, McCombie WR (2015) Oxford Nanopore sequencing, hybrid error correction, and de novo assembly of a

eukaryotic genome. Genome Res 25(11): 1750–1756

8. Schaefer-Ramadan S, Guan Y, Ahmed AA, Aleksic J, Elmagarmid KA, Syed LF, Mohamoud YA, Malek JA (2025) Long-read sequencing to detect full-length protein–protein interactions. Sci Rep 15(1):1–12
9. Shi J, Ye W, Ma D, Yin J, Zhang Z, Wang Y, Qiao Y (2021) Improved whole-genome sequence of Phytophthora capsici generated by long-read sequencing. Mol Plant-Microbe Interact 34(7):866–869
10. Lorenz DA, Sathe S, Einstein JM, Yeo GW (2020) Direct RNA sequencing enables m6A detection in endogenous transcript isoforms at base-specific resolution. RNA 26(1):19–28
11. Hasin Y, Seldin M, Lusis A (2017) Multi-omic approaches to disease. Nat Rev Genet 18:113–120. https://doi.org/10.1038/nrg.2016.135
12. Zhang X et al (2019) Gut microbiome-derived proteins and their relevance to disease. Nat Rev Microbiol 17:345–356. https://doi.org/10.1038/s41579-019-0190-7
13. Lloyd-Price J, Arze C, Ananthakrishnan A et al (2019) Multi-omics of the gut microbiome in inflammatory bowel disease. Nature 569:655–662. https://doi.org/10.1038/s41586-019-1237-9
14. Holmes E, Li J, Marchesi J, Nicholson J (2017) The role of the microbiome in health and disease: an introduction for clinicians. BMJ 357: j831. https://doi.org/10.1136/bmj.j831
15. Zhu W et al (2018) Precision editing of microbiome functions. ISME J 12:2195–2207. https://doi.org/10.1038/s41396-018-0158-3
16. Hughes D, Andersson D (2019) Antibiotic resistance and its cost: antibiotic perturbation of the gut microbiome. Science 364:eaau4673. https://doi.org/10.1126/science.aau4673
17. Wang T, Larson M, Vasan R et al (2011) Metabolomics and cardiovascular risk. N Engl J Med 365:1108–1117. https://doi.org/10.1056/NEJMoa1012
18. Li L, Simopoulos CMA, Mayne J, Ning Z, Zhang X, Hamada M, Butcher J, Serrana JM, Wang L, Cheng K, Qin H, Walker K, Zhang X, Stintzi A, Figeys D (2025) Systematic metaproteomics mapping reveals functional and ecological landscapes of Ex vivo human gut microbiota responses to therapeutic drugs. Nat Commun 16(1):9383. https://doi.org/10.1038/s41467-025-64433-8
19. Wang Z, Klipfell E, Bennett BJ, Koeth R, Levison BS, Dugar B, Feldstein AE, Britt EB, Fu X, Chung YM, Wu Y, Schauer P, Smith JD, Allayee H, Tang WH, DiDonato JA, Lusis AJ, Hazen SL (2011) Gut flora metabolism of phosphatidylcholine promotes cardiovascular disease. Nature 472(7341):57–63. https://doi.org/10.1038/nature09922
20. Gatto L, Lilley K (2012) Proteomics in biology and medicine. J Proteome Res 11:5831–5845. https://doi.org/10.1021/pr3006636
21. Misra B, Langefeld C (2018) Systems biology: integrative analysis of omics data. Nat Rev Genet 19:299–300. https://doi.org/10.1038/s41576-018-0003-0
22. Wang M et al (2016) Sharing and community curation of mass spectrometry data with global natural products social molecular networking. Nat Biotechnol 34:828–837. https://doi.org/10.1038/nbt.3597
23. McKnight D et al (2019) Methods for normalizing microbiome data: compositionality, library size, and beyond. Microbiome 7:90. https://doi.org/10.1186/s40168-019-0709-7
24. DeLong EF, Wickham GS, Pace NR (1989) Phylogenetic stains: ribosomal RNA-based probes for the identification of single cells. Science 243(4896):1360–1363
25. Wang J, Palecek E, Nielsen PE, Rivas G, Cai X, Shiraishi H, Dontha N, Luo D, Farias PA (1996) Peptide nucleic acid probes for sequence-specific DNA biosensors. J Am Chem Soc 118(33):7667–7670
26. Dijk J, Breugelmans P, Philips J, Haest PJ, Smolders E, Springael D (2008) Catalyzed reporter deposition-fluorescent in situ hybridization (CARD-FISH) detection of Dehalococcoides. J Microbiol Methods 73(2):142–147
27. Beliveau BJ, Joyce EF, Apostolopoulos N, Yilmaz F, Fonseka CY, McCole RB, Chang Y, Li JB, Senaratne TN, Williams BR (2012) Versatile design and synthesis platform for visualizing genomes with Oligopaint FISH probes. Proc Natl Acad Sci 109(52):21301–21306
28. Yılmaz LŞ, Noguera DR (2007) Development of thermodynamic models for simulating probe dissociation profiles in fluorescence in situ hybridization. Biotechnol Bioeng 96(2): 349–363
29. Navin N, Grubor V, Hicks J, Leibu E, Thomas E, Troge J, Riggs M, Lundin P, Månér S, Sebat J (2006) PROBER: oligonucleotide FISH probe design software. Bioinformatics 22(19):2437–2438
30. Cambré A, Aertsen A (2020) Bacterial vivisection: how fluorescence-based imaging techniques shed a light on the inner workings of bacteria. Microbiol Mol Biol Rev 84 (4): 10.1128/mmbr. 00008-00020

31. Sheth RU, Cabral V, Chen SP, Wang HH (2016) Manipulating bacterial communities by in situ microbiome engineering. Trends Genet 32(4):189–200
32. Tanna T, Ramachanderan R, Platt RJ (2021) Engineered bacteria to report gut function: technologies and implementation. Curr Opin Microbiol 59:24–33
33. Castaing R, Slodzian G (1962) Microanalyse par émission ionique secondaire. J Microsc 1(1):395–410
34. Reyzer ML, Caprioli RM (2013) Imaging mass spectrometry of proteins and peptides. In: Mass Spectrometry for Drug Discovery and Drug Development, pp 277–302
35. Norris JL, Caprioli RM (2013) Analysis of tissue specimens by matrix-assisted laser desorption/ionization imaging mass spectrometry in biological and clinical research. Chem Rev 113(4):2309–2342
36. Parrot D, Papazian S, Foil D, Tasdemir D (2018) Imaging the unimaginable: desorption electrospray ionization–imaging mass spectrometry (DESI-IMS) in natural product research. Planta Med 84(09/10):584–593
37. Garza KY, Feider CL, Klein DR, Rosenberg JA, Brodbelt JS, Eberlin LS (2018) Desorption electrospray ionization mass spectrometry imaging of proteins directly from biological tissue sections. Anal Chem 90(13):7785–7789
38. Watrous J, Roach P, Heath B, Alexandrov T, Laskin J, Dorrestein PC (2013) Metabolic profiling directly from the Petri dish using nanospray desorption electrospray ionization imaging mass spectrometry. Anal Chem 85(21):10385–10391
39. Anderton CR, Gamble LJ (2016) Secondary ion mass spectrometry imaging of tissues, cells, and microbial systems. Microsc Today 24(2):24–31
40. Lockyer NP, Aoyagi S, Fletcher JS, Gilmore IS, van der Heide PA, Moore KL, Tyler BJ, Weng L-T (2024) Secondary ion mass spectrometry. Nat Rev Method Prim 4(1):32
41. Pozebon D, Scheffler GL, Dressler VL, Nunes MA (2014) Review of the applications of laser ablation inductively coupled plasma mass spectrometry (LA-ICP-MS) to the analysis of biological samples. J Anal At Spectrom 29(12):2204–2228
42. Barger SR, Hoefler BC, Cubillos-Ruiz A, Russell WK, Russell DH, Straight PD (2012) Imaging secondary metabolism of Streptomyces sp. Mg1 during cellular lysis and colony degradation of competing Bacillus subtilis. Antonie Van Leeuwenhoek 102(3):435–445
43. Zaima N, Hayasaka T, Goto-Inoue N, Setou M (2010) Matrix-assisted laser desorption/ionization imaging mass spectrometry. Int J Mol Sci 11(12):5040–5055
44. Boehm DA, Gottlieb PA, Hua SZ (2007) On-chip microfluidic biosensor for bacterial detection and identification. Sensors Actuators B Chem 126(2):508–514. https://doi.org/10.1016/j.snb.2007.03.043
45. Belaïdi FS, Salvagnac L, Souleille SA, Blatché M-C, Bedel-Pereira E, Séguy I, Temple-Boyer-P, Launay J (2020) Accurate physiological monitoring using lab-on-a-chip platform for aquatic micro-organisms growth and optimized culture. Sens Actuators B Chem 321: 128492
46. Mark D, Haeberle S, Roth G, von Stetten F, Zengerle R (2010) Microfluidic lab-on-a-chip platforms: requirements, characteristics and applications. Chem Soc Rev 39:3. https://doi.org/10.1039/b820557b
47. Kaeberlein T, Lewis K, Epstein SS (2002) Isolating "uncultivable" microorganisms in pure culture in a simulated natural environment. Science 296(5570):1127–1129. https://doi.org/10.1126/science.1070633
48. Berdy B, Spoering AL, Ling LL, Epstein SS (2017) In situ cultivation of previously uncultivable microorganisms using the ichip. Nat Protoc 12(10):2232–2242. https://doi.org/10.1038/nprot.2017.074
49. Zengler K, Zaramela LS (2018) The social network of microorganisms — how auxotrophies shape complex communities. Nat Rev Microbiol 16(6):383–390. https://doi.org/10.1038/s41579-018-0004-5
50. Peng H, Darlington AP, South EJ, Chen H-H, Jiang W, Ledesma-Amaro R (2024) A molecular toolkit of cross-feeding strains for engineering synthetic yeast communities. Nat Microbiol 9(3):848–863
51. Fu AY, Spence C, Scherer A, Arnold FH, Quake SR (1999) A microfabricated fluorescence-activated cell sorter. Nat Biotechnol 17(11):1109–1111
52. Müller S, Nebe-von-Caron G (2010) Functional single-cell analyses: flow cytometry and cell sorting of microbial populations and communities. FEMS Microbiol Rev 34(4): 554–587
53. Wang Y, Xue Y, Hong X, Wang H, Zhou C, Wang W, Sheng J, Liang P, Li H, Xu Q (2025) Functionalized microwell laser sorting method for single-cell viable microbial detection. Sens Actuators B Chem 427:137202
54. Hatzenpichler R, Krukenberg V, Spietz RL, Jay ZJ (2020) Next-generation physiology

approaches to study microbiome function at single cell level. Nat Rev Microbiol 18(4): 241–256

55. Cross KL, Campbell JH, Balachandran M, Campbell AG, Cooper CJ, Griffen A, Heaton M, Joshi S, Klingeman D, Leys E (2019) Targeted isolation and cultivation of uncultivated bacteria by reverse genomics. Nat Biotechnol 37(11):1314–1321

Chapter 17

RNA Large Language Models in Virology

Weiwei Shen, Liucun Zhu, Yixue Li, and Tao Huang

Abstract

The viral gene sequence is regarded as a kind of special text, which constitutes a core innovation in the application of large language models to virus research. This idea breaks free from the shackles of the traditional analysis framework and opens up a new way of computing to understand the evolution of the virus. Using this method, researchers can quickly and accurately locate the key mutation sites of the virus. This helps to detect emerging variant strains in time and assess their potential impact on transmissibility and pathogenicity. Especially importantly, this technology can also reconstruct the evolutionary path of the virus and realize real-time tracking of the mutation process. This provides a powerful tool for a systematic understanding of the adaptive evolution of the virus. These profound insights based on the big language model can be used as a theoretical basis for the development of targeted vaccines and drugs. Although the relevant computing technology is in the development stage, it has shown great potential to promote the transformation of virological research to a predictable direction.

Key words Virology, Large language models (LLMs), Artificial intelligence, Genomics

1 Introduction

The genetic sequence of the virus is the material basis and information source of all its biological functions, which fundamentally determines the infectious, pathogenic, and other key characteristics of the virus. This code contains instructions for copying, transmitting, and infecting host cells, as well as long-term evolutionary adaptation mechanisms to ensure their survival [1]. The core challenge facing virology is to connect some key points, and we need to decipher the complex code of the viral genome and clearly understand how it directly affects the key characteristics such as the transmission mode, pathogenicity, and evolutionary path of the virus [2]. However, the three innate characteristics of viruses make this work extremely difficult. They mutate rapidly, are complex in structure, and have a variety of functions [3].

Over the years, the standard toolkit for the study of RNA viruses has always revolved around several key technologies:

Tao Huang (ed.), *Microbial 'Omics: Methods and Protocols*, Methods in Molecular Biology, vol. 3033,
https://doi.org/10.1007/978-1-0716-5264-0_17,

sequence comparison [4], phylogenetic tree construction [5], and machine learning models based on artificially defined characteristics [6]. Although these methods are of great practical value, they have a common flaw. These methods have never been able to effectively capture the long-range interaction in the gene sequence, which limits the ability of the model. At the same time, the neglect of the universal law also affects the breadth of its inference.

The rapid growth of biological data sets is intertwined with major breakthroughs in deep learning, which is making a key transformation of research methods in this field. Among them, LLMs has achieved the most transformative breakthrough, which is redefining the core paradigm of virology research [7].

The breakthrough of LLMs comes from the redefinition of genetic code as a computational language unit. Through self-supervised pretraining, the model can learn the statistical laws within the sequence. Furthermore, it can gradually reveal the deeper evolutionary logic behind the structure and function of the virus [8]. An important advance of the self-attention mechanism is that it can model long-range dependencies in the sequence. Accurate capture of such nonlinear interactions is an indispensable prerequisite for building a reliable RNA model [9], and because of this, the model can exhibit a sequence understanding consistent with biological intuition. Thanks to this basic cognition, the model can perform reliably in some tasks such as predicting structures, analyzing functions, and sorting out evolutionary paths. These abilities are a natural extension of the cognition without additional complicated steps. In practical applications, this shift makes the field no longer overly dependent on those large data sets that need to be carefully sorted out. At the same time, it also provides researchers with a fundamentally new computing resource [10].

2 Technical Principles and Development Overview of Large Language Models

2.1 Data Sources and Preprocessing

When constructing the LLMs, the first task is to obtain data that is both highly reliable and sufficient. Such data researchers can be obtained from authoritative public databases, which store large-scale multimodal sequence resources [11]. NCBI Virus provides data covering hundreds of thousands of viral genomes worldwide [12]. It includes host information and sampling date and can be precisely filtered based on criteria such as classification, host, and integrity [13]. GISAID played an immeasurable and central role in the COVID-19 pandemic, and the scope of data it collected rapidly expanded from the initial influenza field to cover the entire virology community [14, 15]. As of February 2023, the platform had collected over 15 million genome sequences, becoming a core platform for real-time sharing of viral genomes. These databases integrate various forms of biological data, including nucleotide sequences, protein modification, epigenetic markers, and host-pathogen interaction records. With its integrity and high-quality annotations, this resource is not only the basic material for model pretraining but also provides a key biological basis for us to interpret the output results of the model.

The training effect of the model depends to a large extent on strict data quality control. To this end, we need to carry out systematic preprocessing of the original data, such as noise reduction, standardization, and data verification, to eliminate the interference caused by technical differences and ensure that subsequent analysis can draw effective conclusions [16]. The high-efficiency tool Fastp is used to process the link of FASTA files. With its parallel processing ability, the tool can complete joint trimming, mass filtering, and base identification in one stop, which improves the efficiency of data standardization significantly [17]. In the face of the computing bottlenecks brought about by large-scale sequence data sets, these tools efficiently complete data quality control by calling multi-threaded resources, providing key technical support for solving such problems. At the same time, its de-weighting step uses CD-HIT technology to cluster highly similar sequences and only retains the most representative sequences in each cluster. This effectively eliminates redundant data and saves a lot of computing resources for downstream analysis [18]. Another tool, MMseqs2, also stands out for its highly efficient clustering performance. Compared to CD-HIT, it processes ultra-large datasets extremely quickly and with high memory efficiency, making it more suitable for datasets containing tens of millions of viral genomes. As data volumes grow exponentially, MMseqs2 often outperforms CD-HIT in processing speed and memory consumption when handling ultra-large datasets [19]. Ultimately, after completing the entire preprocessing workflow, a high-quality, low-redundancy, and uniformly formatted virus sequence dataset will be obtained. This will lay a solid data foundation for subsequent pretraining of large language models.

2.2 Model Architecture and Training Strategy

Most mainstream large-scale language models are built on the Transformer architecture [20]. The core mechanism of the standard Transformer model is its self-attention mechanism, primarily composed of stacked encoders and decoders. As shown in Fig. 1, the input nucleotide sequence is first mapped to an embedding vector via one-hot encoding [21] or word2vec [22] representation methods and then combined with positional encoding to capture sequence order. To preserve the order of bases in the sequence, we then incorporate positional encoding information into these embeddings. The model architecture employs multiple encoders stacked together, with a multi-head self-attention mechanism at its core. What's particularly clever is how this setup lets every nucleotide in a sequence directly discover all the others, allowing it to

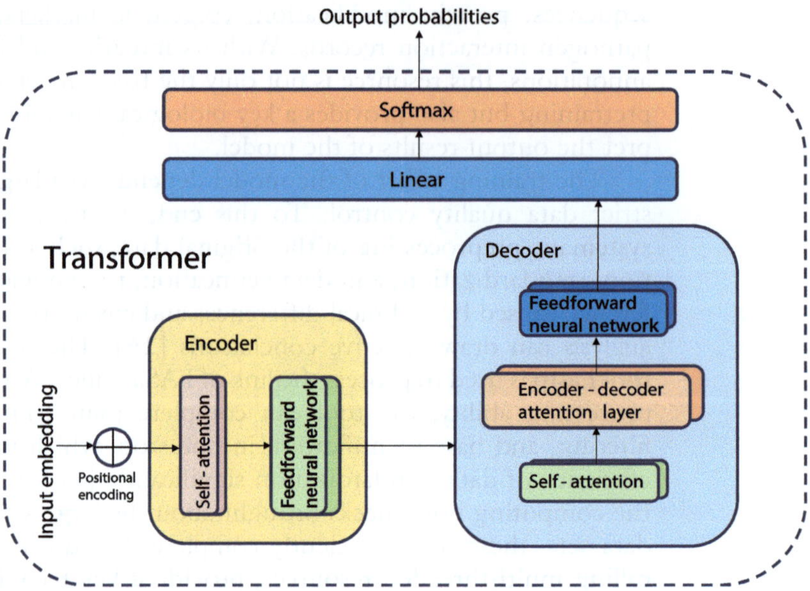

Fig. 1 This diagram illustrates the core architecture of the Transformer model. The model consists of a stacked encoder and decoder. Input sequences undergo embedding and positional encoding before the encoder computes context-aware representations through self-attention mechanisms. The decoder then utilizes the encoder's output and the partially generated sequence to compute the probability distribution of the next token via masked self-attention layers and encoder-decoder attention layers

capture rich contextual relationships across different levels. Residual connections help maintain stability during training by passing each layer's output directly to the next. This mechanism is like adding a built-in stabilizer to the deep neural network. It can make the training process more stable and accelerate the convergence speed. More importantly, it can dynamically calculate the correlation weight between nucleotide pairs and accurately judge the intensity of their interaction. This feature allows it to capture the long-distance dependence in the sequence, which is particularly useful whether it is to study the RNA folding law, locate the remote regulatory element, or mine the evolutionary signals retained in the molecular spectrum [23].

In the pretraining stage, self-supervising language modeling is very important. Its function is to let the model learn the rules from the sequence by itself. Specifically, in the commonly used mask language model method, when pretraining, some bases in the input sequence will be covered, and then the model will try to predict the covered part according to the contextual information on both sides. In this way, the model can slowly learn how to restore the complete sequence, so as to master the internal structure of the sequence [24].

The model components we mentioned above enable the context to be coded in both directions, thus integrating more comprehensive sequence information and forming a holistic characterization beyond local linear dependencies. Because the operating mechanism of biological sequences is different from that of human language, researchers often combine multilayer models with the next token prediction target. In this way, the model can also learn the follow-up content of the prediction sequence, so as to deepen the grasp of the natural organization and generation laws of biological sequences. In fact, the latest study found that when the long-sequence Transformer model is added to an output layer focusing on the next token prediction, it can significantly improve its processing efficiency in actual biological tasks [25]. This two-stage training method allows us to solve the original disease. We can first use large public data sets to build a solid foundation and then quickly adapt the model through targeted adjustment to solve specific problems in virology. The final model will gradually evolve from a broad understanding of the sequence to an accurate special tool that can make reliable predictions about the behavior of the virus.

2.3 Model Evaluation

After the model training is completed, the verification of the model has become the core problem in the new stage, and we will make it subject to rigorous real-world testing. At this stage, the question of this study shifted to how to strictly verify whether the model really insights the inherent laws of biology or only fits the surface statistical model in the training data [26]. In order to effectively evaluate the real ability of the model, we put it in a benchmark test of zero samples or few samples. By observing the performance of the model in untrained tasks such as protein function classification and RNA structure prediction, its generalization performance and the validity of the knowledge learned can be objectively measured. Simultaneously, we must conduct in-depth analysis of the model's internal representations to investigate whether its embedding space clusters functionally similar viral proteins or evolutionarily closely related sequences together in geometric space, thereby forming a biologically meaningful topological structure. This semantically rich embedding space serves as the solid foundation enabling the model to support subsequent diverse application studies, with its quality directly determining the model's lower bound and potential.

3 Applications of Large Language Models in Virology Research

LLMs have demonstrated breakthrough application value in multiple critical areas of virology. In Table 1, RNA-FM, through pretraining on tens of millions of unlabeled RNA sequences, learns representations that effectively capture secondary structure

Table 1
Detailed information on LLMs

Models	Parameter scale	Architecture	Primary tasks
RNABERT	−0.5 M	BERT	Prediction of secondary structure, functional classification
RNA-FM	−100 M	Transformer	Structure prediction, functional annotation
RNA-MSM	−100 M	Transformer	Secondary structure prediction, evolutionary information extraction
RNAErnie	−100 M	Motif-aware transformer	Prediction of secondary structure, functional classification
Nucleotide Transformer	500 M – 2.5 B	Transformer	Regulatory function prediction, species classification, prediction of viral hosts and pathogenicity

Abbreviations: This table presents the parameter scale, network architecture, training data volume, and corresponding core tasks for major RNA large language models as of 2025

information and functional features, enabling functional inference for unknown viral RNAs [27]. It learns energy functions through isothermal graph neural networks and enables multistate modeling of RNA's dynamic conformational ensembles. In the field of RNA three-dimensional structure prediction, the RNA-MSM model represents the latest major breakthrough [28]. Secondly, models specifically designed for functional annotation can classify viral protein families, predict RNA-binding protein interactions, and estimate mRNA vaccine expression levels under zero-shot or few-shot conditions. Examples include models such as RNABERT [29], BERT-RBP [30], and CodonBERT [31], which provide efficient computational tools for rapid functional annotation of novel viruses and screening of vaccine candidate antigens [8]. In recent studies, the Nucleotide Transformer models systematically explore the role of model scale and data breadth, with parameters ranging from 500 million to 2.5 billion [32]. What we aim to achieve is to learn from the genomic data of various organisms and develop a universal "language model" for gene sequences. By doing so, we can systematically examine and evaluate a specific virus sequence within the broader genetic context of the entire living world.

4 Discussion

Although the scale of public databases has reached unprecedented levels, RNA sequencing data that truly combines both high quality and comprehensive functional annotations remains scarce and difficult to find [33].

To improve the interpretability of the model, it is necessary to break through the existing technical limitations. Although the Transformer model has excellent predictive accuracy, how to analyze the biological relationship between its attention weight and embedded vector characterization has yet to be solved. The lack of understanding of this mechanism makes it difficult to trace the model reasoning process, and it is urgent to develop new explanatory algorithms to establish a credible biomedical reasoning chain [34]. Using attention heat maps for visual analysis is an effective way to enhance the interpretability of language models in the field of biomedicine. The value of this method is that it can accurately locate the key sequences that affect the decision-making of the model, so that researchers can intuitively understand its reasoning logic, so as to verify whether the judgment of the model follows known biological laws rather than the whole process being completely in a black box state [35]. With the help of advanced attribution analysis technologies such as Grad-CAM, we can explore the decision-making logic inside the model more deeply [36]. Through this technology, we can locate those embedded representations that make key contributions to model decision-making, so as to understand their internal working mechanisms deeply. In addition to this attribution analysis, the robustness of the model in the face of specific sequence disturbances can also be evaluated by constructing confrontational samples. By locating potential functional sites and providing explanations for model decision-making accurately, the method can establish a testable relationship between computational prediction and experimental verification.

A crucial evaluation criterion is the adaptability of the model to unknown RNA viruses. The system must be able to make effective inferences under the condition of a small number of samples or even zero samples to ensure a rapid and reliable response to emerging pathogen threats [37]. The ideal pretraining model should be embedded in the core principle of virus evolution, so that it can really have practical value when encountering new pathogens. They should be able to quickly give reliable sequence analysis results, predict potential adaptive mutations, and outline evolutionary trajectories. This forward-looking analysis ability will be directly transformed into strategic advantages in key stages such as vaccine research and development and drug development [38].

Acknowledgments

This work was supported by Major Project of Guangzhou National Laboratory (GZNL2024A01003), National Key R&D Program of China (2022YFF1203202), and Strategic Priority Research Program of Chinese Academy of Sciences (XDB38050200, XDB38040202, XDA26040304).

Conflicts of Interest The authors have no conflicts of interest to declare that are relevant to the content of this chapter.

References

1. Retel C, Märkle H, Becks L, Feulner PGD (2019) Ecological and evolutionary processes shaping viral genetic diversity. Viruses 11(3): 220
2. Belshaw R, Gardner A, Rambaut A, Pybus OG (2008) Pacing a small cage: mutation and RNA viruses. Trends Ecol Evol 23(4):188–193. https://doi.org/10.1016/j.tree.2007.11.010
3. Zwart MP, Kupczok A, Iranzo J (2023) Editorial: predicting virus evolution: from genome evolution to epidemiological trends. Front Virol 3. https://doi.org/10.3389/fviro.2023.1215709
4. Lewitter F, Notredame C (2007) Recent evolutions of multiple sequence alignment algorithms. PLoS Comput Biol 3(8). https://doi.org/10.1371/journal.pcbi.0030123
5. Holmes EC, Grenfell BT (2009) Discovering the Phylodynamics of RNA viruses. PLoS Comput Biol 5(10):e1000505. https://doi.org/10.1371/journal.pcbi.1000505
6. Shen W, Li Y, Zhu L, Huang T (2025) Predict SARS-CoV-2 genome interactions based on RNA language models. Biosafety Health 7(5): 333–343. https://doi.org/10.1016/j.bsheal.2025.09.006
7. Sarumi OA, Heider D (2024) Large language models and their applications in bioinformatics. Comput Struct Biotechnol J 23:3498–3505. https://doi.org/10.1016/j.csbj.2024.09.031
8. Rives A, Meier J, Sercu T, Goyal S, Lin Z, Liu J, Guo D, Ott M, Zitnick CL, Ma J, Fergus R (2021) Biological structure and function emerge from scaling unsupervised learning to 250 million protein sequences. Proc Natl Acad Sci U S A 118(15):e2016239118. https://doi.org/10.1073/pnas.2016239118
9. Gong T, Bu D (2024) Language models enable zero-shot prediction of RNA secondary structure including pseudoknots. bioRxiv:2024.2001.2027.577533. https://doi.org/10.1101/2024.01.27.577533
10. Greener JG, Kandathil SM, Moffat L, Jones DT (2022) A guide to machine learning for biologists. Nat Rev Mol Cell Biol 23(1): 40–55. https://doi.org/10.1038/s41580-021-00407-0
11. Kodama Y, Shumway M, Leinonen R (2012) The sequence read archive: explosive growth of sequencing data. Nucleic Acids Res 40(Database issue):D54–D56. https://doi.org/10.1093/nar/gkr854
12. Federhen S (2012) The NCBI taxonomy database. Nucleic Acids Res 40(Database issue): D136–D143. https://doi.org/10.1093/nar/gkr1178
13. Zhao M, Chen J, Wang Q, Lu Z, Jia Z (2022) A landscape analysis on virus: based on NCBI database. China CDC Wkly 4(7):120–125. https://doi.org/10.46234/ccdcw2022.019
14. Sheehan N, Leonelli S, Botta F (2023) From collection to analysis: a comparison of GISAID and the Covid-19 data portal. https://doi.org/10.1101/2023.05.13.540634
15. Shu Y, McCauley J (2017) GISAID: global initiative on sharing all influenza data—from vision to reality. Euro Surveill 22(13). https://doi.org/10.2807/1560-7917.Es.2017.22.13.30494
16. Raza K (2024) Preprocessing and quality control. In: Raza K (ed) Machine learning in single-cell RNA-seq data analysis. Springer Nature Singapore, Singapore, pp 17–30. https://doi.org/10.1007/978-981-97-6703-8_2
17. Chen S, Zhou Y, Chen Y, Gu J (2018) Fastp: an ultra-fast all-in-one FASTQ preprocessor. Bioinformatics 34(17):i884–i890. https://doi.org/10.1093/bioinformatics/bty560
18. Fu L, Niu B, Zhu Z, Wu S, Li W (2012) CD-HIT: accelerated for clustering the next-generation sequencing data. Bioinformatics 28(23):3150–3152. https://doi.org/10.1093/bioinformatics/bts565
19. Steinegger M, Söding J (2017) MMseqs2 enables sensitive protein sequence searching for the analysis of massive data sets. Nat Biotechnol 35(11):1026–1028. https://doi.org/10.1038/nbt.3988
20. Vaswani A, Shazeer N, Parmar N, Uszkoreit J, Jones L, Gomez AN, Kaiser Ł, Polosukhin I (2017) Attention is all you need. Paper presented at the Proceedings of the 31st International Conference on Neural Information Processing Systems, Long Beach, California, USA
21. Alipanahi B, Delong A, Weirauch MT, Frey BJ (2015) Predicting the sequence specificities of DNA- and RNA-binding proteins by deep

learning. Nat Biotechnol 33(8):831–838. https://doi.org/10.1038/nbt.3300

22. Asgari E, Mofrad MR (2015) Continuous distributed representation of biological sequences for deep proteomics and genomics. PLoS One 10(11):e0141287. https://doi.org/10.1371/journal.pone.0141287
23. Xu Y, Wei H, Lin M, Deng Y, Sheng K, Zhang M, Tang F, Dong W, Huang F, Xu C (2022) Transformers in computational visual media: a survey. Comput Vis Media 8(1): 33–62. https://doi.org/10.1007/s41095-021-0247-3
24. Sajun AR, Zualkernan I, Sankalpa D (2024) A historical survey of advances in transformer architectures. Appl Sci 14(10):4316
25. Han X, Zhang Z, Ding N, Gu Y, Liu X, Huo Y, Qiu J, Yao Y, Zhang A, Zhang L, Han W, Huang M, Jin Q, Lan Y, Liu Y, Liu Z, Lu Z, Qiu X, Song R, Tang J, Wen J-R, Yuan J, Zhao WX, Zhu J (2021) Pre-trained models: past, present and future. AI Open 2:225–250. https://doi.org/10.1016/j.aiopen.2021.08.002
26. AlQuraishi M (2021) Machine learning in protein structure prediction. Curr Opin Chem Biol 65:1–8. https://doi.org/10.1016/j.cbpa.2021.04.005
27. Chen J, Hu Z, Sun S, Tan Q, Wang Y, Yu Q, Zong L, Hong L, Xiao J, King I, Li Y (2022) Interpretable RNA foundation model from unannotated data for highly accurate RNA structure and function predictions. bioRxiv
28. Zhang Y, Lang M, Jiang J, Gao Z, Xu F, Litfin T, Chen K, Singh J, Huang X, Song G, Tian Y, Zhan J, Chen J, Zhou Y (2023) Multiple sequence alignment-based RNA language model and its application to structural inference. Nucleic Acids Res 52(1):e3–e3. https://doi.org/10.1093/nar/gkad1031
29. Akiyama M, Sakakibara Y (2022) Informative RNA base embedding for RNA structural alignment and clustering by deep representation learning. NAR Genomics and Bioinformatics 4(1). https://doi.org/10.1093/nargab/lqac012
30. Yamada K, Hamada M (2022) Prediction of RNA-protein interactions using a nucleotide language model. Bioinform Adv 2(1): vbac023. https://doi.org/10.1093/bioadv/vbac023
31. Li S, Moayedpour S, Li R, Bailey M, Riahi S, Miladi M, Miner J, Zheng D, Wang J, Balsubramani A, Tran K, Zacharia M, Wu M, Gu X, Clinton R, Asquith C, Skalesk J, Boeglin L, Chivukula S, Dias A, Montoya FU, Agarwal V, Bar-Joseph Z, Jager S (2023) CodonBERT: Large Language Models for mRNA design and optimization. bioRxiv:2023.2009.2009.556981. https://doi.org/10.1101/2023.09.09.556981
32. Dalla-Torre H, Gonzalez L, Mendoza-Revilla J, Lopez Carranza N, Grzywaczewski AH, Oteri F, Dallago C, Trop E, de Almeida BP, Sirelkhatim H, Richard G, Skwark M, Beguir K, Lopez M, Pierrot T (2025) Nucleotide transformer: building and evaluating robust foundation models for human genomics. Nat Methods 22(2):287–297. https://doi.org/10.1038/s41592-024-02523-z
33. Bileschi ML, Belanger D, Bryant DH, Sanderson T, Carter B, Sculley D, Bateman A, DePristo MA, Colwell LJ (2022) Using deep learning to annotate the protein universe. Nat Biotechnol 40(6):932–937. https://doi.org/10.1038/s41587-021-01179-w
34. Yue T, Wang Y, Zhang L, Gu C, Xue H, Wang W, Lyu Q, Dun Y (2023) Deep learning for genomics: from early neural nets to modern large language models. Int J Mol Sci 24(21): 15858
35. Rao R, Meier J, Sercu T, Ovchinnikov S, Rives A (2020) Transformer protein language models are unsupervised structure learners. bioRxiv:2020.2012.2015.422761. https://doi.org/10.1101/2020.12.15.422761
36. Selvaraju RR, Cogswell M, Das A, Vedantam R, Parikh D, Batra D Grad-CAM: Visual Explanations from Deep Networks via Gradient-Based Localization. In: 2017 IEEE International Conference on Computer Vision (ICCV), 22–29 Oct. 2017. pp 618–626. https://doi.org/10.1109/ICCV.2017.74
37. Hie B, Zhong ED, Berger B, Bryson B (2021) Learning the language of viral evolution and escape. Science 371(6526):284–288. https://doi.org/10.1126/science.abd7331
38. Zvyagin M, Brace A, Hippe K, Deng Y, Zhang B, Bohorquez CO, Clyde A, Kale B, Perez-Rivera D, Ma H, Mann CM, Irvin M, Pauloski JG, Ward L, Hayot-Sasson V, Emani M, Foreman S, Xie Z, Lin D, Shukla M, Nie W, Romero J, Dallago C, Vahdat A, Xiao C, Gibbs T, Foster I, Davis JJ, Papka ME, Brettin T, Stevens R, Anandkumar A, Vishwanath V, Ramanathan A (2022) GenSLMs: genome-scale language models reveal SARS-CoV-2 evolutionary dynamics. bioRxiv. https://doi.org/10.1101/2022.10.10.511571

Chapter 18

Protein Language Models in Virology: A Review of Advances and Applications

Lingxin Luo, Yixue Li, and Tao Huang

Abstract

Protein language models (PLMs) enable functional analysis of divergent viral sequences without homology alignment. This review covers PLM architectures from sequence encoders through structure-aware architectures to generative models and assesses their application to orphan protein structure resolution, virosphere-wide functional classification, host factor identification, and therapeutic antibody optimization. Finally, the limitations of the current model in terms of interpretability and insufficient data representation are discussed while exploring future trends toward multimodal integration and the "dry-wet" experimental loop to accelerate the adoption of artificial intelligence in precision virology.

Key words Protein language models, Virology, Viral dark matter, Deep learning, Virus-host interactions, Antiviral strategy

1 Introduction

Viruses, especially RNA viruses, as a unique form of life, strictly rely on host cells for replication (i.e., obligate parasitism). With their high mutation rate and cross-species transmission capacity, viruses continue to pose a serious threat to global public health [1–3]. Throughout the viral life cycle, proteins play a central role, regulating key processes such as host invasion, genetic material replication, and immune escape [4]. In view of the high dependence of the virus on the host cellular machinery, which increases the difficulty of screening selective drug targets, analyzing the structure and function of viral proteins has become the cornerstone of breaking through the bottleneck of drug design and accelerating the research and development of vaccines and diagnostic tools.

However, metagenomics has revealed a large number of highly divergent and unknown viral sequences (known as "viral dark matter"), which are difficult to annotate using traditional homology alignment due to the lack of homologous sequences [5]. In

Tao Huang (ed.), *Microbial 'Omics: Methods and Protocols*, Methods in Molecular Biology, vol. 3033,
https://doi.org/10.1007/978-1-0716-5264-0_18,

addition, limited by low throughput, high cost, and strict biosafety requirements, traditional experimental methods are difficult to keep up with the explosive growth of sequence data, leaving the functions of a large number of viral proteins uncharacterized [6].

To address these challenges, computational biology methods have become a crucial solution. Early machine learning algorithms based on feature engineering improved the efficiency of viral protein function prediction to a certain extent [7]. However, these methods usually require manual design and extraction of sequence or structural features related to specific functions. This process is not only time-consuming and labor-intensive but also the features obtained often lack universality, thus limiting the generalization ability and prediction accuracy of the model when applied to highly diverse or novel viral proteins.

Inspired by natural language processing (NLP), protein language models (PLMs) offer a powerful approach for overcoming these obstacles. Treating amino acid sequences as a biological language, PLMs use self-supervised training on massive sequence datasets to capture deep evolutionary and structural information. This capability enables accurate protein function prediction, even in zero-shot or few-shot settings [8, 9]. This alignment-free characteristic enables PLMs to effectively extract intrinsic patterns from unannotated data [10]. Therefore, PLMs show great potential in analyzing highly divergent sequences such as "viral dark matter." The protein embeddings generated by these models have been successfully applied in many fields, including viral protein function annotation, structure prediction, mutational effect analysis, and rational design [11, 12].

2 Protein Language Models: From Sequence to Multimodal Information

2.1 Evolution of Self-Supervised Learning Paradigms: From Sequence Representation to Multimodal Generation

Protein language models (PLMs) are based on self-supervised learning. Their core concept is to analogize amino acid sequences to natural language, so that deep neural networks can acquire the "biological grammar" of protein sequences through pretraining on a large number of unlabeled data sets. The process mainly adopts a masked language modeling (MLM) strategy similar to BERT, which aims to reconstruct the masked residues based on contextual information, so as to enable the model to capture complex physicochemical dependencies and evolutionary constraints between amino acids [8, 13, 14]. With the evolution of deep learning architectures, this field has expanded from single sequence modeling to covering a variety of technical paths, including structural representation, sequence translation, and generative design.

In the field of representation learning, Transformer encoder-only architectures represented by the ESM series have established the paradigm of "sequence is structure." Rives et al. (2021)

demonstrated that when training ESM-1b on the UniRef50 database, its deep attention weight matrices, when examined through sparsification analysis, exhibited significant topological consistency with protein three-dimensional contact maps [8]. On this basis, Lin et al. (2023) launched an ESM-2 model with up to 15 billion parameters and established the Scaling Law that correlates the performance of language models with biological accuracy. In terms of architecture, ESM-2 introduced Rotary Position Embeddings (RoPE) to replace the traditional absolute positional encodings, which significantly enhances the model's ability to generalize and extrapolate to long sequences. The model also uses a Pre-LayerNorm structure to improve numerical stability, enabling atomic-level structure prediction without needing multiple sequence alignment (MSA) [15].

Purely sequence-based models like ESM-2 implicitly capture some structural features, but using explicit 3D geometric information can improve representation quality, which is critical for viral proteins, because they often maintain their structure despite high sequence variability. Su et al. (2023) introduced Saprot to address this [16]. Unlike models that only rely on amino acids, Saprot employs the Foldseek algorithm to convert 3D structures into 3Di tokens, constructing a hybrid Structure-Aware Vocabulary [16]. This approach allows the model to learn sequence and structure distributions simultaneously. Experiments showed that Saprot outperforms ESM-2 in zero-shot mutation prediction [16]. Therefore, for predicting viral antigenic drift and functional mutations, it can serve as a powerful tool.

However, Saprot requires three-dimensional structural input (or 3Di sequences derived from Foldseek) during inference, which limits efficiency when processing massive viral sequencing datasets or new variants lacking structural data. Ouyang-Zhang et al. (2025) proposed the Implicit Structure Model (ISM) and introduced Structure Distillation paradigm to address this issue [17]. ISM adopts a microenvironment atomic autoencoder to discretize local 3D atomic coordinates into structural tokens. During the training process, the model uses the structure-tuning to enable the ESM-2-based sequence encoder to predict these tokens, effectively distilling the 3D structural knowledge directly into the parameters of the sequence model. Therefore, ISM can generate deep representations rich in structural information solely from single-sequence inputs, thus eliminating the dependence on explicit structural data in the inference process [17].

In order to further solve the intrinsic bottleneck of completely compressing the huge evolutionary context into the model parameters, Jain et al. (2025) proposed Profluent-E1, a class of retrieval-augmented protein encoder models [18]. Unlike traditional architectures, E1 explicitly incorporates homologous sequences as context during the inference phase. By coordinating intra-sequence attention and block causal multi-sequence attention, the model directly utilizes retrieved evolutionary information without fine-tuning [18]. E1 is trained on the massive Profluent Protein Atlas (containing 4 trillion tokens) and outperforms ESM-2 and comparable baselines in terms of zero-shot fitness prediction and unsupervised contact prediction. These findings prove that the explicit integration of evolutionary context is an effective strategy to enhance the representation of rare or highly divergent protein families [18].

Distinct from encoder-only architectures, which focus on sequence feature extraction, the progress of generative models aims to solve the challenges of de novo design and multimodal unification. Elnaggar et al. (2021) developed the ProtTrans series and built the ProtT5 model using the Encoder-Decoder T5 architecture. By incorporating the Big Fantastic Database (BFD) to cover the highly diverse metagenomic sequence space, ProtTrans maintains unique advantages in sequence translation tasks [19]. Meanwhile, ESM3, released by Hayes et al. (2024), has further achieved native multimodal unification. ESM3 is a generative masked language model built on a bidirectional Transformer architecture. It unifies three modalities of sequence, structure, and function through discretized tokenization [20]. Experimental results indicate that by incorporating a "Chain of Thought" reasoning mechanism, ESM3 has the potential to generate across vast evolutionary timescales and can create new proteins with extremely low homology with known natural sequences but still retain complete biological activity [20].

To solve the inherent conflict between high-quality representation and generation ability, Hallee et al. (2025) recently proposed the Diffusion Sequence Model (DSM), establishing a new paradigm based on Masked Diffusion [21]. Standard MLMs (such as the ESM series) are typically limited to denoising only 15–20% of tokens during training, which restricts their ability to generate coherent full-length sequences. In contrast, although autoregressive models excel in generation, their representation quality is often compromised by the lack of bidirectional context. Using the framework of Large Language Diffusion Models (LLaDA), DSM implements a reverse diffusion process in the ESM-2 architecture, progressing from complete noise (high masking rates) to authentic sequences [21, 22]. This mechanism allows the model to perform robust denoising reconstruction even at very high corruption rates, so it can acquire global generative capabilities while keeping the strong structural representation power of the ESM series. Experiments show that DSM not only generates natural-like sequences with natural protein statistical characteristics and functions, but its fine-tuned variant (DSM_{ppi}) also shows affinity prediction performance better than known natural binders in binder design tasks. This provides a computational framework integrating deep representation learning and controllable sequence generation for the de novo design of viral neutralizing antibodies and blockers [21].

In summary, the development of this field has evolved from the extensive coverage and translation ability of the metagenomic sequence space of the ProtTrans series to the depth of single-sequence structure representation established by ESM-1b/ESM-2. It then developed into Profluent-E1, which introduced explicit evolutionary context through retrieval augmentation, ESM3 which realized multimodal unified generation of sequence-structure-function, and the global diffusion denoising paradigm established by DSM. Collectively, these protein language models are driving the field toward higher accuracy, enhanced controllable generation, and deeper biological understanding. To visually clarify the differences between these state-of-the-art models in terms of core architecture, training data and key advantages, Table 1 makes a systematic comparison of the representative models mentioned above.

2.2 Multimodal Representation Enhancement: Prediction Frameworks

2.2.1 Integrating Orthogonal Biological Features

Although the foundation models described in Subheading 2.1 have made a major breakthrough in general representation, biological systems are inherently multidimensional and dynamic. The function of a protein is not only determined by the amino acid sequence but also by the geometric features of the three-dimensional conformation, the distribution of the surface electrostatic potential, and its topological positioning in the protein-protein interaction (PPI) network. In the context of virology, a single sequence modality is often unable to fully capture the complex host adaptability and immune evasion mechanisms unique to viral infection. Therefore, extending feature extraction from one-dimensional sequences to a multimodal fusion framework that includes structure, surface physicochemical properties, functional text descriptions, and interaction networks uses the complementary advantages of these data types, improving the predictive performance of the model [23, 24].

In developing TransFactor, a SARS-CoV-2 host factor prediction model, An et al. (2025) pointed out that although fine-tuned ESM-2 can effectively identify key host proteins, models relying only on sequence information show limitations when identifying proteins with low homology to the training set, which indicates integrating orthogonal information, such as 3D structure, post-translational modifications, and interaction networks, is important to break through the current limit of predictive accuracy [25].

In order to address the limitations of using only sequence data, researchers began to explicitly integrate physicochemical properties and structural features into the language model. Wang et al. (2025) proposed the DeepHVI framework, which features a multimodal fusion module that jointly encodes the sequence features extracted by ESM-2 or LucaProt and the physicochemical properties of

Table 1
Characteristics of mainstream protein language models

Model	Architecture and paradigm	Training data	Key advantages
ProtT5 [19]	Encoder-decoder (T5) and Seq2Seq generation	BFD + UniRef	Broad-spectrum coverage and generative capability
ESM-1b/ ESM-2 [8, 15]	Transformer encoder and pure sequence representation	UniRef 50/90	Structure emergence enables alignment-free, atomic-level structure prediction directly from single sequences via scaling
Saprot [16]	Structure-aware encoder and explicit structure tokenization	AlphaFoldDB (Sequence +3Di structure)	Explicit structure encoding uses Foldseek structure tokens to improve accuracy in zero-shot mutation prediction
ISM [17]	Implicit structure encoder and structure distillation	UniRef + PDB	Implicit structure distillation distills 3D structural knowledge into the encoder to allow high precision using only single sequences
Profluent-E1 [18]	Retrieval-augmented encoder and context-aware representation	Profluent Protein Atlas + UniRef	Retrieval augmentation uses evolutionary context to better represent rare families without fine-tuning
ESM-3 [20]	Bidirectional Transformer; generative masked modeling	2.78B proteins (3.15B sequences + 236 M structures + 539 M functional annotations)	Multimodal unification simulates evolution across sequence, structure, and function via "chain of thought" reasoning
DSM [21]	Masked diffusion Transformer and global denoising generation	OMGprot50	Global denoising reconciles the representation-generation trade-off to enable controllable de novo design

amino acids derived from AAindex (such as hydrophobicity and charge distribution). Experiments show that this representation, which combines sequence semantics and explicit biochemical properties, is significantly better than the unimodal model in predicting unknown human-virus protein-protein interactions (HV-PPI). It can more accurately capture the electrostatic interactions and complementary pairing mechanisms between viral proteins and host receptors [24]. On this basis, Flöge et al. (2025) introduced OneProt, further expanding the dimension to three-dimensional. Through cross-modal latent space alignment, the model maps sequences, binding pockets, structure graphs, and biomedical text annotations into a shared semantic space [23]. For viral research, this ability to capture surface pocket features is valuable. It directly helps to identify key binding sites between viral proteins (such as spike proteins) and host receptors, and these geometric features are often difficult to capture in pure sequence models.

Apart from structural and physicochemical features, macroscopic biological networks provide global contextual information. In the process of infection, viral proteins often function by hijacking existing host signaling pathways or protein complexes. Therefore, the topological position and connectivity of proteins in the global human interactome are key criteria for determining the potential of host factors [26]. Hu et al. (2025), leveraging the STRING database for the SPACE project, demonstrated that highly complementary feature representations can be achieved by aligning existing cross-species PPI network embeddings with sequence embeddings, without training a large-scale generation model from scratch [27, 28]. This strategy allows researchers to directly use precomputed network features to accurately locate potential virus targets. It avoids the computing overhead required to construct networks de novo and at the same time provides a systems biology perspective beyond single-sequence analysis to analyze the viral perturbation of host systems.

3 Applications of Protein Language Models in Basic Virology

3.1 Structure and Function of Viral Proteins

Viral proteins are frequently characterized by extreme sequence variability and conformational dynamics. These characteristics make it difficult to use traditional structural prediction methods that rely on multiple sequence alignment (MSA) when applied to orphan viral proteins that lack detectable homologous sequences [29, 30]. Protein language models (PLMs) offer a new way to analyze "viral dark matter" by using evolutionary patterns acquired from massive data.

Lin et al. (2023) utilized the ESM-2 language model and the ESMFold module to fold hundreds of millions of sequences from metagenomic databases. This analysis identified a large number of high-confidence protein structures that are different from existing knowledge, many of which lack detectable sequence homologs. This finding proves that language models can achieve precise, atomic-level structure prediction with only single-sequence inputs, without relying on MSAs, thus providing an efficient computational tool for characterizing viral dark matter [15].

In the field of virus discovery and functional annotation, traditional alignment-based methods (such as BLAST) often struggle to identify highly divergent viral proteins. To solve this problem, Hou et al. (2024) developed LucaProt, a deep learning model based on the Transformer architecture. By using protein language models

(PLMs) to capture deep sequence motifs and structural characteristics of viral core enzymes, they successfully mined more than 160,000 novel RNA viruses from global ecosystems, significantly expanding the known virosphere. Experimental results show that LucaProt can accurately identify functional viral proteins even under conditions of extremely low sequence homology [31].

In addition, PLM-based zero-shot inference has been widely used to evaluate the functional adaptability of viral proteins. Hie et al. (2021) proved that pretrained language models can acquire implicit biophysical rules (i.e., biological grammar) through self-supervised learning, providing a method for the rapid screening of high-risk viral variants. The embeddings generated by these models not only cluster viral proteins with similar functions, but the calculated sequence probability scores also reflect stability and infectivity [32].

3.2 Elucidating Virus-Host Protein-Protein Interactions

Understanding virus-host protein-protein interactions (VH-PPIs) is fundamental to learning how viruses cause disease and developing antiviral drugs. It is difficult for traditional experiments to scale to match the explosion of metagenomic data, so computational approaches have become the primary alternative. Early models like DeepViral (2021) attempted to use CNNs or RNNs to combine sequence and phenotype data, but these architectures always struggled to capture long-range dependencies between residues [33].

Early Siamese network architectures process viral and host protein sequences independently, failing to capture binding-induced conformational adaptation effect during physical contact. PLM-interact (2025) adopts a joint encoding strategy that takes viral and host sequences as a single input, learning interaction patterns directly in the embedding space and enabling detection of single-point mutations affecting binding affinity [34].

TransFactor (2025) incorporates low-rank adaptation (LoRA) into ESM-2 to address labeled data scarcity and quantifies per-residue contributions through computational alanine scanning. This approach identifies key domains such as COMM and PX involved in viral hijacking of host pathways [25]. DeepHVI (2025) integrates sequence embeddings with physicochemical properties in a multimodal framework and adds a conditional sequence generation task to generate potential host target sequences from viral proteins [24]. Ablation experiments in One-Prot (2025) demonstrate that incorporating structural features and text annotations improves predictive performance [23]. Table 2 summarizes the key features of these models.

Table 2
Comparison of computational models for virus-host interaction analysis

Model (year)	Core task	Architecture	Input
DeepViral (2021) [33]	Prediction of novel virus-host PPIs	CNN + joint neural network	Sequence + phenotype
PLM-interact (2025) [34]	Cross-species PPI prediction and mutational effect assessment	Transformer (ESM-2)	Sequence-only (concatenated viral and host sequences)
TransFactor (2025) [25]	Host factor screening and prioritization	Transformer (ESM-2)	Sequence-only (host protein sequences)
DeepHVI (2025) [24]	Human-virus PPI prediction and target generation	Multimodal transformer (fusion: TMO-net module)	Sequence + physicochemical

3.3 Tracking Viral Adaptation and Evolutionary Trajectories

Host immune pressure drives continuous viral evolution, which can lead to reduced vaccine efficacy and drug resistance [35, 36]. Traditional phylogenetic analysis reconstructs evolutionary history but struggles to predict future mutation trajectories.

Hie et al. (2021) proposed that viral genomes follow a "biological grammar" analogous to natural language. Their framework computes a "grammaticality" score for mutant sequences using pretrained language models—the probability assigned to a given sequence by the model. High scores indicate retained infectivity; low scores suggest functional loss. This zero-shot approach enables rapid fitness assessment without experimental data [32].

Viral escape, however, requires antigenic variation while maintaining fitness to evade neutralizing antibodies, a phenomenon termed "semantic drift." The constrained semantic change search (CSCS) algorithm quantifies this semantic shift in PLM embedding space, searching for mutants with high fitness but large antigenic differences. Using sequence data alone, this method predicted immune escape pathways for influenza, HIV, and SARS-CoV-2 [32].

The Evo-velocity framework (2022), built on ESM-1b, introduces "evolutionary velocity" to track viral temporal evolution: Sequence embeddings construct a K-nearest neighbor network, and a masked language model calculates likelihood changes between sequences to quantify evolutionary direction and rate. By adding direction to the network, it creates a global evolutionary vector field. Evo-velocity finds evolutionary roots and orders sequences by time and successfully reconstructed decades of influenza history. This proves PLMs can map global evolutionary dynamics using local clues [37].

4 Emerging Advances in PLM-Driven Antiviral Strategies

4.1 Discovery and Design of Antiviral Therapeutic Targets

Protein language models (PLMs) are becoming essential tools for experimental science, which are especially useful during the early stages of viral outbreaks. By using models to screen targets or design molecules, PLMs facilitate the search space for wet labs. This significantly reduces trial-and-error costs.

When new viruses appear, high-resolution crystal structures are often missing. This lack of data limits traditional structure-based drug design. However, sequence-based deep learning can predict drug-target interactions without 3D structures. Beck et al. (2020) created the MT-DTI framework for drug repurposing early in the pandemic. This model uses self-attention mechanisms to analyze drug molecular sequences, known as SMILES. It performed large-scale screening against SARS-CoV-2 proteins like the 3CL protease. The model successfully predicted high binding affinities ($K_d <$ 200 nM) for drugs like remdesivir, atazanavir, and efavirenz. It did this without relying on any 3D structural information. These predictions matched later in vitro experiments and clinical results [38]. This proves that sequence-based models function as rapid response tools for drug repurposing when structural data is absent.

Hie et al. (2023) proposed a zero-shot antibody evolution strategy using pretrained language models to navigate the fitness landscape without antigen-antibody complex structures. Evaluated on SARS-CoV-2 and Ebolavirus, affinity-matured variants showed improved binding and broadened neutralization breadth [39].

4.2 Viral Vaccine Design

PLMs contribute to vaccine development through epitope prediction, antigen optimization, and sequence design [40–42].

4.2.1 B-cell Epitope Prediction and Antibody Design

For B-cell epitope identification, homology-based methods fail to capture conformational information. Ivanisenko et al. (2024) introduced the SEMA 2.0 platform to address this problem. SEMA 2.0 combines sequence-level (SEMA-1D) and structure-level (SEMA-3D) predictions, outperforming existing state-of-the-art methods such as DiscoTope-3.0 and BepiPred-3.0 on both masked and unmasked benchmark datasets [43–45]. The platform incorporates N-glycosylation site prediction, which is relevant for glycan-shielded epitopes such as those on the SARS-CoV-2 Spike protein, and cross-strain epitope conservation analysis for broad-spectrum vaccine design [43].

4.2.2 mRNA Vaccine Sequence Optimization

During the COVID-19 pandemic, mRNA vaccines showed great potential because of their safety and fast production. But optimizing mRNA sequences is still a difficult problem. Traditional methods usually focus only on host codon bias. They often overlook how local secondary structures affect translation efficiency and stability [40, 46, 47].

Li et al. (2024) developed CodonBERT, a large language model applying the BERT architecture to mRNA sequences [40]. This model is trained with masked language modeling and taxonomy prediction on codon sequences, predicting influenza vaccine expression levels (Spearman $\rho = 0.81$, approximately 10% improvement over the next best method) and SARS-CoV-2 mRNA degradation rates, which shows the rewarding potential of capturing the link between mRNA sequences and protein expression and optimizing vaccine design [40].

4.2.3 Complementarity Between PLMs and Structure Prediction

PLMs implicitly learn residue contact patterns but lack explicit 3D coordinates for validating surface exposure and steric feasibility. Therefore, researchers still rely on explicit structure prediction to verify if specific mutations might disrupt global protein folding [48, 49].

Jumper et al. (2021) released AlphaFold2, enabling atomic-level structure prediction from amino acid sequences with unprecedented accuracy [50]. Gema et al. (2023) used Vaxformer to generate SARS-CoV-2 Spike protein variants, filtering candidates through AlphaFold2 to retain sequences with native-like folding and thermodynamic stability [42].

5 Conclusion

PLMs extract evolutionary and structural patterns directly from sequences without requiring alignment, enabling functional annotation of uncharacterized viral proteins and interaction prediction. Current limitations include difficulty in interpreting learned representations for mechanistic insight and training data bias toward well-characterized viral families, potentially reducing accuracy on divergent lineages. Multimodal architectures integrating sequence embeddings with predicted structures and physicochemical descriptors offer one direction for improvement. Platforms such as ColabFold and SaprotHub provide pretrained models and inference pipelines accessible without local GPU infrastructure [51]. Virologists can thus build custom models without extensive programming skills. This capability accelerates the feedback loop between computational prediction and experimental validation. Deep learning methods thereby become practical tools for routine virological analysis.

Acknowledgments

This work was supported by Major Project of Guangzhou National Laboratory (GZNL2024A01003), National Key R&D Program of China (2022YFF1203202), and Strategic Priority Research

Program of Chinese Academy of Sciences (XDB38050200, XDB38040202, XDA26040304).

Conflict of Interest This work was funded by Major Project of Guangzhou National Laboratory (GZNL2024A01003). The authors have no conflicts of interest to declare that are relevant to the content of this chapter.

References

1. Miller CM, Selvam S, Fuchs G (2021) Fatal attraction: the roles of ribosomal proteins in the viral life cycle. Wiley Interdiscip Rev: RNA 12:e1613. https://doi.org/10.1002/wrna.1613
2. Avula K, Singh B, Kumar PV, Syed GH (2021) Role of lipid transfer proteins (LTPs) in the viral life cycle. Front Microbiol 12. https://doi.org/10.3389/fmicb.2021.673509
3. Holmes EC, Krammer F, Goodrum FD (2024) Virology—the next fifty years. Cell 187:5128–5145. https://doi.org/10.1016/j.cell.2024.07.025
4. Villanueva RA, Rouillé Y, Dubuisson J (2005) Interactions between virus proteins and host cell membranes during the viral life cycle. Int Rev Cytol 245:171–244. https://doi.org/10.1016/S0074-7696(05)45006-8
5. Kosmopoulos JC, Anantharaman K (2025) Viral dark matter: illuminating protein function, ecology, and biotechnological promises. Biochemistry. https://doi.org/10.1021/acs.biochem.5c00349
6. Artika IM, Ma'roef CN (2017) Laboratory biosafety for handling emerging viruses. Asian Pac J Trop Biomed 7:483–491. https://doi.org/10.1016/j.apjtb.2017.01.020
7. Auslander N, Gussow AB, Koonin EV (2021) Incorporating machine learning into established bioinformatics frameworks. IJMS 22:2903. https://doi.org/10.3390/ijms22062903
8. Rives A, Meier J, Sercu T, Goyal S, Lin Z, Liu J, Guo D, Ott M, Zitnick CL, Ma J, Fergus R (2021) Biological structure and function emerge from scaling unsupervised learning to 250 million protein sequences. Proc Natl Acad Sci USA 118:e2016239118. https://doi.org/10.1073/pnas.2016239118
9. Lv L, Lin Z, Li H, Liu Y, Cui J, Yu-Chian Chen C, Yuan L, Tian Y (2024) ProLLaMA: a protein large language model for multi-task protein language processing. arXiv e-prints arXiv:2402.16445. https://doi.org/10.48550/arXiv.2402.16445
10. Chen J-Y, Wang J-F, Hu Y, Li X-H, Qian Y-R, Song C-L (2025) Evaluating the advancements in protein language models for encoding strategies in protein function prediction: a comprehensive review. Front Bioeng Biotechnol 13:1506508. https://doi.org/10.3389/fbioe.2025.1506508
11. Sawhney R, Ferrell B, Dejean T, Schreiber Z, Harrigan W, Polson SW, Wommack KE, Belcaid M (2025) Fine-Tuning Protein Language Models Unlocks the Potential of Underrepresented Viral Proteomes bioRxiv 2025.04.17.649224. https://doi.org/10.1101/2025.04.17.649224
12. Ito J, Strange A, Liu W, Joas G, Lytras S, Sato K (2025) A protein language model for exploring viral fitness landscapes. Nat Commun 16:4236. https://doi.org/10.1038/s41467-025-59422-w
13. Ferruz N, Schmidt S, Höcker B (2022) ProtGPT2 is a deep unsupervised language model for protein design. Nat Commun 13:4348. https://doi.org/10.1038/s41467-022-32007-7
14. Devlin J, Chang M-W, Lee K, Toutanova K (2019) BERT: pre-training of deep bidirectional transformers for language understanding. In: Burstein J, Doran C, Solorio T (eds) Proceedings of the 2019 conference of the North American chapter of the Association for Computational Linguistics: human language technologies, volume 1 (Long and short papers). Association for Computational Linguistics, Minneapolis, pp 4171–4186
15. Lin Z, Akin H, Rao R, Hie B, Zhu Z, Lu W, Smetanin N, Verkuil R, Kabeli O, Shmueli Y, dos Santos CA, Fazel-Zarandi M, Sercu T, Candido S, Rives A (2023) Evolutionary-scale prediction of atomic-level protein structure with a language model. Science 379:1123–1130. https://doi.org/10.1126/science.ade2574
16. Su J, Han C, Zhou Y, Shan J, Zhou X, Yuan F (2024) SaProt: protein language modeling with structure-aware vocabulary. In: The

twelfth international conference on learning representations, ICLR 2024, Vienna, Austria, May 7–11, 2024. OpenReview.net

17. Ouyang-Zhang J, Gong C, Zhao Y, Krähenbühl P, Klivans AR, Diaz DJ (2025) Distilling structural representations into protein sequence models. bioRxiv 2024.11.08.622579. https://doi.org/10.1101/2024.11.08.622579
18. Jain S, Beazer J, Ruffolo JA, Bhatnagar A, Madani A (2025) E1: retrieval-augmented protein encoder models. bioRxiv 2025.11.12.688125. https://doi.org/10.1101/2025.11.12.688125
19. Elnaggar A, Heinzinger M, Dallago C, Rehawi G, Wang Y, Jones L, Gibbs T, Feher T, Angerer C, Steinegger M, Bhowmik D, Rost B (2022) ProtTrans: toward understanding the language of life through self-supervised learning. IEEE Trans Pattern Anal Mach Intell 44:7112–7127. https://doi.org/10.1109/TPAMI.2021.3095381
20. Hayes T, Rao R, Akin H, Sofroniew NJ, Oktay D, Lin Z, Verkuil R, Tran VQ, Deaton J, Wiggert M, Badkundri R, Shafkat I, Gong J, Derry A, Molina RS, Thomas N, Khan YA, Mishra C, Kim C, Bartie LJ, Nemeth M, Hsu PD, Sercu T, Candido S, Rives A (2024) Simulating 500 million years of evolution with a language model. bioRxiv 2024.07.01.600583. https://doi.org/10.1101/2024.07.01.600583
21. Hallee L, Rafailidis N, Bichara DB, Gleghorn JP (2025) Diffusion sequence models for enhanced protein representation and generation. arXiv e-prints arXiv:2506.08293. https://doi.org/10.48550/arXiv.2506.08293
22. Nie S, Zhu F, You Z, Zhang X, Ou J, Hu J, Zhou J, Lin Y, Wen J-R, Li C (2025) Large language diffusion models. arXiv e-prints arXiv:2502.09992. https://doi.org/10.48550/arXiv.2502.09992
23. Flöge K, Udayakumar S, Sommer J, Piraud M, Kesselheim S, Fortuin V, Günnemann S, Van Der Weg KJ, Gohlke H, Merdivan E, Bazarova A (2025) OneProt: towards multi-modal protein foundation models via latent space alignment of sequence, structure, binding sites and text encoders. PLoS Comput Biol 21: e1013679. https://doi.org/10.1371/journal.pcbi.1013679
24. Wang X, Luo J, Cai X, Liu R, Li Y, Hon C (2025) DeepHVI: a multimodal deep learning framework for predicting human-virus protein-protein interactions using protein language models. Biosaf Health 7:257–266. https://doi.org/10.1016/j.bsheal.2025.07.005
25. An Y, Bergant V, Firmani S, Grünke C, Bonnal B, Henrici A, Pichlmair A, Schubert B, Marsico A (2025) TransFactor-prediction of pro-viral SARS-CoV-2 host factors using a protein language model. Bioinformatics 41:btaf491. https://doi.org/10.1093/bioinformatics/btaf491
26. Ray S, Alberuni S, Schönhuth A (2025) A graph neural network-based approach for predicting SARS-CoV-2–human protein interactions from multiview data. PLoS One 20: e0332794. https://doi.org/10.1371/journal.pone.0332794
27. Hu D, Szklarczyk D, von Mering C, Jensen LJ (2025) SPACE: STRING proteins as complementary embeddings. Bioinformatics 41: btaf496. https://doi.org/10.1093/bioinformatics/btaf496
28. Szklarczyk D, Kirsch R, Koutrouli M, Nastou K, Mehryary F, Hachilif R, Gable AL, Fang T, Doncheva NT, Pyysalo S, Bork P, Jensen LJ, von Mering C (2023) The STRING database in 2023: protein-protein association networks and functional enrichment analyses for any sequenced genome of interest. Nucleic Acids Res 51:D638–D646. https://doi.org/10.1093/nar/gkac1000
29. Fischer D, Eisenberg D (1999) Finding families for genomic ORFans. Bioinformatics 15:759–762. https://doi.org/10.1093/bioinformatics/15.9.759
30. Krishnamurthy SR, Wang D (2017) Origins and challenges of viral dark matter. Virus Res 239:136–142. https://doi.org/10.1016/j.virusres.2017.02.002
31. Hou X, He Y, Fang P, Mei S-Q, Xu Z, Wu W-C, Tian J-H, Zhang S, Zeng Z-Y, Gou Q-Y, Xin G-Y, Le S-J, Xia Y-Y, Zhou Y-L, Hui F-M, Pan Y-F, Eden J-S, Yang Z-H, Han C, Shu Y-L, Guo D, Li J, Holmes EC, Li Z-R, Shi M (2024) Using artificial intelligence to document the hidden RNA virosphere. Cell 187:6929–6942.e16. https://doi.org/10.1016/j.cell.2024.09.027
32. Hie B, Zhong ED, Berger B, Bryson B (2021) Learning the language of viral evolution and escape. Science 371:284–288. https://doi.org/10.1126/science.abd7331
33. Liu-Wei W, Kafkas Ş, Chen J, Dimonaco NJ, Tegnér J, Hoehndorf R (2021) DeepViral: prediction of novel virus–host interactions from protein sequences and infectious disease phenotypes. Bioinformatics 37:2722–2729. https://doi.org/10.1093/bioinformatics/btab147
34. Liu D, Young F, Lamb KD, Claudio Quiros A, Pancheva A, Miller CJ, Macdonald C, Robertson DL, Yuan K (2025) PLM-interact: extending protein language models to predict protein-protein interactions. Nat Commun

16:9012. https://doi.org/10.1038/s41467-025-64512-w

35. Domingo E, Sheldon J, Perales C (2012) Viral quasispecies evolution. Microbiol Mol Biol Rev 76:159–216. https://doi.org/10.1128/MMBR.05023-11
36. Harvey WT, Carabelli AM, Jackson B, Gupta RK, Thomson EC, Harrison EM, Ludden C, Reeve R, Rambaut A, COVID-19 Genomics UK (COG-UK) Consortium, Peacock SJ, Robertson DL (2021) SARS-CoV-2 variants, spike mutations and immune escape. Nat Rev Microbiol 19:409–424. doi:https://doi.org/10.1038/s41579-021-00573-0
37. Hie BL, Yang KK, Kim PS (2022) Evolutionary velocity with protein language models predicts evolutionary dynamics of diverse proteins. Cell Syst 13:274–285.e6. https://doi.org/10.1016/j.cels.2022.01.003
38. Beck BR, Shin B, Choi Y, Park S, Kang K (2020) Predicting commercially available antiviral drugs that may act on the novel coronavirus (SARS-CoV-2) through a drug-target interaction deep learning model. Comput Struct Biotechnol J 18:784–790. https://doi.org/10.1016/j.csbj.2020.03.025
39. Hie BL, Shanker VR, Xu D, Bruun TUJ, Weidenbacher PA, Tang S, Wu W, Pak JE, Kim PS (2024) Efficient evolution of human antibodies from general protein language models. Nat Biotechnol 42:275–283. https://doi.org/10.1038/s41587-023-01763-2
40. Li S, Moayedpour S, Li R, Bailey M, Riahi S, Kogler-Anele L, Miladi M, Miner J, Pertuy F, Zheng D, Wang J, Balsubramani A, Tran K, Zacharia M, Wu M, Gu X, Clinton R, Asquith C, Skaleski J, Boeglin L, Chivukula S, Dias A, Strugnell T, Montoya FU, Agarwal V, Bar-Joseph Z, Jager S (2024) CodonBERT large language model for mRNA vaccines. Genome Res 34:1027–1035. https://doi.org/10.1101/gr.278870.123
41. da Silva MK, Campos DM, Akash S, Akter S, Yee LC, Fulco UL, Oliveira JIN (2023) Advances of reverse vaccinology for mRNA vaccine design against SARS-CoV-2: a review of methods and tools. Viruses 15:2130. https://doi.org/10.3390/v15102130
42. Gema AP, Kobiela M, Fraisse A, Rajan A, Oyarzún DA, Alfaro JA (2023) Vaxformer: antigenicity-controlled transformer for vaccine design against SARS-CoV-2. arXiv e-prints arXiv:2305.11194. https://doi.org/10.48550/arXiv.2305.11194
43. Ivanisenko NV, Shashkova TI, Shevtsov A, Sindeeva M, Umerenkov D, Kardymon O (2024) SEMA 2.0: web-platform for B-cell conformational epitopes prediction using artificial intelligence. Nucleic Acids Res 52: W533–W539. https://doi.org/10.1093/nar/gkae386
44. Clifford JN, Høie MH, Deleuran S, Peters B, Nielsen M, Marcatili P (2022) BepiPred-3.0: improved B-cell epitope prediction using protein language models. Protein Sci 31:e4497. https://doi.org/10.1002/pro.4497
45. Høie MH, Gade FS, Johansen JM, Würtzen C, Winther O, Nielsen M, Marcatili P (2024) DiscoTope-3.0: improved B-cell epitope prediction using inverse folding latent representations. Front Immunol 15:1322712. https://doi.org/10.3389/fimmu.2024.1322712
46. Jin L, Zhou Y, Zhang S, Chen S-J (2024) mRNA vaccine sequence and structure design and optimization: advances and challenges. J Biol Chem 301:108015. https://doi.org/10.1016/j.jbc.2024.108015
47. Li S, Noroozizadeh S, Moayedpour S, Kogler-Anele L, Xue Z, Zheng D, Montoya FU, Agarwal V, Bar-Joseph Z, Jager S (2025) mRNA-LM: full-length integrated SLM for mRNA analysis. Nucleic Acids Res 53: gkaf044. https://doi.org/10.1093/nar/gkaf044
48. Rao R, Meier J, Sercu T, Ovchinnikov S, Rives A (2020) Transformer protein language models are unsupervised structure learners bioRxiv 2020.12.15.422761. https://doi.org/10.1101/2020.12.15.422761
49. Qin Z, Wu S, Sun J, Li S (2024) Single-sequence protein structure prediction by integrating protein language models. Proc Natl Acad Sci USA 121(16):e2308788121. https://doi.org/10.1073/pnas.2308788121
50. Jumper J, Evans R, Pritzel A, Green T, Figurnov M, Ronneberger O, Tunyasuvunakool K, Bates R, Žídek A, Potapenko A, Bridgland A, Meyer C, Kohl SAA, Ballard AJ, Cowie A, Romera-Paredes B, Nikolov S, Jain R, Adler J, Back T, Petersen S, Reiman D, Clancy E, Zielinski M, Steinegger M, Pacholska M, Berghammer T, Bodenstein S, Silver D, Vinyals O, Senior AW, Kavukcuoglu K, Kohli P, Hassabis D (2021) Highly accurate protein structure prediction with AlphaFold. Nature 596:583–589. https://doi.org/10.1038/s41586-021-03819-2
51. Su J, Li Z, Tao T, Han C, He Y, Dai F, Yuan Q, Gao Y, Si T, Zhang X, Zhou Y, Shan J, Zhou X, Chang X, Jiang S, Ma D, The OPMC, Steinegger M, Ovchinnikov S, Yuan F (2025) Democratizing protein language model training, sharing and collaboration. Nat Biotechnol. https://doi.org/10.1038/s41587-025-02859-7

Index

Tao Huang (ed.), *Microbial 'Omics: Methods and Protocols*, Methods in Molecular Biology, vol. 3033,
https://doi.org/10.1007/978-1-0716-5264-0,

MIX
Papier aus verantwortungsvollen Quellen
Paper from responsible sources
FSC® C105338

If you have any concerns about our products,
you can contact us on
ProductSafety@springernature.com

In case Publisher is established outside the EU,
the EU authorized representative is:
Springer Nature Customer Service Center GmbH
Europaplatz 3, 69115 Heidelberg, Germany

Printed by Libri Plureos GmbH
in Hamburg, Germany